Universitext

Universitext

Universitext is a series of textbooks that presents material from a wide variety of mathematical disciplines at master's level and beyond. The books, often well class-tested by their author, may have an informal, personal, even experimental approach to their subject matter. Some of the most successful and established books in the series have evolved through several editions, always following the evolution of teaching curricula, into very polished texts.

Thus as research topics trickle down into graduate-level teaching, first textbooks written for new, cutting-edge courses may make their way into *Universitext*.

For further volumes:
http://www.springer.com/series/223

Fumio Hiai · Dénes Petz

Introduction to Matrix
Analysis and Applications

Fumio Hiai
Graduate School of Information Sciences
Tohoku University
Sendai
Japan

Dénes Petz
Alfréd Rényi Institute of Mathematics
Budapest
Hungary

Department for Mathematical Analysis
Budapest University of Technology
  and Economics
Budapest
Hungary

A co-publication with the Hindustan Book Agency, New Delhi, licensed for sale in all countries outside of India. Sold and distributed within India by the Hindustan Book Agency, P 19 Green Park Extn., New Delhi 110 016, India.
HBA ISBN 978-93-80250-60-1

ISSN 0172-5939   ISSN 2191-6675   (electronic)
ISBN 978-3-319-04149-0   ISBN 978-3-319-04150-6   (eBook)
DOI 10.1007/978-3-319-04150-6
Springer Cham Heidelberg New York Dordrecht London

Library of Congress Control Number: 2013957552

Mathematics Subject Classification: 15A99, 15A60

Printed on acid-free paper

Springer is part of Springer Science+Business Media (www.springer.com)

Preface

The material of this book is partly based on the lectures of the authors given at the Graduate School of Information Sciences of Tohoku University and at the Budapest University of Technology and Economics. The aim of the lectures was to explain certain important topics in matrix analysis from the point of view of functional analysis. The concept of Hilbert space appears many times, but only finite-dimensional spaces are used. The book treats some aspects of analysis related to matrices including such topics as matrix monotone functions, matrix means, majorization, entropies, quantum Markov triplets, and so on. There are several popular matrix applications in quantum theory.

The book is organized into seven chapters. Chapters 1–3 form an introductory part of the book and could be used as a textbook for an advanced undergraduate special topics course. The word "matrix" was first introduced in 1848 and applications subsequently appeared in many different areas. Chapters 4–7 contain a number of more advanced and less well-known topics. This material could be used for an advanced specialized graduate-level course aimed at students who wish to specialize in quantum information. But the best use for this part is as a reference for active researchers in the field of quantum information theory. Researchers in statistics, engineering, and economics may also find this book useful.

Chapter 1 introduces the basic notions of matrix analysis. We prefer the Hilbert space concepts, so complex numbers are used. The spectrum and eigenvalues are important, and the determinant and trace are used later in several applications. The final section covers tensor products and their symmetric and antisymmetric subspaces. The chapter concludes with a selection of exercises. We point out that in this book "positive" means ≥ 0; we shall not use the term "non-negative."

Chapter 2 covers block matrices, partial ordering, and an elementary theory of von Neumann algebras in the finite-dimensional setting. The Hilbert space concept requires projections, i.e., matrices P satisfying $P = P^2 = P^*$. Self-adjoint matrices are linear combinations of projections. Not only are single matrices required, but subalgebras of matrices are also used. This material includes Kadison's inequality and completely positive mappings.

Chapter 3 details matrix functional calculus. Functional calculus provides a new matrix $f(A)$ when a matrix A and a function f are given. This is an essential tool in matrix theory as well as in operator theory. A typical example is the exponential function $e^A = \sum_{n=0}^{\infty} A^n/n!$ If f is sufficiently smooth, then $f(A)$ is also smooth and we have a useful Fréchet differential formula.

Chapter 4 covers matrix monotone functions. A real function defined on an interval is matrix monotone if $A \leq B$ implies $f(A) \leq f(B)$ for Hermitian matrices A and B whose eigenvalues are in the domain of f. We have a beautiful theory of such functions, initiated by Löwner in 1934. A highlight is the integral expression of such functions. Matrix convex functions are also considered. Graduate students in mathematics and in information theory will benefit from having all of this material collected into a single source.

Chapter 5 covers matrix (operator) means for positive matrices. Matrix extensions of the arithmetic mean $(a + b)/2$ and the harmonic mean

$$\left(\frac{a^{-1} + b^{-1}}{2} \right)^{-1}$$

are rather trivial, however it is nontrivial to define matrix version of the geometric mean $\sqrt{ab}$. This was first done by Pusz and Woronowicz. A general theory of matrix means developed by Kubo and Ando is closely related to operator monotone functions on $(0, \infty)$. There are also more complicated means. The mean transformation $M(A, B) := m(\mathbb{L}_A, \mathbb{R}_B)$ is a mean of the left-multiplication $\mathbb{L}_A$ and the right-multiplication $\mathbb{R}_B$, recently studied by Hiai and Kosaki. Another useful concept is a multivariable extension of two-variable matrix means.

Chapter 6 discusses majorizations for eigenvalues and singular values of matrices. Majorization is a certain order relation between two real vectors. Section 6.1 recalls classical material that can be found in other sources. There are several famous majorizations for matrices which have strong applications to matrix norm inequalities in symmetric norms. For instance, an extremely useful inequality is the Lidskii–Wielandt theorem.

The last chapter contains topics related to quantum applications. Positive matrices with trace 1, also called density matrices, are the states in quantum theory. The relative entropy appeared in 1962, and matrix theory has found many applications in the quantum formalism. The unknown quantum states can be described via the use of positive operators $F(x)$ with $\sum_x F(x) = I$. This is called a POVM and a few mathematical results are shown, but in quantum theory there are much more relevant subjects. These subjects are close to the authors' interests, and there are some very recent results.

The authors thank several colleagues for useful communications. They are particularly grateful to Prof. Tsuyoshi Ando for insightful comments and to Prof. Rajendra Bhatia for valuable advice.

April 2013   Fumio Hiai
                                                                                Dénes Petz

Contents

Chapter 1
Fundamentals of Operators and Matrices

A linear mapping is essentially a matrix if the vector space is finite-dimensional. In this book the vector space is typically a finite-dimensional complex Hilbert space. The first chapter collects introductory materials on matrices and operators. Section 1.2 is a concise exposition of Hilbert spaces. The polar and spectral decompositions useful in studying operators on Hilbert spaces are also essential for matrices. A finer decomposition for matrices is the Jordan canonical form described in Sect. 1.3. Among the most basic notions for matrices are eigenvalues, singular values, trace and determinant, included in the subsequent sections. A less elementary but important subject is tensor products, discussed in the last section.

1.1 Basics on Matrices

For $n, m \in \mathbb{N}$, $\mathbb{M}_{n \times m} = \mathbb{M}_{n \times m}(\mathbb{C})$ denotes the space of all $n \times m$ complex matrices. A matrix $M \in \mathbb{M}_{n \times m}$ is a mapping $\{1, 2, \ldots, n\} \times \{1, 2, \ldots, m\} \to \mathbb{C}$. It is represented as an array with n rows and m columns:

$$
M = \begin{bmatrix} m_{11} & m_{12} & \cdots & m_{1m} \\ m_{21} & m_{22} & \cdots & m_{2m} \\ \vdots & \vdots & \ddots & \vdots \\ m_{n1} & m_{n2} & \cdots & m_{nm} \end{bmatrix},
$$

where m_{ij} is the intersection of the ith row and the jth column. If the matrix is denoted by M, then this entry is denoted by M_{ij}. If $n = m$, then we write $\mathbb{M}_n$ instead of $\mathbb{M}_{n \times n}$. A simple example is the **identity matrix** $I_n \in \mathbb{M}_n$ defined as $m_{ij} = \delta_{i,j}$, or

F. Hiai and D. Petz, *Introduction to Matrix Analysis and Applications*,
Universitext, DOI: 10.1007/978-3-319-04150-6_1,
© Hindustan Book Agency 2014

$$I_n = \begin{bmatrix} 1 & 0 & \cdots & 0 \\ 0 & 1 & \cdots & 0 \\ \vdots & \vdots & \ddots & \vdots \\ 0 & 0 & \cdots & 1 \end{bmatrix}.$$

$\mathbb{M}_{n \times m}$ is a complex vector space of dimension nm. The linear operations are defined as follows:

$$[\lambda A]_{ij} := \lambda A_{ij}, \qquad [A + B]_{ij} := A_{ij} + B_{ij},$$

where λ is a complex number and $A, B \in \mathbb{M}_{n \times m}$.

Example 1.1 For $i, j = 1, \ldots, n$ let $E(ij)$ be the $n \times n$ matrix such that the (i, j)-entry is equal to one and all other entries are equal to zero. Then $E(ij)$ are called the **matrix units** and form a basis of $\mathbb{M}_n$:

$$A = \sum_{i,j=1}^{n} A_{ij} E(ij).$$

In particular,

$$I_n = \sum_{i=1}^{n} E(ii).$$

If $A \in \mathbb{M}_{n \times m}$ and $B \in \mathbb{M}_{m \times k}$, then the **product** AB of A and B is defined by

$$[AB]_{ij} = \sum_{\ell=1}^{m} A_{i\ell} B_{\ell j},$$

where $1 \le i \le n$ and $1 \le j \le k$. Hence $AB \in \mathbb{M}_{n \times k}$. So $\mathbb{M}_n$ becomes an algebra. The most significant feature of matrices is the non-commutativity of the product $AB \ne BA$. For example,

$$\begin{bmatrix} 0 & 1 \\ 0 & 0 \end{bmatrix} \begin{bmatrix} 0 & 0 \\ 1 & 0 \end{bmatrix} = \begin{bmatrix} 1 & 0 \\ 0 & 0 \end{bmatrix}, \qquad \begin{bmatrix} 0 & 0 \\ 1 & 0 \end{bmatrix} \begin{bmatrix} 0 & 1 \\ 0 & 0 \end{bmatrix} = \begin{bmatrix} 0 & 0 \\ 0 & 1 \end{bmatrix}.$$

In the matrix algebra $\mathbb{M}_n$, the identity matrix I_n behaves as a unit: $I_n A = A I_n = A$ for every $A \in \mathbb{M}_n$. The matrix $A \in \mathbb{M}_n$ is **invertible** if there is a $B \in \mathbb{M}_n$ such that $AB = BA = I_n$. This B is called the **inverse** of A, and is denoted by A^{-1}.   □

Example 1.2 The linear equations

$$ax + by = u$$

$$cx + dy = v$$

can be written in a matrix formalism:

$$\begin{bmatrix} a & b \\ c & d \end{bmatrix} \begin{bmatrix} x \\ y \end{bmatrix} = \begin{bmatrix} u \\ v \end{bmatrix}.$$

If x and y are the unknown parameters and the coefficient matrix is invertible, then the solution is

$$\begin{bmatrix} x \\ y \end{bmatrix} = \begin{bmatrix} a & b \\ c & d \end{bmatrix}^{-1} \begin{bmatrix} u \\ v \end{bmatrix}.$$

So the solution of linear equations is based on the inverse matrix, which is formulated in Theorem 1.33.   □

The **transpose** A^t of the matrix $A \in \mathbb{M}_{n \times m}$ is an $m \times n$ matrix,

$$[A^t]_{ij} = A_{ji} \qquad (1 \le i \le m, 1 \le j \le n).$$

It is easy to see that if the product AB is defined, then $(AB)^t = B^t A^t$. The **adjoint matrix** A^* is the complex conjugate of the transpose A^t. The space $\mathbb{M}_n$ is a *-algebra:

$$(AB)C = A(BC), \quad (A+B)C = AC + BC, \quad A(B+C) = AB + AC,$$
$$(A+B)^* = A^* + B^*, \quad (\lambda A)^* = \bar{\lambda} A^*, \quad (A^*)^* = A, \quad (AB)^* = B^* A^*.$$

Let $A \in \mathbb{M}_n$. The **trace** of A is the sum of the diagonal entries:

$$\operatorname{Tr} A := \sum_{i=1}^{n} A_{ii}.$$

It is easy to show that $\operatorname{Tr} AB = \operatorname{Tr} BA$, see Theorem 1.28.

The **determinant** of $A \in \mathbb{M}_n$ is slightly more complicated:

$$\det A := \sum_{\pi} (-1)^{\sigma(\pi)} A_{1\pi(1)} A_{2\pi(2)} \ldots A_{n\pi(n)}, \tag{1.1}$$

where the sum is over all permutations π of the set $\{1, 2, \ldots, n\}$ and $\sigma(\pi)$ is the parity of the permutation π. Therefore

$$\det \begin{bmatrix} a & b \\ c & d \end{bmatrix} = ad - bc,$$

and another example is the following:

$$\det \begin{bmatrix} A_{11} & A_{12} & A_{13} \\ A_{21} & A_{22} & A_{23} \\ A_{31} & A_{32} & A_{33} \end{bmatrix}$$

$$= A_{11} \det \begin{bmatrix} A_{22} & A_{23} \\ A_{32} & A_{33} \end{bmatrix} - A_{12} \det \begin{bmatrix} A_{21} & A_{23} \\ A_{31} & A_{33} \end{bmatrix} + A_{13} \det \begin{bmatrix} A_{21} & A_{22} \\ A_{31} & A_{33} \end{bmatrix}.$$

It can be proven that

$$\det(AB) = (\det A)(\det B).$$

1.2 Hilbert Space

Let $\mathcal{H}$ be a complex vector space. A functional $\langle \cdot, \cdot \rangle : \mathcal{H} \times \mathcal{H} \to \mathbb{C}$ of two variables is called an **inner product** if it satisfies:

(1) $\langle x + y, z \rangle = \langle x, z \rangle + \langle y, z \rangle$ $(x, y, z \in \mathcal{H})$;
(2) $\langle \lambda x, y \rangle = \overline{\lambda} \langle x, y \rangle$ $(\lambda \in \mathbb{C},\ x, y \in \mathcal{H})$;
(3) $\langle x, y \rangle = \overline{\langle y, x \rangle}$ $(x, y \in \mathcal{H})$;
(4) $\langle x, x \rangle \geq 0$ for every $x \in \mathcal{H}$ and $\langle x, x \rangle = 0$ only for $x = 0$.

Condition (2) states that the inner product is conjugate linear in the first variable (and it is linear in the second variable). The **Schwarz inequality**

$$|\langle x, y \rangle|^2 \leq \langle x, x \rangle \langle y, y \rangle \tag{1.2}$$

holds. The inner product determines a **norm** for the vectors:

$$\|x\| := \sqrt{\langle x, x \rangle}.$$

This has the properties

$$\|x + y\| \leq \|x\| + \|y\| \quad \text{and} \quad |\langle x, y \rangle| \leq \|x\| \cdot \|y\|.$$

$\|x\|$ is interpreted as the length of the vector x. A further requirement in the definition of a Hilbert space is that every Cauchy sequence must be convergent, that is, the space is **complete**. (In the finite-dimensional case, completeness always holds.)

The linear space $\mathbb{C}^n$ of all n-tuples of complex numbers becomes a Hilbert space with the inner product

$$\langle x, y \rangle = \sum_{i=1}^{n} \bar{x}_i y_i = [\bar{x}_1, \bar{x}_2, \ldots, \bar{x}_n] \begin{bmatrix} y_1 \\ y_2 \\ \vdots \\ y_n \end{bmatrix},$$

where $\bar{z}$ denotes the complex conjugate of the complex number $z \in \mathbb{C}$. Another example is the space of square integrable complex-valued functions on the real Euclidean space $\mathbb{R}^n$. If f and g are such functions then

$$\langle f, g \rangle = \int_{\mathbb{R}^n} \overline{f(x)}\, g(x)\, dx$$

gives the inner product. The latter space is denoted by $L^2(\mathbb{R}^n)$ and, in contrast to the n-dimensional space $\mathbb{C}^n$, it is infinite-dimensional. Below we are mostly concerned with finite-dimensional spaces.

If $\langle x, y \rangle = 0$ for vectors x and y of a Hilbert space, then x and y are called **orthogonal**, denoted $x \perp y$. When $H \subset \mathcal{H}$, $H^\perp := \{x \in \mathcal{H} : x \perp h \text{ for every } h \in H\}$ is called the **orthogonal complement** of H. For any subset $H \subset \mathcal{H}$, $H^\perp$ is a closed subspace.

A family $\{e_i\}$ of vectors is called **orthonormal** if $\langle e_i, e_i \rangle = 1$ and $\langle e_i, e_j \rangle = 0$ if $i \neq j$. A maximal orthonormal system is called a **basis** or orthonormal basis. The cardinality of a basis is called the dimension of the Hilbert space. (The cardinality of any two bases is the same.)

In the space $\mathbb{C}^n$, the standard orthonormal basis consists of the vectors

$$\delta_1 = (1, 0, \ldots, 0), \quad \delta_2 = (0, 1, 0, \ldots, 0), \quad \ldots, \quad \delta_n = (0, 0, \ldots, 0, 1); \quad (1.3)$$

each vector has 0 coordinate $n - 1$ times and one coordinate equals 1.

Example 1.3 The space $\mathbb{M}_n$ of matrices becomes a Hilbert space with the inner product

$$\langle A, B \rangle = \operatorname{Tr} A^* B$$

which is called **Hilbert–Schmidt inner product**. The matrix units $E(ij)$ ($1 \le i, j \le n$) form an orthonormal basis.

It follows that the **Hilbert–Schmidt norm**

$$\|A\|_2 := \sqrt{\langle A, A \rangle} = \sqrt{\operatorname{Tr} A^* A} = \left(\sum_{i,j=1}^{n} |A_{ij}|^2 \right)^{1/2} \qquad (1.4)$$

is a norm for the matrices. □

Assume that in an n-dimensional Hilbert space, linearly independent vectors $v_1, v_2, \ldots, v_n$ are given. By the **Gram–Schmidt procedure** an orthonormal basis can be obtained by linear combinations:

$$e_1 := \frac{1}{\|v_1\|} v_1,$$

$$e_2 := \frac{1}{\|w_2\|} w_2 \quad \text{with} \quad w_2 := v_2 - \langle e_1, v_2 \rangle e_1,$$

$$e_3 := \frac{1}{\|w_3\|} w_3 \quad \text{with} \quad w_3 := v_3 - \langle e_1, v_3 \rangle e_1 - \langle e_2, v_3 \rangle e_2,$$

$$\vdots$$

$$e_n := \frac{1}{\|w_n\|} w_n \quad \text{with} \quad w_n := v_n - \langle e_1, v_n \rangle e_1 - \cdots - \langle e_{n-1}, v_n \rangle e_{n-1}.$$

The next theorem tells us that any vector has a unique **Fourier expansion**.

Theorem 1.4 *Let $e_1, e_2, \ldots$ be a basis in a Hilbert space $\mathcal{H}$. Then for any vector $x \in \mathcal{H}$ the expansion*

$$x = \sum_n \langle e_n, x \rangle e_n$$

holds. Moreover,

$$\|x\|^2 = \sum_n |\langle e_n, x \rangle|^2.$$

Let $\mathcal{H}$ and $\mathcal{K}$ be Hilbert spaces. A mapping $A : \mathcal{H} \to \mathcal{K}$ is called linear if it preserves linear combinations:

$$A(\lambda f + \mu g) = \lambda A f + \mu A g \quad (f, g \in \mathcal{H}, \quad \lambda, \mu \in \mathbb{C}).$$

The **kernel** and the **range** of A are

$$\ker A := \{x \in \mathcal{H} : Ax = 0\}, \quad \operatorname{ran} A := \{Ax \in \mathcal{K} : x \in \mathcal{H}\}.$$

The dimension formula familiar in linear algebra is

$$\dim \mathcal{H} = \dim(\ker A) + \dim(\operatorname{ran} A).$$

The quantity $\dim(\operatorname{ran} A)$ is called the **rank** of A and is denoted by rank A. It is easy to see that rank $A \leq \dim \mathcal{H}$, $\dim \mathcal{K}$.

Let $e_1, e_2, \ldots, e_n$ be a basis of the Hilbert space $\mathcal{H}$ and $f_1, f_2, \ldots, f_m$ be a basis of $\mathcal{K}$. The linear mapping $A : \mathcal{H} \to \mathcal{K}$ is determined by the vectors $Ae_j, j = 1, 2, \ldots, n$.

Furthermore, the vector Ae_j is determined by its coordinates:

$$Ae_j = c_{1,j}f_1 + c_{2,j}f_2 + \cdots + c_{m,j}f_m.$$

The numbers $c_{i,j}$, $1 \le i \le m$, $1 \le j \le n$, form an $m \times n$ matrix, which is called the **matrix** of the linear transformation A with respect to the bases $(e_1, e_2, \ldots, e_n)$ and $(f_1, f_2, \ldots, f_m)$. If we want to distinguish the linear operator A from its matrix, then the latter will be denoted by $[A]$. We have

$$[A]_{ij} = \langle f_i, Ae_j \rangle \qquad (1 \le i \le m, \quad 1 \le j \le n).$$

Note that the order of the basis vectors is important. We shall mostly consider linear operators of a Hilbert space into itself. Then only one basis is needed and the matrix of the operator has the form of a square. So a linear transformation and a basis yield a matrix. If an $n \times n$ matrix is given, then it can be always considered as a linear transformation of the space $\mathbb{C}^n$ endowed with the standard basis (1.3).

The inner product of the vectors $|x\rangle$ and $|y\rangle$ will often be denoted as $\langle x|y \rangle$. This notation, sometimes called **bra and ket**, is popular in physics. On the other hand, $|x\rangle\langle y|$ is a linear operator which acts on the vector $|z\rangle$ as

$$\big(|x\rangle\langle y|\big)\, |z\rangle := |x\rangle \, \langle y|z\rangle \equiv \langle y|z\rangle \, |x\rangle.$$

Therefore,

$$|x\rangle\langle y| = \begin{bmatrix} x_1 \\ x_2 \\ \cdot \\ \cdot \\ x_n \end{bmatrix} \begin{bmatrix} \bar{y}_1, \bar{y}_2, \ldots, \bar{y}_n \end{bmatrix}$$

is conjugate linear in $|y\rangle$, while $\langle x|y \rangle$ is linear in $|y\rangle$.

The next example shows the possible use of the bra and ket.

Example 1.5 If $X, Y \in \mathbb{M}_n(\mathbb{C})$, then

$$\sum_{i,j=1}^{n} \operatorname{Tr} E(ij)XE(ji)Y = (\operatorname{Tr} X)(\operatorname{Tr} Y).$$

Since both sides are bilinear in the variables X and Y, it is enough to check the case $X = E(ab)$ and $Y = E(cd)$. Simple computation gives that the left-hand side is $\delta_{ab}\delta_{cd}$ and this is the same as the right-hand side.

Another possibility is to use the formula $E(ij) = |e_i\rangle\langle e_j|$. So

$$\sum_{i,j} \operatorname{Tr} E(ij)XE(ji)Y = \sum_{i,j} \operatorname{Tr} |e_i\rangle \langle e_j|X|e_j\rangle \langle e_i|Y = \sum_{i,j} \langle e_j|X|e_j\rangle \langle e_i|Y|e_i\rangle$$

$$= \sum_j \langle e_j|X|e_j\rangle \sum_i \langle e_i|Y|e_i\rangle$$

and the last expression is $(\operatorname{Tr} X)(\operatorname{Tr} Y)$.   □

Example 1.6 Fix a natural number n and let $\mathcal{H}$ be the space of polynomials of degree at most n. Assume that the variable of these polynomials is t and the coefficients are complex numbers. The typical elements are

$$p(t) = \sum_{i=0}^{n} u_i t^i \quad \text{and} \quad q(t) = \sum_{i=0}^{n} v_i t^i.$$

If their inner product is defined as

$$\langle p(t), q(t) \rangle := \sum_{i=0}^{n} \bar{u}_i v_i,$$

then $\{1, t, t^2, \ldots, t^n\}$ is an orthonormal basis.

Differentiation is a linear operator on $\mathcal{H}$:

$$\sum_{k=0}^{n} u_k t^k \mapsto \sum_{k=1}^{n} k u_k t^{k-1}.$$

With respect to the above basis, its matrix is

$$\begin{bmatrix} 0 & 1 & 0 & \ldots & 0 & 0 \\ 0 & 0 & 2 & \ldots & 0 & 0 \\ 0 & 0 & 0 & \ldots & 0 & 0 \\ \vdots & \vdots & \vdots & \ddots & \vdots & 0 \\ 0 & 0 & 0 & \ldots & 0 & n \\ 0 & 0 & 0 & \ldots & 0 & 0 \end{bmatrix}.$$

This is an **upper triangular matrix**; the (i, j) entry is 0 if $i > j$.   □

Let $\mathcal{H}_1$, $\mathcal{H}_2$ and $\mathcal{H}_3$ be Hilbert spaces and fix a basis in each of them. If $B : \mathcal{H}_1 \to \mathcal{H}_2$ and $A : \mathcal{H}_2 \to \mathcal{H}_3$ are linear mappings, then the composition

$$f \mapsto A(Bf) \in \mathcal{H}_3 \quad (f \in \mathcal{H}_1)$$

is linear as well and it is denoted by AB. The matrix $[AB]$ of the composition AB can be computed from the matrices $[A]$ and $[B]$ as follows:

$$[AB]_{ij} = \sum_k [A]_{ik}[B]_{kj}.$$

The right-hand side is defined to be the product $[A][B]$ of the matrices $[A]$ and $[B]$, that is, $[AB] = [A][B]$ holds. It is obvious that for an $\ell \times m$ matrix $[A]$ and an $m \times n$ matrix $[B]$, their product $[A][B]$ is an $\ell \times n$ matrix.

Let $\mathcal{H}_1$ and $\mathcal{H}_2$ be Hilbert spaces and fix a basis in each of them. If $A, B : \mathcal{H}_1 \to \mathcal{H}_2$ are linear mappings, then their linear combination

$$(\lambda A + \mu B)f \mapsto \lambda(Af) + \mu(Bf)$$

is a linear mapping and

$$[\lambda A + \mu B]_{ij} = \lambda[A]_{ij} + \mu[B]_{ij}.$$

Let $\mathcal{H}$ be a Hilbert space. The linear operators $\mathcal{H} \to \mathcal{H}$ form an algebra. This algebra $B(\mathcal{H})$ has a unit, the identity operator denoted by I, and the product is non-commutative. Assume that $\mathcal{H}$ is n-dimensional and fix a basis. Then to each linear operator $A \in B(\mathcal{H})$ an $n \times n$ matrix A is associated. The correspondence $A \mapsto [A]$ is an algebraic isomorphism from $B(\mathcal{H})$ to the algebra $\mathbb{M}_n(\mathbb{C})$ of $n \times n$ matrices. This isomorphism shows that the theory of linear operators on an n-dimensional Hilbert space is the same as the theory of $n \times n$ matrices.

Theorem 1.7 (Riesz–Fischer theorem) *Let $\phi : \mathcal{H} \to \mathbb{C}$ be a linear mapping on a finite-dimensional Hilbert space $\mathcal{H}$. Then there is a unique vector $v \in \mathcal{H}$ such that $\phi(x) = \langle v, x \rangle$ for every vector $x \in \mathcal{H}$.*

Proof: Let $e_1, e_2, \ldots, e_n$ be an orthonormal basis in $\mathcal{H}$. Then we need a vector $v \in \mathcal{H}$ such that $\phi(e_i) = \langle v, e_i \rangle$. The vector

$$v = \sum_i \overline{\phi(e_i)} e_i$$

will satisfy the condition. $\square$

The linear mappings $\phi : \mathcal{H} \to \mathbb{C}$ are called functionals. If the Hilbert space is not finite-dimensional, then in the previous theorem the boundedness condition $|\phi(x)| \le c\|x\|$ should be added, where c is a positive number.

Let $\mathcal{H}$ and $\mathcal{K}$ be finite-dimensional Hilbert spaces. The **operator norm** of a linear operator $A : \mathcal{H} \to \mathcal{K}$ is defined as

$$\|A\| := \sup\{\|Ax\| : x \in \mathcal{H}, \|x\| = 1\}.$$

It can be shown that $\|A\|$ is finite. In addition to the common properties $\|A + B\| \le \|A\| + \|B\|$ and $\|\lambda A\| = |\lambda|\|A\|$, submultiplicativity

$$\|AB\| \le \|A\| \, \|B\|$$

also holds.

If $\|A\| \le 1$, then the operator A is called a **contraction**.

The set of linear operators $A : \mathcal{H} \to \mathcal{H}$ is denoted by $B(\mathcal{H})$. The convergence $A_n \to A$ means $\|A_n - A\| \to 0$ in terms of the operator norm defined above. But in the case of a finite-dimensional Hilbert space, the Hilbert–Schmidt norm can also be used. Unlike the Hilbert–Schmidt norm, the operator norm of a matrix is not expressed explicitly by the matrix entries.

Example 1.8 Let $A \in B(\mathcal{H})$ and $\|A\| < 1$. Then $I - A$ is invertible and

$$(I - A)^{-1} = \sum_{n=0}^{\infty} A^n.$$

Since

$$(I - A) \sum_{n=0}^{N} A^n = I - A^{N+1} \quad \text{and} \quad \|A^{N+1}\| \le \|A\|^{N+1},$$

we can see that the limit of the first equation is

$$(I - A) \sum_{n=0}^{\infty} A^n = I.$$

This proves the statement, the formula of which is called a **Neumann series**.   □

Let $\mathcal{H}$ and $\mathcal{K}$ be Hilbert spaces. If $T : \mathcal{H} \to \mathcal{K}$ is a linear operator, then its **adjoint** $T^* : \mathcal{K} \to \mathcal{H}$ is determined by the formula

$$\langle x, Ty \rangle_{\mathcal{K}} = \langle T^*x, y \rangle_{\mathcal{H}} \quad (x \in \mathcal{K}, y \in \mathcal{H}).$$

An operator $T \in B(\mathcal{H})$ is called **self-adjoint** if $T^* = T$. An operator T is self-adjoint if and only if $\langle x, Tx \rangle$ is a real number for every vector $x \in \mathcal{H}$. For the self-adjoint operators on $\mathcal{H}$ and the self-adjoint $n \times n$ matrices the notations $B(\mathcal{H})^{sa}$ and $\mathbb{M}_n^{sa}$ are used.

Theorem 1.9 *The properties of the adjoint are:*

(1) $(A + B)^* = A^* + B^*$, $(\lambda A)^* = \bar{\lambda} A^*$ $(\lambda \in \mathbb{C})$;
(2) $(A^*)^* = A$, $(AB)^* = B^* A^*$;
(3) $(A^{-1})^* = (A^*)^{-1}$ *if A is invertible;*
(4) $\|A\| = \|A^*\|$, $\|A^*A\| = \|A\|^2$.

Example 1.10 Let $A : \mathcal{H} \to \mathcal{H}$ be a linear operator and $e_1, e_2, \ldots, e_n$ be a basis in the Hilbert space $\mathcal{H}$. The (i, j) element of the matrix of A is $\langle e_i, Ae_j \rangle$. Since

$$\langle e_i, Ae_j \rangle = \overline{\langle e_j, A^*e_i \rangle},$$

this is the complex conjugate of the (j, i) element of the matrix of A^*.

If A is self-adjoint, then the (i, j) element of the matrix of A is the conjugate of the (j, i) element. In particular, all diagonal entries are real. The self-adjoint matrices are also called **Hermitian** matrices. □

Theorem 1.11 (Projection theorem) *Let $\mathcal{M}$ be a closed subspace of a Hilbert space $\mathcal{H}$. Any vector $x \in \mathcal{H}$ can be written in a unique way in the form $x = x_0 + y$, where $x_0 \in \mathcal{M}$ and $y \perp \mathcal{M}$.*

Note that a subspace of a finite-dimensional Hilbert space is always closed. The mapping $P : x \mapsto x_0$ defined in the context of the previous theorem is called the **orthogonal projection** onto the subspace $\mathcal{M}$. This mapping is linear:

$$P(\lambda x + \mu y) = \lambda Px + \mu Py.$$

Moreover, $P^2 = P = P^*$. The converse is also true: If $P^2 = P = P^*$, then P is an orthogonal projection (onto its range).

Example 1.12 A matrix $A \in \mathbb{M}_n$ is self-adjoint if $A_{ji} = \overline{A_{ij}}$. A particular example is a **Toeplitz matrix**:

$$\begin{bmatrix} a_1 & a_2 & a_3 & \ldots & a_{n-1} & a_n \\ \overline{a_2} & a_1 & a_2 & \ldots & a_{n-2} & a_{n-1} \\ \overline{a_3} & \overline{a_2} & a_1 & \ldots & a_{n-3} & a_{n-2} \\ \vdots & \vdots & \vdots & \ddots & \vdots & \vdots \\ \overline{a_{n-1}} & \overline{a_{n-2}} & \overline{a_{n-3}} & \ldots & a_1 & a_2 \\ \overline{a_n} & \overline{a_{n-1}} & \overline{a_{n-2}} & \ldots & \overline{a_2} & a_1 \end{bmatrix},$$

where $a_1 \in \mathbb{R}$. □

An operator $U \in B(\mathcal{H})$ is called a **unitary** if U^* is the inverse of U. Then $U^*U = I$ and

$$\langle x, y \rangle = \langle U^*Ux, y \rangle = \langle Ux, Uy \rangle$$

for any vectors $x, y \in \mathcal{H}$. Therefore the unitary operators preserve the inner product. In particular, orthogonal unit vectors are mapped by a unitary operator onto orthogonal unit vectors.

Example 1.13 The **permutation matrices** are simple unitaries. Let π be a permutation of the set $\{1, 2, \ldots, n\}$. The $A_{i,\pi(i)}$ entries of $A \in \mathbb{M}_n(\mathbb{C})$ are 1 and all others are 0. Every row and every column contain exactly one 1 entry. If such a matrix A is applied to a vector, it permutes the coordinates:

$$\begin{bmatrix} 0 & 1 & 0 \\ 0 & 0 & 1 \\ 1 & 0 & 0 \end{bmatrix} \begin{bmatrix} x_1 \\ x_2 \\ x_3 \end{bmatrix} = \begin{bmatrix} x_2 \\ x_3 \\ x_1 \end{bmatrix}.$$

This shows the reason behind the terminology. Another possible formalism is $A(x_1, x_2, x_3) = (x_2, x_3, x_1)$.   □

An operator $A \in B(\mathcal{H})$ is called **normal** if $AA^* = A^*A$. It immediately follows that

$$\|Ax\| = \|A^*x\|$$

for any vector $x \in \mathcal{H}$. Self-adjoint and unitary operators are normal.

The operators we need are mostly linear, but sometimes **conjugate-linear** operators appear. $\Lambda : \mathcal{H} \to \mathcal{K}$ is conjugate-linear if

$$\Lambda(\lambda x + \mu y) = \overline{\lambda}\, \Lambda x + \overline{\mu}\, \Lambda y$$

for any complex numbers λ and μ and for any vectors $x, y \in \mathcal{H}$. The adjoint Λ^* of a conjugate-linear operator Λ is determined by the equation

$$\langle x, \Lambda y \rangle_{\mathcal{K}} = \langle y, \Lambda^* x \rangle_{\mathcal{H}} \qquad (x \in \mathcal{K}, \ y \in \mathcal{H}). \tag{1.5}$$

A mapping $\phi : \mathcal{H} \times \mathcal{H} \to \mathbb{C}$ is called a **complex bilinear form** if ϕ is linear in the second variable and conjugate linear in the first variable. The inner product is a particular example.

Theorem 1.14 *On a finite-dimensional Hilbert space there is a one-to-one correspondence*

$$\phi(x, y) = \langle Ax, y \rangle$$

between the complex bilinear forms $\phi : \mathcal{H} \times \mathcal{H} \to \mathbb{C}$ and the linear operators $A : \mathcal{H} \to \mathcal{H}$.

Proof: Fix $x \in \mathcal{H}$. Then $y \mapsto \phi(x, y)$ is a linear functional. By the Riesz–Fischer theorem, $\phi(x, y) = \langle z, y \rangle$ for a vector $z \in \mathcal{H}$. We set $Ax = z$.   □

The **polarization identity**

$$\begin{aligned} 4\phi(x, y) = {}& \phi(x + y, x + y) + i\phi(x + iy, x + iy) \\ & - \phi(x - y, x - y) - i\phi(x - iy, x - iy) \end{aligned} \tag{1.6}$$

shows that a complex bilinear form ϕ is determined by its so-called quadratic form $x \mapsto \phi(x, x)$.

1.3 Jordan Canonical Form

A **Jordan block** is a matrix

$$
J_k(a) = \begin{bmatrix}
a & 1 & 0 & \cdots & 0 \\
0 & a & 1 & \cdots & 0 \\
0 & 0 & a & \cdots & 0 \\
\vdots & \vdots & \vdots & \ddots & \vdots \\
0 & 0 & 0 & \cdots & a
\end{bmatrix},
$$

where $a \in \mathbb{C}$. This is an upper triangular matrix $J_k(a) \in \mathbb{M}_k$. We also use the notation $J_k := J_k(0)$. Then

$$
J_k(a) = aI_k + J_k
$$

and the sum consists of commuting matrices.

Example 1.15 The matrix J_k is

$$
(J_k)_{ij} = \begin{cases} 1 & \text{if } j = i + 1, \\ 0 & \text{otherwise.} \end{cases}
$$

Therefore

$$
(J_k)_{ij}(J_k)_{jk} = \begin{cases} 1 & \text{if } j = i + 1 \text{ and } k = i + 2, \\ 0 & \text{otherwise.} \end{cases}
$$

It follows that

$$
(J_k^2)_{ij} = \begin{cases} 1 & \text{if } j = i + 2, \\ 0 & \text{otherwise.} \end{cases}
$$

We observe that when taking the powers of J_k, the line of the 1 entries moves upward, in particular $J_k^k = 0$. The matrices J_k^m ($0 \le m \le k - 1$) are linearly independent.

If $a \ne 0$, then $\det J_k(a) \ne 0$ and $J_k(a)$ is invertible. We can search for the inverse via the equation

$$
(aI_k + J_k) \left(\sum_{j=0}^{k-1} c_j J_k^j \right) = I_k.
$$

Rewriting this equation we get

$$ac_0 I_k + \sum_{j=1}^{k-1}(ac_j + c_{j-1})J_k^j = I_k.$$

The solution is

$$c_j = -(-a)^{-j-1} \qquad (0 \le j \le k-1).$$

In particular,

$$\begin{bmatrix} a & 1 & 0 \\ 0 & a & 1 \\ 0 & 0 & a \end{bmatrix}^{-1} = \begin{bmatrix} a^{-1} & -a^{-2} & a^{-3} \\ 0 & a^{-1} & -a^{-2} \\ 0 & 0 & a^{-1} \end{bmatrix}.$$

Computation with Jordan blocks is convenient.   □

The **Jordan canonical form theorem** is the following:

Theorem 1.16 *Given a matrix $X \in \mathbb{M}_n$, there is an invertible matrix $S \in \mathbb{M}_n$ such that*

$$X = S \begin{bmatrix} J_{k_1}(\lambda_1) & 0 & \cdots & 0 \\ 0 & J_{k_2}(\lambda_2) & \cdots & 0 \\ \vdots & \vdots & \ddots & \vdots \\ 0 & 0 & \cdots & J_{k_m}(\lambda_m) \end{bmatrix} S^{-1} = SJS^{-1},$$

where $k_1 + k_2 + \cdots + k_m = n$. The Jordan matrix J is uniquely determined (up to a permutation of the Jordan blocks in the diagonal).

Note that the numbers $\lambda_1, \lambda_2, \ldots, \lambda_m$ are not necessarily different. The theorem is about complex matrices. Example 1.15 showed that it is rather easy to handle a Jordan block. If the Jordan canonical decomposition is known, then the inverse can be computed.

Example 1.17 An essential application concerns the determinant. Since $\det X = \det(SJS^{-1}) = \det J$, it is enough to compute the determinant of the upper-triangular Jordan matrix J. Therefore

$$\det X = \prod_{j=1}^{m} \lambda_j^{k_j}. \tag{1.7}$$

The **characteristic polynomial** of $X \in \mathbb{M}_n$ is defined as

$$p(x) := \det(xI_n - X).$$

From the computation (1.7) we have

$$p(x) = \prod_{j=1}^{m}(x - \lambda_j)^{k_j} = x^n - \left(\sum_{j=1}^{m} k_j\lambda_j\right) x^{n-1} + \cdots + (-1)^n \prod_{j=1}^{m} \lambda_j^{k_j}. \quad (1.8)$$

The numbers λ_j are the roots of the characteristic polynomial. □

The powers of a matrix $X \in \mathbb{M}_n$ are well-defined. For a polynomial $p(x) = \sum_{k=0}^{m} c_k x^k$ the matrix $p(X)$ is

$$\sum_{k=0}^{m} c_k X^k.$$

A polynomial q is said to annihilate a matrix $X \in \mathbb{M}_n$ if $q(X) = 0$.

The next result is the **Cayley–Hamilton theorem**.

Theorem 1.18 *If p is the characteristic polynomial of $X \in \mathbb{M}_n$, then $p(X) = 0$.*

1.4 Spectrum and Eigenvalues

Let $\mathcal{H}$ be a Hilbert space. For $A \in B(\mathcal{H})$ and $\lambda \in \mathbb{C}$, we say that λ is an **eigenvalue** of A if there is a non-zero vector $v \in \mathcal{H}$ such that $Av = \lambda v$. Such a vector v is called an **eigenvector** of A for the eigenvalue λ. If $\mathcal{H}$ is finite-dimensional, then $\lambda \in \mathbb{C}$ is an eigenvalue of A if and only if $A - \lambda I$ is not invertible.

Generally, the **spectrum** $\sigma(A)$ of $A \in B(\mathcal{H})$ consists of the numbers $\lambda \in \mathbb{C}$ such that $A - \lambda I$ is not invertible. Therefore in the finite-dimensional case the spectrum is the set of eigenvalues.

Example 1.19 We show that $\sigma(AB) = \sigma(BA)$ for $A, B \in \mathbb{M}_n$. It is enough to prove that $\det(\lambda I - AB) = \det(\lambda I - BA)$. Assume first that A is invertible. We then have

$$\det(\lambda I - AB) = \det(A^{-1}(\lambda I - AB)A) = \det(\lambda I - BA)$$

and hence $\sigma(AB) = \sigma(BA)$.

When A is not invertible, choose a sequence $\varepsilon_k \in \mathbb{C} \setminus \sigma(A)$ with $\varepsilon_k \to 0$ and set $A_k := A - \varepsilon_k I$. Then

$$\det(\lambda I - AB) = \lim_{k\to\infty} \det(\lambda I - A_k B) = \lim_{k\to\infty} \det(\lambda I - BA_k) = \det(\lambda I - BA).$$

(Another argument appears in Exercise 3 of Chap. 2.)   □

Example 1.20 In the history of matrix theory the particular matrix

$$\begin{bmatrix} 0 & 1 & 0 & \ldots & 0 & 0 \\ 1 & 0 & 1 & \ldots & 0 & 0 \\ 0 & 1 & 0 & \ldots & 0 & 0 \\ \vdots & \vdots & \vdots & \ddots & \vdots & \vdots \\ 0 & 0 & 0 & \ldots & 0 & 1 \\ 0 & 0 & 0 & \ldots & 1 & 0 \end{bmatrix} \tag{1.9}$$

has importance. Its eigenvalues were computed by **Joseph Louis Lagrange** in 1759. He found that the eigenvalues are $2\cos j\pi/(n+1)$ $(j = 1, 2, \ldots, n)$.   □

The matrix (1.9) is **tridiagonal**. This means that $A_{ij} = 0$ if $|i - j| > 1$.

Example 1.21 Let $\lambda \in \mathbb{R}$ and consider the matrix

$$J_3(\lambda) = \begin{bmatrix} \lambda & 1 & 0 \\ 0 & \lambda & 1 \\ 0 & 0 & \lambda \end{bmatrix}.$$

Now λ is the only eigenvalue and $(1, 0, 0)$ is the only eigenvector up to a constant multiple. The situation is similar in the $k \times k$ generalization $J_k(\lambda)$: λ is the eigenvalue of $SJ_k(\lambda)S^{-1}$ for an arbitrary invertible S and there is one eigenvector (up to a constant multiple).

If X has the Jordan form as in Theorem 1.16, then all λ_j's are eigenvalues. Therefore the roots of the characteristic polynomial are eigenvalues. When λ is an eigenvalue of X, $\ker(X - \lambda I) = \{v \in \mathbb{C}^n : (X - \lambda I)v = 0\}$ is called the **eigenspace** of X for λ. Note that the dimension of $\ker(X - \lambda I)$ is the number of j such that $\lambda_j = \lambda$, which is called the **geometric multiplicity** of λ; on the other hand, the multiplicity of λ as a root of the characteristic polynomial is called the **algebraic multiplicity**.

For the above $J_3(\lambda)$ we can see that

$$J_3(\lambda)(0, 0, 1) = (0, 1, \lambda), \qquad J_3(\lambda)^2(0, 0, 1) = (1, 2\lambda, \lambda^2).$$

Therefore $(0, 0, 1)$ and these two vectors linearly span the whole space $\mathbb{C}^3$. The vector $(0, 0, 1)$ is called a **cyclic vector**.

Assume that a matrix $X \in \mathbb{M}_n$ has a cyclic vector $v \in \mathbb{C}^n$ which means that the set $\{v, Xv, X^2v, \ldots, X^{n-1}v\}$ spans $\mathbb{C}^n$. Then $X = SJ_n(\lambda)S^{-1}$ with some invertible matrix S, so the Jordan canonical form consists of a single block.   □

Theorem 1.22 *Assume that* $A \in B(\mathcal{H})$ *is normal. Then there exist* $\lambda_1, \ldots, \lambda_n \in \mathbb{C}$ *and* $u_1, \ldots, u_n \in \mathcal{H}$ *such that* $\{u_1, \ldots, u_n\}$ *is an orthonormal basis of* $\mathcal{H}$ *and* $Au_i = \lambda_i u_i$ *for all* $1 \le i \le n$.

Proof: Let us prove the theorem by induction on $n = \dim \mathcal{H}$. The case $n = 1$ trivially holds. Suppose the assertion holds for dimension $n-1$. Assume that $\dim \mathcal{H} = n$ and $A \in B(\mathcal{H})$ is normal. Choose a root λ_1 of $\det(\lambda I - A) = 0$. As explained before the theorem, λ_1 is an eigenvalue of A so that there is an eigenvector u_1 with $Au_1 = \lambda_1 u_1$. One may assume that u_1 is a unit vector, i.e., $\|u_1\| = 1$. Since A is normal, we have

$$
\begin{aligned}
(A - \lambda_1 I)^*(A - \lambda_1 I) &= (A^* - \overline{\lambda}_1 I)(A - \lambda_1 I) \\
&= A^*A - \overline{\lambda}_1 A - \lambda_1 A^* + \lambda_1 \overline{\lambda}_1 I \\
&= AA^* - \overline{\lambda}_1 A - \lambda_1 A^* + \lambda_1 \overline{\lambda}_1 I \\
&= (A - \lambda_1 I)(A - \lambda_1 I)^*,
\end{aligned}
$$

that is, $A - \lambda_1 I$ is also normal. Therefore,

$$
\|(A^* - \overline{\lambda}_1 I)u_1\| = \|(A - \lambda_1 I)^* u_1\| = \|(A - \lambda_1 I)u_1\| = 0
$$

so that $A^* u_1 = \overline{\lambda}_1 u_1$. Let $\mathcal{H}_1 := \{u_1\}^\perp$, the orthogonal complement of $\{u_1\}$. If $x \in \mathcal{H}_1$ then

$$
\langle Ax, u_1 \rangle = \langle x, A^* u_1 \rangle = \langle x, \overline{\lambda}_1 u_1 \rangle = \overline{\lambda}_1 \langle x, u_1 \rangle = 0,
$$
$$
\langle A^* x, u_1 \rangle = \langle x, Au_1 \rangle = \langle x, \lambda_1 u_1 \rangle = \lambda_1 \langle x, u_1 \rangle = 0
$$

so that $Ax, A^*x \in \mathcal{H}_1$. Hence we have $A\mathcal{H}_1 \subset \mathcal{H}_1$ and $A^*\mathcal{H}_1 \subset \mathcal{H}_1$. So one can define $A_1 := A|_{\mathcal{H}_1} \in B(\mathcal{H}_1)$. Then $A_1^* = A^*|_{\mathcal{H}_1}$, which implies that A_1 is also normal. Since $\dim \mathcal{H}_1 = n - 1$, the induction hypothesis can be applied to obtain $\lambda_2, \ldots, \lambda_n \in \mathbb{C}$ and $u_2, \ldots, u_n \in \mathcal{H}_1$ such that $\{u_2, \ldots, u_n\}$ is an orthonormal basis of $\mathcal{H}_1$ and $A_1 u_i = \lambda_i u_i$ for all $i = 2, \ldots, n$. Then $\{u_1, u_2, \ldots, u_n\}$ is an orthonormal basis of $\mathcal{H}$ and $Au_i = \lambda_i u_i$ for all $i = 1, 2, \ldots, n$. Thus the assertion holds for dimension n as well. $\qquad\square$

It is an important consequence that the matrix of a normal operator is diagonal with respect to an appropriate orthonormal basis and the trace is the sum of the eigenvalues.

Theorem 1.23 *Assume that $A \in B(\mathcal{H})$ is self-adjoint. If $Av = \lambda v$ and $Aw = \mu w$ with non-zero eigenvectors v, w and the eigenvalues λ and μ are different, then $v \perp w$ and $\lambda, \mu \in \mathbb{R}$.*

Proof: First we show that the eigenvalues are real:

$$
\lambda \langle v, v \rangle = \langle v, \lambda v \rangle = \langle v, Av \rangle = \langle Av, v \rangle = \langle \lambda v, v \rangle = \overline{\lambda} \langle v, v \rangle.
$$

The orthogonality $\langle v, w \rangle = 0$ comes similarly:

$$
\mu \langle v, w \rangle = \langle v, \mu w \rangle = \langle v, Aw \rangle = \langle Av, w \rangle = \langle \lambda v, w \rangle = \lambda \langle v, w \rangle. \qquad\square
$$

If A is a self-adjoint operator on an n-dimensional Hilbert space, then from the eigenvectors we can find an orthonormal basis $v_1, v_2, \ldots, v_n$. If $Av_i = \lambda_i v_i$, then

$$A = \sum_{i=1}^{n} \lambda_i |v_i\rangle\langle v_i| \qquad (1.10)$$

which is called the **Schmidt decomposition**. The Schmidt decomposition is unique if all the eigenvalues are different, otherwise not. Another useful decomposition is the **spectral decomposition**. Assume that a self-adjoint operator A has eigenvalues $\mu_1 > \mu_2 > \cdots > \mu_k$. Then

$$A = \sum_{j=1}^{k} \mu_j P_j, \qquad (1.11)$$

where P_j is the orthogonal projection onto the eigenspace for the eigenvalue μ_j. (From the Schmidt decomposition (1.10),

$$P_j = \sum_{i} |v_i\rangle\langle v_i|,$$

where the summation is over all i such that $\lambda_i = \mu_j$.) This decomposition is always unique. Actually, the Schmidt decomposition and the spectral decomposition exist for all normal operators.

If $\lambda_i \geq 0$ in (1.10), then we can set $|x_i\rangle := \sqrt{\lambda_i}|v_i\rangle$ and we have

$$A = \sum_{i=1}^{n} |x_i\rangle\langle x_i|.$$

If the orthogonality of the vectors $|x_i\rangle$ is not assumed, then there are several similar decompositions, but they are connected by a unitary. The next lemma and its proof is a good exercise for the bra and ket formalism. (The result and the proof is due to **Schrödinger** [78].)

Lemma 1.24 *If*

$$A = \sum_{j=1}^{n} |x_j\rangle\langle x_j| = \sum_{i=1}^{n} |y_i\rangle\langle y_i|,$$

then there exists a unitary matrix $[U_{ij}]_{i,j=1}^{n}$ *such that*

$$\sum_{j=1}^{n} U_{ij}|x_j\rangle = |y_i\rangle \qquad (1 \leq i \leq n). \qquad (1.12)$$

Proof: Assume first that the vectors $|x_j\rangle$ are orthogonal. Typically they are not unit vectors and several of them can be 0. Assume that $|x_1\rangle, |x_2\rangle, \ldots, |x_k\rangle$ are not 0 and $|x_{k+1}\rangle = \cdots = |x_n\rangle = 0$. Then the vectors $|y_i\rangle$ are in the linear span of $\{|x_j\rangle : 1 \le j \le k\}$. Therefore

$$|y_i\rangle = \sum_{j=1}^{k} \frac{\langle x_j|y_i\rangle}{\langle x_j|x_j\rangle} |x_j\rangle$$

is the orthogonal expansion. We can define $[U_{ij}]$ by the formula

$$U_{ij} = \frac{\langle x_j|y_i\rangle}{\langle x_j|x_j\rangle} \qquad (1 \le i \le n, 1 \le j \le k).$$

We easily compute that

$$\sum_{i=1}^{k} U_{it} U_{iu}^* = \sum_{i=1}^{k} \frac{\langle x_t|y_i\rangle}{\langle x_t|x_t\rangle} \frac{\langle y_i|x_u\rangle}{\langle x_u|x_u\rangle}$$
$$= \frac{\langle x_t|A|x_u\rangle}{\langle x_u|x_u\rangle \langle x_t|x_t\rangle} = \delta_{t,u},$$

and this relation shows that the k column vectors of the matrix $[U_{ij}]$ are orthonormal. If $k < n$, then we can append further columns to get an $n \times n$ unitary, see Exercise 37. (One can see in (1.12) that if $|x_j\rangle = 0$, then U_{ij} does not play any role.)

In the general case

$$A = \sum_{j=1}^{n} |z_j\rangle\langle z_j| = \sum_{i=1}^{n} |y_i\rangle\langle y_i|,$$

we can find a unitary U from an orthogonal family to the $|y_i\rangle$'s and a unitary V from the same orthogonal family to the $|z_i\rangle$'s. Then UV^* maps from the $|z_i\rangle$'s to the $|y_i\rangle$'s. □

Example 1.25 Let $A \in B(\mathcal{H})$ be a self-adjoint operator with eigenvalues $\lambda_1 \ge \lambda_2 \ge \cdots \ge \lambda_n$ (counted with multiplicity). Then

$$\lambda_1 = \max\{\langle v, Av\rangle : v \in \mathcal{H}, \|v\| = 1\}. \tag{1.13}$$

We can take the Schmidt decomposition (1.10). Assume that

$$\max\{\langle v, Av\rangle : v \in \mathcal{H}, \|v\| = 1\} = \langle w, Aw\rangle$$

for a unit vector w. This vector has the expansion

$$w = \sum_{i=1}^{n} c_i |v_i\rangle$$

and we have

$$\langle w, Aw \rangle = \sum_{i=1}^{n} |c_i|^2 \lambda_i \leq \lambda_1.$$

Equality holds if and only if $\lambda_i < \lambda_1$ implies $c_i = 0$. The maximizer should be an eigenvector for the eigenvalue λ_1.

Similarly,

$$\lambda_n = \min\{\langle v, Av \rangle : v \in \mathcal{H}, \|v\| = 1\}. \tag{1.14}$$

The formulas (1.13) and (1.14) will be extended below.   □

Theorem 1.26 (Poincaré's inequality) *Let $A \in B(\mathcal{H})$ be a self-adjoint operator with eigenvalues $\lambda_1 \geq \lambda_2 \geq \cdots \geq \lambda_n$ (counted with multiplicity) and let $\mathcal{K}$ be a k-dimensional subspace of $\mathcal{H}$. Then there are unit vectors $x, y \in \mathcal{K}$ such that*

$$\langle x, Ax \rangle \leq \lambda_k \quad and \quad \langle y, Ay \rangle \geq \lambda_k.$$

Proof: Let $v_k, \ldots, v_n$ be orthonormal eigenvectors corresponding to the eigenvalues $\lambda_k, \ldots, \lambda_n$. They span a subspace $\mathcal{M}$ of dimension $n - k + 1$ which must have intersection with $\mathcal{K}$. Take a unit vector $x \in \mathcal{K} \cap \mathcal{M}$ which has the expansion

$$x = \sum_{i=k}^{n} c_i v_i.$$

This vector x has the required property:

$$\langle x, Ax \rangle = \sum_{i=k}^{n} |c_i|^2 \lambda_i \leq \lambda_k \sum_{i=k}^{n} |c_i|^2 = \lambda_k.$$

To find the other vector y, the same argument can be used with the matrix $-A$. □

The next result is a **minimax principle**.

Theorem 1.27 *Let $A \in B(\mathcal{H})$ be a self-adjoint operator with eigenvalues $\lambda_1 \geq \lambda_2 \geq \cdots \geq \lambda_n$ (counted with multiplicity). Then*

$$\lambda_k = \min\left\{\max\{\langle v, Av \rangle : v \in \mathcal{K}, \|v\| = 1\} : \mathcal{K} \subset \mathcal{H}, \dim \mathcal{K} = n + 1 - k\right\}.$$

Proof: Let $v_k, \ldots, v_n$ be orthonormal eigenvectors corresponding to the eigenvalues $\lambda_k, \ldots, \lambda_n$. They span a subspace $\mathcal{K}$ of dimension $n + 1 - k$. According to
(1.13) we have

$$\lambda_k = \max\{\langle v, Av \rangle : v \in \mathcal{K}\}$$

and it follows that in the statement of the theorem $\geq$ is true.

To complete the proof we have to show that for any subspace $\mathcal{K}$ of dimension
$n + 1 - k$ there is a unit vector v such that $\lambda_k \leq \langle v, Av \rangle$, or $-\lambda_k \geq \langle v, (-A)v \rangle$.
The decreasing eigenvalues of $-A$ are $-\lambda_n \geq -\lambda_{n-1} \geq \cdots \geq -\lambda_1$ where the ℓth
is $-\lambda_{n+1-\ell}$. The existence of a unit vector v is guaranteed by Poincaré's inequality,
where we take $\ell = n + 1 - k$.   □

1.5 Trace and Determinant

When $\{e_1, \ldots, e_n\}$ is an orthonormal basis of $\mathcal{H}$, the **trace** $\operatorname{Tr} A$ of $A \in B(\mathcal{H})$ is
defined as

$$\operatorname{Tr} A := \sum_{i=1}^{n} \langle e_i, Ae_i \rangle. \tag{1.15}$$

Theorem 1.28 *The definition* (1.15) *is independent of the choice of an orthonormal
basis* $\{e_1, \ldots, e_n\}$ *and* $\operatorname{Tr} AB = \operatorname{Tr} BA$ *for all* $A, B \in B(\mathcal{H})$.

Proof: We have

$$\operatorname{Tr} AB = \sum_{i=1}^{n} \langle e_i, ABe_i \rangle = \sum_{i=1}^{n} \langle A^* e_i, Be_i \rangle = \sum_{i=1}^{n} \sum_{j=1}^{n} \overline{\langle e_j, A^* e_i \rangle} \langle e_j, Be_i \rangle$$

$$= \sum_{j=1}^{n} \sum_{i=1}^{n} \overline{\langle e_i, B^* e_j \rangle} \langle e_i, Ae_j \rangle = \sum_{j=1}^{n} \langle e_j, BAe_j \rangle = \operatorname{Tr} BA.$$

Now let $\{f_1, \ldots, f_n\}$ be another orthonormal basis of $\mathcal{H}$. Then a unitary U is
defined by $Ue_i = f_i$, $1 \leq i \leq n$, and we have

$$\sum_{i=1}^{n} \langle f_i, Af_i \rangle = \sum_{i=1}^{n} \langle Ue_i, AUe_i \rangle = \operatorname{Tr} U^* AU = \operatorname{Tr} AUU^* = \operatorname{Tr} A,$$

which says that the definition of $\operatorname{Tr} A$ is actually independent of the choice of an
orthonormal basis.   □

When $A \in \mathbb{M}_n$, the trace of A is nothing but the sum of the principal diagonal entries of A:

$$\mathrm{Tr}\, A = A_{11} + A_{22} + \cdots + A_{nn}.$$

The trace is the sum of the eigenvalues.

Computation of the trace is very simple, however the case of the determinant (1.1) is very different. In terms of the Jordan canonical form described in Theorem 1.16, we have

$$\mathrm{Tr}\, X = \sum_{j=1}^{m} k_j \lambda_j \quad \text{and} \quad \det X = \prod_{j=1}^{m} \lambda_j^{k_j}.$$

Formula (1.8) shows that trace and determinant are certain coefficients of the characteristic polynomial.

The next example concerns the determinant of a special linear mapping.

Example 1.29 On the linear space $\mathbb{M}_n$ we can define a linear mapping $\alpha : \mathbb{M}_n \to \mathbb{M}_n$ as $\alpha(A) = VAV^*$, where $V \in \mathbb{M}_n$ is a fixed matrix. We are interested in $\det \alpha$.

Let $V = SJS^{-1}$ be the canonical Jordan decomposition and set

$$\alpha_1(A) = S^{-1}A(S^{-1})^*, \qquad \alpha_2(B) = JBJ^*, \qquad \alpha_3(C) = SCS^*.$$

Then $\alpha = \alpha_3 \circ \alpha_2 \circ \alpha_1$ and $\det \alpha = \det \alpha_3 \times \det \alpha_2 \times \det \alpha_1$. Since $\alpha_1 = \alpha_3^{-1}$, we have $\det \alpha = \det \alpha_2$, so only the Jordan block part has influence on the determinant.

The following example helps to understand the situation. Let

$$J = \begin{bmatrix} \lambda_1 & x \\ 0 & \lambda_2 \end{bmatrix}$$

and

$$A_1 = \begin{bmatrix} 1 & 0 \\ 0 & 0 \end{bmatrix}, \quad A_2 = \begin{bmatrix} 0 & 1 \\ 0 & 0 \end{bmatrix}, \quad A_3 = \begin{bmatrix} 0 & 0 \\ 1 & 0 \end{bmatrix}, \quad A_4 = \begin{bmatrix} 0 & 0 \\ 0 & 1 \end{bmatrix}.$$

Then $\{A_1, A_2, A_3, A_4\}$ is a basis in $\mathbb{M}_2$. If $\alpha(A) = JAJ^*$, then from the data

$$\alpha(A_1) = \lambda_1 \overline{\lambda_1} A_1, \qquad \alpha(A_2) = \lambda_1 \overline{x} A_1 + \lambda_1 \overline{\lambda_2} A_2,$$

$$\alpha(A_3) = \overline{\lambda_1} x A_1 + \overline{\lambda_1} \lambda_2 A_3, \qquad \alpha(A_4) = x\overline{x} A_1 + \overline{\lambda_2} x A_2 + \lambda_2 \overline{x} A_3 + \lambda_2 \overline{\lambda_2} A_4$$

we can observe that the matrix of α is upper triangular:

$$\begin{bmatrix} \lambda_1\overline{\lambda_1} & \lambda_1\overline{x} & \overline{\lambda_1}x & x\overline{x} \\ 0 & \lambda_1\lambda_2 & 0 & \overline{\lambda_2}x \\ 0 & 0 & \overline{\lambda_1}\lambda_2 & \lambda_2\overline{x} \\ 0 & 0 & 0 & \lambda_2\overline{\lambda_2} \end{bmatrix}.$$

So its determinant is the product of the diagonal entries:

$$\lambda_1\overline{\lambda_1} \cdot \lambda_1\overline{\lambda_2} \cdot \overline{\lambda_1}\lambda_2 \cdot \lambda_2\overline{\lambda_2} = |\lambda_1\lambda_2|^4 = |\det J|^4.$$

Now let $J \in \mathbb{M}_n$ and assume that only the entries J_{ii} and $J_{i,i+1}$ can be non-zero. In $\mathbb{M}_n$ we choose the basis of matrix units,

$$E(1, 1), E(1, 2), \ldots, E(1, n), E(2, 1), \ldots, E(2, n), \ldots, E(3, 1), \ldots, E(n, n).$$

We want to show that the matrix of α is upper triangular.

From the computation

$$JE(j, k)J^* = J_{j-1,j}\overline{J_{k-1,k}}\, E(j - 1, k - 1) + J_{j-1,j}\overline{J_{k,k}}\, E(j - 1, k)$$

$$+ J_{jj}\overline{J_{k-1,k}}\, E(j, k - 1) + J_{jj}\overline{J_{k,k}}\, E(j, k)$$

we can see that the matrix of the mapping $A \mapsto JAJ^*$ is upper triangular. (In the lexicographical order of the matrix units $E(j - 1, k - 1), E(j - 1, k), E(j, k - 1)$ precede $E(j, k)$.) The determinant is the product of the diagonal entries:

$$\prod_{j,k=1}^{n} J_{jj}\overline{J_{kk}} = \prod_{k=1}^{m} (\det J)\overline{J_{kk}}^{n} = (\det J)^n \overline{\det J}^n.$$

Hence the determinant of $\alpha(A) = VAV^*$ is $(\det V)^n \overline{\det V}^n = |\det V|^{2n}$, since the determinant of V is equal to the determinant of its Jordan block J. If $\beta(A) = VAV^t$, then the argument is similar, $\det \beta = (\det V)^{2n}$, thus only the conjugate is missing.

Next we consider the space $\mathcal{M}$ of real symmetric $n \times n$ matrices. Let V be a real matrix and set $\gamma : \mathcal{M} \to \mathcal{M}$, $\gamma(A) = VAV^t$. Then γ is a symmetric operator on the real Hilbert space $\mathcal{M}$. The Jordan blocks in Theorem 1.16 for the real V are generally non-real. However, the real form of the Jordan canonical decomposition holds in such a way that there is a real invertible matrix S such that

$$V = SJS^{-1}, \qquad J = \begin{bmatrix} J_1 & 0 & \cdots & 0 \\ 0 & J_2 & \cdots & 0 \\ \vdots & \vdots & \ddots & \vdots \\ 0 & 0 & \cdots & J_m \end{bmatrix}$$

and each block J_i is either the Jordan block $J_k(\lambda)$ with real λ or a matrix of the form

$$
\begin{bmatrix}
C & I & 0 & \cdots & 0 \\
0 & C & I & \cdots & 0 \\
\vdots & \vdots & \ddots & \ddots & \vdots \\
\vdots & \vdots & & \ddots & I \\
0 & 0 & \cdots & \cdots & C
\end{bmatrix}, \quad
C = \begin{bmatrix} a & b \\ -b & a \end{bmatrix} \text{ with real } a, b, \quad
I = \begin{bmatrix} 1 & 0 \\ 0 & 1 \end{bmatrix}.
$$

Similarly to the above argument, $\det \gamma$ is equal to the determinant of $X \mapsto JXJ^t$. Since the computation by using the above real Jordan decomposition and a basis $\{E(j,k) + E(k,j) : 1 \leq j \leq k \leq n\}$ in $\mathcal{M}$ is rather complicated, we shall be satisfied with the computation for the special case:

$$
J = \begin{bmatrix}
\alpha & \beta & 0 & 0 \\
-\beta & \alpha & 0 & 0 \\
0 & 0 & \lambda & 1 \\
0 & 0 & 0 & \lambda
\end{bmatrix}.
$$

For a 4×4 real matrix

$$
X = \begin{bmatrix} X_{11} & X_{12} \\ X_{12}^t & X_{22} \end{bmatrix}
$$

with

$$
X_{11} = \begin{bmatrix} x_{11} & x_{12} \\ x_{12} & x_{22} \end{bmatrix}, \quad
X_{12} = \begin{bmatrix} x_{13} & x_{14} \\ x_{23} & x_{24} \end{bmatrix}, \quad
X_{22} = \begin{bmatrix} x_{33} & x_{34} \\ x_{34} & x_{44} \end{bmatrix},
$$

a direct computation gives

$$
JXJ^t = \begin{bmatrix} Y_{11} & Y_{12} \\ Y_{12}^t & Y_{22} \end{bmatrix}
$$

with

$$
Y_{11} = \begin{bmatrix}
\alpha^2 x_{11} + 2\alpha\beta x_{12} + \beta^2 x_{22} & -\alpha\beta x_{11} + (\alpha^2 - \beta^2)x_{12} + \alpha\beta x_{22} \\
-\alpha\beta x_{11} + (\alpha^2 - \beta^2)x_{12} + \alpha\beta x_{22} & \beta^2 x_{11} - 2\alpha\beta x_{12} + \alpha^2 x_{22}
\end{bmatrix},
$$

$$
Y_{12} = \begin{bmatrix}
\alpha\lambda x_{13} + \alpha x_{14} + \beta\lambda x_{23} + \beta x_{24} & \alpha\lambda x_{14} + \beta\lambda x_{24} \\
-\beta\lambda x_{13} - \beta x_{14} + \alpha\lambda x_{23} + \alpha x_{24} & -\beta\lambda x_{14} + \alpha\lambda x_{24}
\end{bmatrix},
$$

$$
Y_{22} = \begin{bmatrix}
\lambda^2 x_{33} + 2\lambda x_{34} + x_{44} & \lambda^2 x_{34} + \lambda x_{44} \\
\lambda^2 x_{34} + \lambda x_{44} & \lambda^2 x_{44}
\end{bmatrix}.
$$

Therefore, the matrix of $X \mapsto JXJ^t$ is the direct sum of

$$
\begin{bmatrix} \alpha^2 & 2\alpha\beta & \beta^2 \\ -\alpha\beta & \alpha^2 - \beta^2 & \alpha\beta \\ \beta^2 & -2\alpha\beta & \alpha^2 \end{bmatrix},
\quad
\begin{bmatrix} \alpha\lambda & \alpha & \beta\lambda & \beta \\ 0 & \alpha\lambda & 0 & \beta\lambda \\ -\beta\lambda & -\beta & \alpha\lambda & \lambda \\ 0 & -\beta\lambda & 0 & \alpha\lambda \end{bmatrix},
\quad
\begin{bmatrix} \lambda^2 & 2\lambda & 1 \\ 0 & \lambda^2 & \lambda \\ 0 & 0 & \lambda^2 \end{bmatrix}.
$$

The determinant can be computed as the product of the determinants of the above three matrices and it is

$$
(\alpha^2 + \beta^2)^5 \lambda^{10} = (\det J)^5 = (\det V)^5.
$$

For a general $n \times n$ real V we have $\det \gamma = (\det V)^{n+1}$.   □

Theorem 1.30 *The determinant of a positive matrix $A \in \mathbb{M}_n$ does not exceed the product of the diagonal entries:*

$$
\det A \le \prod_{i=1}^{n} A_{ii}.
$$

This is a consequence of the concavity of the log function, see Example 4.18 (or Example 1.43).

If $A \in \mathbb{M}_n$ and $1 \le i, j \le n$, then in the following theorems $[A]^{ij}$ denotes the $(n-1) \times (n-1)$ matrix which is obtained from A by striking out the ith row and the jth column.

Theorem 1.31 *Let $A \in \mathbb{M}_n$ and $1 \le j \le n$. Then*

$$
\det A = \sum_{i=1}^{n} (-1)^{i+j} A_{ij} \det([A]^{ij}).
$$

Example 1.32 Here is a simple computation using the row version of the previous theorem.

$$
\det \begin{bmatrix} 1 & 2 & 0 \\ 3 & 0 & 4 \\ 0 & 5 & 6 \end{bmatrix} = 1 \cdot (0 \cdot 6 - 5 \cdot 4) - 2 \cdot (3 \cdot 6 - 0 \cdot 4) + 0 \cdot (3 \cdot 5 - 0 \cdot 0).
$$

This theorem is useful if the matrix has several 0 entries.   □

The determinant has an important role in the computation of the **inverse**.

Theorem 1.33 *Let $A \in \mathbb{M}_n$ be invertible. Then*

$$[A^{-1}]_{ki} = (-1)^{i+k} \frac{\det([A]^{ik})}{\det A}$$

for $1 \leq i, k \leq n.$

Example 1.34 A standard formula is

$$\begin{bmatrix} a & b \\ c & d \end{bmatrix}^{-1} = \frac{1}{ad - bc} \begin{bmatrix} d & -b \\ -c & a \end{bmatrix}$$

when the determinant $ad - bc$ is not 0.   □

The next example concerns the Haar measure on some group of matrices. Mathematical analysis is essential here.

Example 1.35 $\mathcal{G}$ denotes the set of invertible real 2×2 matrices. $\mathcal{G}$ is a (noncommutative) group and $\mathcal{G} \subset M_2(\mathbb{R}) \cong \mathbb{R}^4$ is an open set. Therefore it is a locally compact topological group.

The **Haar measure** μ is defined by the left-invariance property:

$$\mu(H) = \mu(\{BA : A \in H\}) \qquad (B \in G)$$

($H \subset \mathcal{G}$ is measurable). We assume that

$$\mu(H) = \int_H p(A) \, dA,$$

where $p : \mathcal{G} \to \mathbb{R}^+$ is a function and dA is the Lebesgue measure in $\mathbb{R}^4$:

$$A = \begin{bmatrix} x & y \\ z & w \end{bmatrix}, \qquad dA = dx \, dy \, dz \, dw.$$

The left-invariance is equivalent to the condition

$$\int f(A) p(A) \, dA = \int f(BA) p(A) \, dA$$

for all continuous functions $f : \mathcal{G} \to \mathbb{R}$ and for every $B \in \mathcal{G}$. The integral can be changed:

$$\int f(BA) p(A) \, dA = \int f(A') p(B^{-1}A') \left| \frac{\partial A}{\partial A'} \right| dA',$$

where BA is replaced with A'. If

$$B = \begin{bmatrix} a & b \\ c & d \end{bmatrix}$$

then

$$A' := BA = \begin{bmatrix} ax + bz & ay + bw \\ cx + dz & cy + dw \end{bmatrix}$$

and the Jacobi matrix is

$$\frac{\partial A'}{\partial A} = \begin{bmatrix} a & 0 & b & 0 \\ 0 & a & 0 & b \\ c & 0 & d & 0 \\ 0 & c & 0 & d \end{bmatrix} = B \otimes I_2 .$$

We have

$$\left| \frac{\partial A}{\partial A'} \right| := \left| \det \left[\frac{\partial A}{\partial A'} \right] \right| = \frac{1}{|\det(B \otimes I_2)|} = \frac{1}{(\det B)^2}$$

and

$$\int f(A)p(A) \, dA = \int f(A) \frac{p(B^{-1}A)}{(\det B)^2} \, dA.$$

So the condition for the invariance of the measure is

$$p(A) = \frac{p(B^{-1}A)}{(\det B)^2} .$$

The solution is

$$p(A) = \frac{1}{(\det A)^2} .$$

This defines the left invariant Haar measure, but it is actually also right invariant. For $n \times n$ matrices the computation is similar; then

$$p(A) = \frac{1}{(\det A)^n} .$$

(Another example appears in Exercise 61.)   □

1.6 Positivity and Absolute Value

Let $\mathcal{H}$ be a Hilbert space and $T : \mathcal{H} \to \mathcal{H}$ be a bounded linear operator. T is called a **positive operator** (or a positive semidefinite matrix) if $\langle x, Tx \rangle \geq 0$ for every vector $x \in \mathcal{H}$, denoted $T \geq 0$. It follows from the definition that a positive operator is self-adjoint. Moreover, if T_1 and T_2 are positive operators, then $T_1 + T_2$ is positive as well.

Theorem 1.36 *Let $T \in B(\mathcal{H})$ be an operator. The following conditions are equivalent.*

(1) *T is positive.*
(2) *$T = T^*$ and the spectrum of T lies in $\mathbb{R}^+ = [0, \infty)$.*
(3) *T is of the form A^*A for some operator $A \in B(\mathcal{H})$.*

An operator T is positive if and only if UTU^* is positive for a unitary U.

We can reformulate positivity for a matrix $T \in \mathbb{M}_n$. For $(a_1, a_2, \ldots, a_n) \in \mathbb{C}^n$ the inequality

$$\sum_i \sum_j \overline{a_i} T_{ij} a_j \geq 0 \tag{1.16}$$

should be true. It is easy to see that if $T \geq 0$, then $T_{ii} \geq 0$ for all $1 \leq i \leq n$. For a special unitary U the matrix UTU^* can be diagonal $\mathrm{Diag}(\lambda_1, \lambda_2, \ldots, \lambda_n)$ where the λ_i's are the eigenvalues. So the positivity of T means that it is Hermitian and the eigenvalues are positive (condition (2) above).

Example 1.37 If the matrix

$$A = \begin{bmatrix} a & b & c \\ \overline{b} & d & e \\ \overline{c} & \overline{e} & f \end{bmatrix}$$

is positive, then the matrices

$$B = \begin{bmatrix} a & b \\ \overline{b} & d \end{bmatrix}, \quad C = \begin{bmatrix} a & c \\ \overline{c} & f \end{bmatrix}$$

are positive as well. (We take the positivity condition (1.16) for A and the choice $a_3 = 0$ gives the positivity of B. A similar argument with $a_2 = 0$ shows that C is positive.) $\qquad\square$

Theorem 1.38 *Let T be a positive operator. Then there is a unique positive operator B such that $B^2 = T$. If a self-adjoint operator A commutes with T, then it commutes with B as well.*

Proof: We restrict ourselves to the finite-dimensional case. In this case it is enough to find the eigenvalues and the eigenvectors. If $Bx = \lambda x$, then x is an eigenvector of T with eigenvalue λ^2. This determines B uniquely; T and B have the same eigenvectors.

$AB = BA$ holds if for any eigenvector x of B the vector Ax is an eigenvector of B, too. If $TA = AT$, then this follows. $\qquad\square$

B is called the **square root** of T and is denoted by $T^{1/2}$ or $\sqrt{T}$. It follows from the theorem that the product of commuting positive operators T and A is positive. Indeed,

$$TA = T^{1/2}T^{1/2}A = T^{1/2}AT^{1/2} = (A^{1/2}T^{1/2})^*A^{1/2}T^{1/2}.$$

For each $A \in B(\mathcal{H})$, we have $A^*A \geq 0$. We define the **absolute value** of A to be $|A| := (A^*A)^{1/2}$. The mapping

$$|A|x \mapsto Ax$$

is norm preserving:

$$\| |A|x\|^2 = \langle|A|x, |A|x\rangle = \langle x, |A|^2x\rangle = \langle x, A^*Ax\rangle = \langle Ax, Ax\rangle = \|Ax\|^2.$$

This mapping can be extended to a unitary U. So $A = U|A|$ and this is called the **polar decomposition** of A.

$|A| := (A^*A)^{1/2}$ makes sense if $A : \mathcal{H}_1 \to \mathcal{H}_2$. Then $|A| \in B(\mathcal{H}_1)$. The above argument tells us that $|A|x \mapsto Ax$ is norm preserving, but it is not always true that it can be extended to a unitary. If $\dim \mathcal{H}_1 \leq \dim \mathcal{H}_2$, then $|A|x \mapsto Ax$ can be extended to an isometry $V : \mathcal{H}_1 \to \mathcal{H}_2$. Then $A = V|A|$, where $V^*V = I$.

The eigenvalues $s_i(A)$ of $|A|$ are called the **singular values** of A. If $A \in \mathbb{M}_n$, then the usual notation is

$$s(A) = (s_1(A), \ldots, s_n(A)), \qquad s_1(A) \geq s_2(A) \geq \cdots \geq s_n(A).$$

Example 1.39 Let T be a positive operator acting on a finite-dimensional Hilbert space such that $\|T\| \leq 1$. We want to show that there is a unitary operator U such that

$$T = \frac{1}{2}(U + U^*).$$

We can choose an orthonormal basis $e_1, e_2, \ldots, e_n$ consisting of eigenvectors of T and with respect to this basis the matrix of T is diagonal, say, $\text{Diag}(t_1, t_2, \ldots, t_n)$, $0 \leq t_j \leq 1$ from the positivity. For any $1 \leq j \leq n$ we can find a real number θ_j such that

$$t_j = \frac{1}{2}(e^{i\theta_j} + e^{-i\theta_j}).$$

Then the unitary operator U with matrix $\mathrm{Diag}(\exp(i\theta_1), \ldots, \exp(i\theta_n))$ will have the desired property.   □

If T acts on a finite-dimensional Hilbert space which has an orthonormal basis $e_1, e_2, \ldots, e_n$, then T is uniquely determined by its matrix

$$[\langle e_i, T e_j \rangle]_{i,j=1}^n.$$

T is positive if and only if its matrix is positive (semidefinite).

Example 1.40 Let

$$A = \begin{bmatrix} \lambda_1 & \lambda_2 & \ldots & \lambda_n \\ 0 & 0 & \ldots & 0 \\ \vdots & \vdots & \ddots & \vdots \\ 0 & 0 & \ldots & 0 \end{bmatrix}.$$

Then

$$[A^*A]_{i,j} = \bar{\lambda}_i \lambda_j \qquad (1 \le i, j \le n)$$

and this matrix is positive:

$$\sum_i \sum_j \bar{a}_i [A^*A]_{i,j} a_j = \sum_i a_i \lambda_i \sum_j a_j \lambda_j \ge 0.$$

Every positive matrix is the sum of matrices of this form. (The minimum number of the summands is the rank of the matrix.)   □

Example 1.41 Take numbers $\lambda_1, \lambda_2, \ldots, \lambda_n > 0$ and define the matrix A by

$$A_{ij} = \frac{1}{\lambda_i + \lambda_j}.$$

(A is called a **Cauchy matrix**.) We have

$$\frac{1}{\lambda_i + \lambda_j} = \int_0^\infty e^{-t\lambda_i} e^{-t\lambda_j} \, dt$$

and the matrix

$$A(t)_{ij} := e^{-t\lambda_i} e^{-t\lambda_j}$$

is positive for every $t \in \mathbb{R}$ by Example 1.40. Therefore

$$A = \int_0^\infty A(t)\, dt$$

is positive as well.

The above argument can be generalized. If $r > 0$, then

$$\frac{1}{(\lambda_i + \lambda_j)^r} = \frac{1}{\Gamma(r)} \int_0^\infty e^{-t\lambda_i} e^{-t\lambda_j} t^{r-1}\, dt.$$

This implies that

$$A_{ij} = \frac{1}{(\lambda_i + \lambda_j)^r} \qquad (r > 0)$$

is positive. □

The Cauchy matrix is an example of an **infinitely divisible matrix**. If A is an entrywise positive matrix, then it is called infinitely divisible if the matrices

$$A(r)_{ij} = (A_{ij})^r$$

are positive for every number $r > 0$.

Theorem 1.42 *Let $T \in B(\mathcal{H})$ be an invertible self-adjoint operator and e_1, $e_2, \ldots, e_n$ be a basis in the Hilbert space $\mathcal{H}$. T is positive if and only if for any $1 \le k \le n$ the determinant of the $k \times k$ matrix*

$$[\langle e_i, Te_j \rangle]_{ij=1}^k$$

is positive (that is, ≥ 0).

An invertible positive matrix is called **positive definite**. Such matrices appear in probability theory in the concept of a **Gaussian distribution**. The work with Gaussian distributions in probability theory requires experience with matrices. (This is described in the next example, but also in Example 2.7.)

Example 1.43 Let M be a positive definite $n \times n$ real matrix and $\mathbf{x} = (x_1, x_2, \ldots, x_n)$. Then

$$f_M(\mathbf{x}) := \sqrt{\frac{\det M}{(2\pi)^n}} \exp\left(-\tfrac{1}{2}\langle \mathbf{x}, M\mathbf{x} \rangle\right) \qquad (1.17)$$

is a multivariate Gaussian probability distribution (with 0 expectation, see, for example, III.6 in [37]). The matrix M will be called the **quadratic matrix** of the Gaussian distribution.

For an $n \times n$ matrix B, the relation

$$\int \langle \mathbf{x}, B\mathbf{x} \rangle f_M(\mathbf{x})\, d\mathbf{x} = \mathrm{Tr}\, BM^{-1} \tag{1.18}$$

holds.

We first note that if (1.18) is true for a matrix M, then

$$\int \langle \mathbf{x}, B\mathbf{x} \rangle f_{U^*MU}(\mathbf{x})\, d\mathbf{x} = \int \langle U^*\mathbf{x}, BU^*\mathbf{x} \rangle f_M(\mathbf{x})\, d\mathbf{x}$$
$$= \mathrm{Tr}\,(UBU^*)M^{-1}$$
$$= \mathrm{Tr}\, B(U^*MU)^{-1}$$

for a unitary U, since the Lebesgue measure on $\mathbb{R}^n$ is invariant under unitary transformation. This means that (1.18) also holds for U^*MU. Therefore to check (1.18), we may assume that M is diagonal. Another reduction concerns B, we may assume that B is a matrix unit E_{ij}. Then the n variable integral reduces to integrals on $\mathbb{R}$ and the known integrals

$$\int_{\mathbb{R}} t \exp\left(-\frac{1}{2}\lambda t^2\right) dt = 0 \quad \text{and} \quad \int_{\mathbb{R}} t^2 \exp\left(-\frac{1}{2}\lambda t^2\right) dt = \frac{\sqrt{2\pi}}{\lambda}$$

can be used.

Formula (1.18) has an important consequence. When the joint distribution of the random variables $(\xi_1, \xi_2, \ldots, \xi_n)$ is given by (1.17), then the **covariance matrix** is M^{-1}.

The **Boltzmann entropy** of a probability density $f(\mathbf{x})$ is defined as

$$h(f) := -\int f(\mathbf{x}) \log f(\mathbf{x})\, d\mathbf{x}$$

if the integral exists. For a Gaussian f_M we have

$$h(f_M) = \frac{n}{2} \log(2\pi e) - \frac{1}{2} \log \det M.$$

Assume that f_M is the joint distribution of (real-valued) random variables $\xi_1, \xi_2, \ldots, \xi_n$. Their joint Boltzmann entropy is

$$h(\xi_1, \xi_2, \ldots, \xi_n) = \frac{n}{2} \log(2\pi e) + \log \det M^{-1}$$

and the Boltzmann entropy of ξ_i is

$$h(\xi_i) = \frac{1}{2}\log(2\pi e) + \frac{1}{2}\log(M^{-1})_{ii}.$$

The subadditivity of the Boltzmann entropy is the inequality

$$h(\xi_1, \xi_2, \ldots, \xi_n) \le h(\xi_1) + h(\xi_2) + \cdots + h(\xi_n),$$

which is

$$\log \det A \le \sum_{i=1}^{n} \log A_{ii}$$

in our particular Gaussian case, $A = M^{-1}$. What we have obtained is the **Hadamard inequality**

$$\det A \le \prod_{i=1}^{n} A_{ii}$$

for a positive definite matrix A, see Theorem 1.30. □

Example 1.44 If the matrix $X \in \mathbb{M}_n$ can be written in the form

$$X = S\mathrm{Diag}(\lambda_1, \lambda_2, \ldots, \lambda_n)S^{-1},$$

with $\lambda_1, \lambda_2, \ldots, \lambda_n > 0$, then X is called **weakly positive**. Such a matrix has n linearly independent eigenvectors with strictly positive eigenvalues. If the eigenvectors are orthogonal, then the matrix is positive definite. Since X has the form

$$\left(S\mathrm{Diag}(\sqrt{\lambda_1}, \sqrt{\lambda_2}, \ldots, \sqrt{\lambda_n})S^*\right)\left((S^*)^{-1}\mathrm{Diag}(\sqrt{\lambda_1}, \sqrt{\lambda_2}, \ldots, \sqrt{\lambda_n})S^{-1}\right),$$

it is the product of two positive definite matrices.

Although this X is not positive, the eigenvalues are strictly positive. Therefore we can define the square root as

$$X^{1/2} = S\mathrm{Diag}(\sqrt{\lambda_1}, \sqrt{\lambda_2}, \ldots, \sqrt{\lambda_n})S^{-1}.$$

(See also Example 3.16.) □

The next result is called the **Wielandt inequality**. In the proof the operator norm will be used.

Theorem 1.45 *Let A be a self-adjoint operator such that for some numbers $a, b > 0$ the inequalities $aI \ge A \ge bI$ hold. Then for orthogonal unit vectors x and y the inequality*

$$|\langle x, Ay \rangle|^2 \leq \left(\frac{a-b}{a+b} \right)^2 \langle x, Ax \rangle \langle y, Ay \rangle$$

holds.

Proof: The conditions imply that A is a positive invertible operator. The following argument holds for any real number α:

$$\langle x, Ay \rangle = \langle x, Ay \rangle - \alpha \langle x, y \rangle = \langle x, (A - \alpha I)y \rangle$$

$$= \langle A^{1/2}x, (I - \alpha A^{-1})A^{1/2}y \rangle$$

and

$$|\langle x, Ay \rangle|^2 \leq \langle x, Ax \rangle \|I - \alpha A^{-1}\|^2 \langle y, Ay \rangle.$$

It is enough to prove that

$$\|I - \alpha A^{-1}\| \leq \frac{a-b}{a+b}$$

for an appropriate α.

Since A is self-adjoint, it is diagonal in a basis, so $A = \mathrm{Diag}(\lambda_1, \lambda_2, \ldots, \lambda_n)$ and

$$I - \alpha A^{-1} = \mathrm{Diag}\left(1 - \frac{\alpha}{\lambda_1}, \ldots, 1 - \frac{\alpha}{\lambda_n} \right).$$

Recall that $b \leq \lambda_i \leq a$. If we choose

$$\alpha = \frac{2ab}{a+b},$$

then it is elementary to check that

$$-\frac{a-b}{a+b} \leq 1 - \frac{\alpha}{\lambda_i} \leq \frac{a-b}{a+b},$$

which gives the proof.   □

The description of the generalized inverse of an $m \times n$ matrix can be described in terms of the **singular value decomposition**.

Let $A \in \mathbb{M}_{m \times n}$ with strictly positive singular values $\sigma_1, \sigma_2, \ldots, \sigma_k$. (Then $k \leq m, n$.) Define a matrix $\Sigma \in \mathbb{M}_{m \times n}$ as

$$\Sigma_{ij} = \begin{cases} \sigma_i & \text{if } i = j < k, \\ 0 & \text{otherwise.} \end{cases}$$

This matrix appears in the singular value decomposition described in the next theorem.

Theorem 1.46 *A matrix $A \in \mathbb{M}_{m \times n}$ has the decomposition*

$$A = U \Sigma V^*, \tag{1.19}$$

where $U \in \mathbb{M}_m$ and $V \in \mathbb{M}_n$ are unitaries and $\Sigma \in \mathbb{M}_{m \times n}$ is defined above.

For the sake of simplicity we consider the case $m = n$. Then A has the polar decomposition $U_0|A|$ and $|A|$ can be diagonalized:

$$|A| = U_1 \mathrm{Diag}(\sigma_1, \sigma_2, \ldots, \sigma_k, 0, \ldots, 0) U_1^*.$$

Therefore, $A = (U_0 U_1) \Sigma U_1^*$, where U_0 and U_1 are unitaries.

Theorem 1.47 *For a matrix $A \in \mathbb{M}_{m \times n}$ there exists a unique matrix $A^\dagger \in \mathbb{M}_{n \times m}$ such that the following four properties hold:*

(1) $AA^\dagger A = A$.
(2) $A^\dagger AA^\dagger = A^\dagger$.
(3) $AA^\dagger$ is self-adjoint.
(4) $A^\dagger A$ is self-adjoint.

It is easy to describe $A^\dagger$ in terms of the singular value decomposition (1.19). Namely, $A^\dagger = V \Sigma^\dagger U^*$, where

$$\Sigma_{ij}^\dagger = \begin{cases} \dfrac{1}{\sigma_i} & \text{if } i = j < k, \\[2mm] 0 & \text{otherwise.} \end{cases}$$

If A is invertible, then $n = m$ and $\Sigma^\dagger = \Sigma^{-1}$. Hence $A^\dagger$ is the inverse of A. Therefore $A^\dagger$ is called the generalized inverse of A or the **Moore–Penrose generalized inverse**. The generalized inverse has the properties

$$(\lambda A)^\dagger = \frac{1}{\lambda} A^\dagger, \quad (A^\dagger)^\dagger = A, \quad (A^\dagger)^* = (A^*)^\dagger.$$

It is worthwhile to note that for a matrix A with real entries $A^\dagger$ has real entries as well. Another important observation is the fact that the generalized inverse of AB is not always $B^\dagger A^\dagger$.

Example 1.48 If $M \in \mathbb{M}_m$ is an invertible matrix and $v \in \mathbb{C}^m$, then the linear system

$$Mx = v$$

has the obvious solution $x = M^{-1}v$. If $M \in \mathbb{M}_{m \times n}$, then the generalized inverse can be used. From property (1) a necessary condition of the solvability of the equation is $MM^\dagger v = v$. If this condition holds, then the solution is

$$x = M^\dagger v + (I_n - M^\dagger M)z$$

with arbitrary $z \in \mathbb{C}^n$. This example justifies the importance of the generalized inverse. $\square$

1.7 Tensor Product

Let $\mathcal{H}$ be the linear space of polynomials in the variable x and with degree less than or equal to n. A natural basis consists of the powers $1, x, x^2, \ldots, x^n$. Similarly, let $\mathcal{K}$ be the space of polynomials in y of degree less than or equal to m. Its basis is $1, y, y^2, \ldots, y^m$. The tensor product of these two spaces is the space of polynomials of two variables with basis $x^i y^j, 0 \le i \le n$ and $0 \le j \le m$. This simple example contains the essential ideas.

Let $\mathcal{H}$ and $\mathcal{K}$ be Hilbert spaces. Their **algebraic tensor product** consists of the formal finite sums

$$\sum_{i,j} x_i \otimes y_j \qquad (x_i \in \mathcal{H}, y_j \in \mathcal{K}).$$

Computing with these sums, one should use the following rules:

$$(x_1 + x_2) \otimes y = x_1 \otimes y + x_2 \otimes y, \quad (\lambda x) \otimes y = \lambda(x \otimes y),$$
$$x \otimes (y_1 + y_2) = x \otimes y_1 + x \otimes y_2, \quad x \otimes (\lambda y) = \lambda(x \otimes y). \qquad (1.20)$$

The inner product is defined as

$$\left\langle \sum_{i,j} x_i \otimes y_j, \sum_{k,l} z_k \otimes w_l \right\rangle = \sum_{i,j,k,l} \langle x_i, z_k \rangle \langle y_j, w_l \rangle.$$

When $\mathcal{H}$ and $\mathcal{K}$ are finite-dimensional spaces, then we arrive at the **tensor product** Hilbert space $\mathcal{H} \otimes \mathcal{K}$; otherwise the algebraic tensor product must be completed in order to get a Hilbert space.

Example 1.49 $L^2[0, 1]$ is the Hilbert space of the square integrable functions on $[0, 1]$. If $f, g \in L^2[0, 1]$, then the elementary tensor $f \otimes g$ can be interpreted as a function of two variables, $f(x)g(y)$ defined on $[0, 1] \times [0, 1]$. The computational rules (1.20) are obvious in this approach. $\square$

The tensor product of finitely many Hilbert spaces is defined similarly.

If $e_1, \ldots, e_n$ and $f_1, \ldots, f_m$ are bases in finite-dimensional $\mathcal{H}$ and $\mathcal{K}$, respectively, then $\{e_i \otimes f_j : i, j\}$ is a basis in the tensor product space. This basis is called the **product basis**. An arbitrary vector $x \in \mathcal{H} \otimes \mathcal{K}$ admits an expansion

$$x = \sum_{i,j} c_{ij}\, e_i \otimes f_j \tag{1.21}$$

for some coefficients c_{ij}, $\sum_{i,j} |c_{ij}|^2 = \|x\|^2$. This kind of expansion is general, but sometimes it is not the best.

Lemma 1.50 *Any unit vector $x \in \mathcal{H} \otimes \mathcal{K}$ can be written in the form*

$$x = \sum_k \sqrt{p_k}\, g_k \otimes h_k, \tag{1.22}$$

where the vectors $g_k \in \mathcal{H}$ and $h_k \in \mathcal{K}$ are orthonormal and (p_k) is a probability distribution.

Proof: We can define a conjugate-linear mapping $\Lambda : \mathcal{H} \to \mathcal{K}$ as

$$\langle \Lambda \alpha, \beta \rangle = \langle x, \alpha \otimes \beta \rangle$$

for every vector $\alpha \in \mathcal{H}$ and $\beta \in \mathcal{K}$. In the computation we can use the bases $(e_i)_i$ in $\mathcal{H}$ and $(f_j)_j$ in $\mathcal{K}$. If x has the expansion (1.21), then

$$\langle \Lambda e_i, f_j \rangle = c_{ij}$$

and the adjoint Λ^* is determined by

$$\langle \Lambda^* f_j, e_i \rangle = \overline{c_{ij}}.$$

(Concerning the adjoint of a conjugate-linear mapping, see (1.5).)

One can compute that the partial trace $\mathrm{Tr}_2 |x\rangle\langle x|$ of the matrix $|x\rangle\langle x|$ is $D := \Lambda^* \Lambda$ (see the definition before Example 1.56). It is enough to check that

$$\langle x, (|e_k\rangle\langle e_\ell| \otimes I_\mathcal{K})x \rangle = \mathrm{Tr}\, \Lambda^* \Lambda |e_k\rangle\langle e_\ell|$$

for every k and ℓ.

Choose now the orthogonal unit vectors g_k such that they are eigenvectors of D with corresponding non-zero eigenvalues p_k, $Dg_k = p_k g_k$. Then

$$h_k := \frac{1}{\sqrt{p_k}}|\Lambda g_k\rangle$$

is a family of pairwise orthogonal unit vectors. Now

$$\langle x, g_k \otimes h_\ell \rangle = \langle \Lambda g_k, h_\ell \rangle = \frac{1}{\sqrt{p_\ell}} \langle \Lambda g_k, \Lambda g_\ell \rangle = \frac{1}{\sqrt{p_\ell}} \langle g_\ell, \Lambda^* \Lambda g_k \rangle = \delta_{k,\ell} \sqrt{p_\ell}$$

and we have arrived at the orthogonal expansion (1.22).   □

The product basis tells us that

$$\dim(\mathcal{H} \otimes \mathcal{K}) = \dim(\mathcal{H}) \times \dim(\mathcal{K}).$$

Example 1.51 In the quantum formalism the orthonormal basis in the two-dimensional Hilbert space $\mathcal{H}$ is denoted by $| \uparrow \rangle$, $| \downarrow \rangle$. Instead of $| \uparrow \rangle \otimes | \downarrow \rangle$, the notation $| \uparrow \downarrow \rangle$ is used. Therefore the product basis is

$$| \uparrow \uparrow \rangle, \quad | \uparrow \downarrow \rangle, \quad | \downarrow \uparrow \rangle, \quad | \downarrow \downarrow \rangle.$$

Sometimes $\downarrow$ is replaced by 0 and $\uparrow$ by 1.
Another basis

$$\frac{1}{\sqrt{2}}(|00\rangle + |11\rangle), \quad \frac{1}{\sqrt{2}}(|01\rangle + |10\rangle), \quad \frac{i}{\sqrt{2}}(|10\rangle - |01\rangle), \quad \frac{1}{\sqrt{2}}(|00\rangle - |11\rangle)$$

is often used, which is called the **Bell basis**.   □

Example 1.52 In the Hilbert space $L^2(\mathbb{R}^2)$ we can get a basis if the space is considered as $L^2(\mathbb{R}) \otimes L^2(\mathbb{R})$. In the space $L^2(\mathbb{R})$ the Hermite functions

$$\varphi_n(x) = \exp(-x^2/2)H_n(x)$$

form a good basis, where $H_n(x)$ is the appropriately normalized Hermite polynomial. Therefore, the two variable Hermite functions

$$\varphi_{nm}(x, y) := e^{-(x^2+y^2)/2}H_n(x)H_m(y) \qquad (n, m = 0, 1, \dots)$$

form a basis in $L^2(\mathbb{R}^2)$.   □

The tensor product of linear transformations can be defined as well. If $A : \mathcal{H}_1 \to \mathcal{K}_1$ and $B : \mathcal{H}_2 \to \mathcal{K}_2$ are linear transformations, then there is a unique linear transformation $A \otimes B : \mathcal{H}_1 \otimes \mathcal{H}_2 \to \mathcal{K}_1 \otimes \mathcal{K}_2$ such that

$$(A \otimes B)(v_1 \otimes v_2) = Av_1 \otimes Bv_2 \qquad (v_1 \in \mathcal{H}_1, \ v_2 \in \mathcal{H}_2).$$

Since the linear mappings (between finite-dimensional Hilbert spaces) are identified with matrices, the tensor product of matrices appears as well.

Example 1.53 Let $\{e_1, e_2, e_3\}$ be a basis in $\mathcal{H}$ and $\{f_1, f_2\}$ be a basis in $\mathcal{K}$. If $[A_{ij}]$ is the matrix of $A \in B(\mathcal{H}_1)$ and $[B_{kl}]$ is the matrix of $B \in B(\mathcal{H}_2)$, then

$$(A \otimes B)(e_j \otimes f_l) = \sum_{i,k} A_{ij} B_{kl} e_i \otimes f_k \,.$$

It is useful to order the tensor product bases lexicographically: $e_1 \otimes f_1, e_1 \otimes f_2, e_2 \otimes f_1, e_2 \otimes f_2, e_3 \otimes f_1, e_3 \otimes f_2$. Fixing this ordering, we can write down the matrix of $A \otimes B$ and we have

$$\begin{bmatrix}
A_{11}B_{11} & A_{11}B_{12} & A_{12}B_{11} & A_{12}B_{12} & A_{13}B_{11} & A_{13}B_{12} \\
A_{11}B_{21} & A_{11}B_{22} & A_{12}B_{21} & A_{12}B_{22} & A_{13}B_{21} & A_{13}B_{22} \\
A_{21}B_{11} & A_{21}B_{12} & A_{22}B_{11} & A_{22}B_{12} & A_{23}B_{11} & A_{23}B_{12} \\
A_{21}B_{21} & A_{21}B_{22} & A_{22}B_{21} & A_{22}B_{22} & A_{23}B_{21} & A_{23}B_{22} \\
A_{31}B_{11} & A_{31}B_{12} & A_{32}B_{11} & A_{32}B_{12} & A_{33}B_{11} & A_{33}B_{12} \\
A_{31}B_{21} & A_{31}B_{22} & A_{32}B_{21} & A_{32}B_{22} & A_{33}B_{21} & A_{33}B_{22}
\end{bmatrix}.$$

In the block matrix formalism we have

$$A \otimes B = \begin{bmatrix}
A_{11}B & A_{12}B & A_{13}B \\
A_{21}B & A_{22}B & A_{23}B \\
A_{31}B & A_{32}B & A_{33}B
\end{bmatrix}, \tag{1.23}$$

see Sect. 2.1. The tensor product of matrices is also called the **Kronecker product**. $\square$

Example 1.54 When $A \in \mathbb{M}_n$ and $B \in \mathbb{M}_m$, the matrix

$$I_m \otimes A + B \otimes I_n \in \mathbb{M}_{nm}$$

is called the **Kronecker sum** of A and B.

If u is an eigenvector of A with eigenvalue λ and v is an eigenvector of B with eigenvalue μ, then

$$(I_m \otimes A + B \otimes I_n)(u \otimes v) = \lambda(u \otimes v) + \mu(u \otimes v) = (\lambda + \mu)(u \otimes v).$$

So $u \otimes v$ is an eigenvector of the Kronecker sum with eigenvalue $\lambda + \mu$. $\square$

The computation rules of the tensor product of Hilbert spaces imply straightforward properties of the tensor product of matrices (or linear operators).

Theorem 1.55 *The following rules hold:*

(1) $(A_1 + A_2) \otimes B = A_1 \otimes B + A_2 \otimes B.$

(2) $B \otimes (A_1 + A_2) = B \otimes A_1 + B \otimes A_2$.
(3) $(\lambda A) \otimes B = A \otimes (\lambda B) = \lambda (A \otimes B)$ $(\lambda \in \mathbb{C})$.
(4) $(A \otimes B)(C \otimes D) = AC \otimes BD$.
(5) $(A \otimes B)^* = A^* \otimes B^*$.
(6) $(A \otimes B)^{-1} = A^{-1} \otimes B^{-1}$ if A and B are invertible.
(7) $\|A \otimes B\| = \|A\| \, \|B\|$.

For example, the tensor product of self-adjoint matrices is self-adjoint and the tensor product of unitaries is unitary.

The linear mapping $\mathbb{M}_n \otimes \mathbb{M}_m \to \mathbb{M}_n$ defined as

$$\mathrm{Tr}_2 : A \otimes B \mapsto (\mathrm{Tr}\, B)A$$

is called a **partial trace**. The other partial trace is

$$\mathrm{Tr}_1 : A \otimes B \mapsto (\mathrm{Tr}\, A)B.$$

Example 1.56 Assume that $A \in \mathbb{M}_n$ and $B \in \mathbb{M}_m$. Then $A \otimes B$ is an $nm \times nm$ matrix. Let $C \in \mathbb{M}_{nm}$. How can we decide if it has the form of $A \otimes B$ for some $A \in \mathbb{M}_n$ and $B \in \mathbb{M}_m$?

First we study how to recognize A and B from $A \otimes B$. (Of course, A and B are not uniquely determined, since $(\lambda A) \otimes (\lambda^{-1} B) = A \otimes B$.) If we take the trace of all entries of (1.23), then we get

$$\begin{bmatrix} A_{11} \mathrm{Tr}\, B & A_{12} \mathrm{Tr}\, B & A_{13} \mathrm{Tr}\, B \\ A_{21} \mathrm{Tr}\, B & A_{22} \mathrm{Tr}\, B & A_{23} \mathrm{Tr}\, B \\ A_{31} \mathrm{Tr}\, B & A_{32} \mathrm{Tr}\, B & A_{33} \mathrm{Tr}\, B \end{bmatrix} = \mathrm{Tr}\, B \begin{bmatrix} A_{11} & A_{12} & A_{13} \\ A_{21} & A_{22} & A_{23} \\ A_{31} & A_{32} & A_{33} \end{bmatrix} = (\mathrm{Tr}\, B)A.$$

The sum of the diagonal entries is

$$A_{11}B + A_{12}B + A_{13}B = (\mathrm{Tr}\, A)B.$$

If $X = A \otimes B$, then

$$(\mathrm{Tr}\, X)X = (\mathrm{Tr}_2 X) \otimes (\mathrm{Tr}_1 X).$$

For example, the matrix

$$X := \begin{bmatrix} 0 & 0 & 0 & 0 \\ 0 & 1 & 1 & 0 \\ 0 & 1 & 1 & 0 \\ 0 & 0 & 0 & 0 \end{bmatrix}$$

in $\mathbb{M}_2 \otimes \mathbb{M}_2$ is not a tensor product. Indeed,

$$\mathrm{Tr}_1 X = \mathrm{Tr}_2 X = \begin{bmatrix} 1 & 0 \\ 0 & 1 \end{bmatrix}$$

and their tensor product is the identity in $\mathbb{M}_4$.   □

Let $\mathcal{H}$ be a Hilbert space. The k-fold tensor product $\mathcal{H} \otimes \cdots \otimes \mathcal{H}$ is called the kth tensor power of $\mathcal{H}$, denoted by $\mathcal{H}^{\otimes k}$. When $A \in B(\mathcal{H})$, then $A^{(1)} \otimes A^{(2)} \cdots \otimes A^{(k)}$ is a linear operator on $\mathcal{H}^{\otimes k}$ and it is denoted by $A^{\otimes k}$. (Here the $A^{(i)}$'s are copies of A.)

$\mathcal{H}^{\otimes k}$ has two important subspaces, the symmetric and the antisymmetric subspaces. If $v_1, v_2, \cdots, v_k \in \mathcal{H}$ are vectors, then their **antisymmetric** tensor product is the linear combination

$$v_1 \wedge v_2 \wedge \cdots \wedge v_k := \frac{1}{\sqrt{k!}} \sum_{\pi} (-1)^{\sigma(\pi)} v_{\pi(1)} \otimes v_{\pi(2)} \otimes \cdots \otimes v_{\pi(k)}$$

where the summation is over all permutations π of the set $\{1, 2, \ldots, k\}$ and $\sigma(\pi)$ is the number of inversions in π. The terminology "antisymmetric" comes from the property that an antisymmetric tensor changes its sign if two elements are exchanged. In particular, $v_1 \wedge v_2 \wedge \cdots \wedge v_k = 0$ if $v_i = v_j$ for different i and j.

The computational rules for the antisymmetric tensors are similar to (1.20):

$$\lambda(v_1 \wedge v_2 \wedge \cdots \wedge v_k) = v_1 \wedge v_2 \wedge \cdots \wedge v_{\ell-1} \wedge (\lambda v_\ell) \wedge v_{\ell+1} \wedge \cdots \wedge v_k$$

for every ℓ and

$$(v_1 \wedge v_2 \wedge \cdots \wedge v_{\ell-1} \wedge v \wedge v_{\ell+1} \wedge \cdots \wedge v_k)$$
$$+ (v_1 \wedge v_2 \wedge \cdots \wedge v_{\ell-1} \wedge v' \wedge v_{\ell+1} \wedge \cdots \wedge v_k)$$
$$= v_1 \wedge v_2 \wedge \cdots \wedge v_{\ell-1} \wedge (v + v') \wedge v_{\ell+1} \wedge \cdots \wedge v_k .$$

Lemma 1.57 *The inner product of $v_1 \wedge v_2 \wedge \cdots \wedge v_k$ and $w_1 \wedge w_2 \wedge \cdots \wedge w_k$ is the determinant of the $k \times k$ matrix whose (i, j) entry is $\langle v_i, w_j \rangle$.*

Proof: The inner product is

$$\frac{1}{k!} \sum_{\pi} \sum_{\kappa} (-1)^{\sigma(\pi)} (-1)^{\sigma(\kappa)} \langle v_{\pi(1)}, w_{\kappa(1)} \rangle \langle v_{\pi(2)}, w_{\kappa(2)} \rangle \ldots \langle v_{\pi(k)}, w_{\kappa(k)} \rangle$$

$$= \frac{1}{k!} \sum_{\pi} \sum_{\kappa} (-1)^{\sigma(\pi)} (-1)^{\sigma(\kappa)} \langle v_1, w_{\pi^{-1}\kappa(1)} \rangle \langle v_2, w_{\pi^{-1}\kappa(2)} \rangle \ldots \langle v_k, w_{\pi^{-1}\kappa(k)} \rangle$$

$$= \frac{1}{k!} \sum_{\pi} \sum_{\kappa} (-1)^{\sigma(\pi^{-1}\kappa)} \langle v_1, w_{\pi^{-1}\kappa(1)} \rangle \langle v_2, w_{\pi^{-1}\kappa(2)} \rangle \ldots \langle v_k, w_{\pi^{-1}\kappa(k)} \rangle$$

$$= \sum_{\pi} (-1)^{\sigma(\pi)} \langle v_1, w_{\pi(1)} \rangle \langle v_2, w_{\pi(2)} \rangle \ldots \langle v_k, w_{\pi(k)} \rangle .$$

This is the determinant.   □

It follows from the previous lemma that $v_1 \wedge v_2 \wedge \cdots \wedge v_k \neq 0$ if and only if the vectors $v_1, v_2, \cdots v_k$ are linearly independent. The subspace spanned by the vectors $v_1 \wedge v_2 \wedge \cdots \wedge v_k$ is called the kth antisymmetric tensor power of $\mathcal{H}$, denoted by $\mathcal{H}^{\wedge k}$. So $\mathcal{H}^{\wedge k} \subset \mathcal{H}^{\otimes k}$.

Lemma 1.58 *The linear extension of the map*

$$x_1 \otimes \cdots \otimes x_k \mapsto \frac{1}{\sqrt{k!}} x_1 \wedge \cdots \wedge x_k$$

is the projection of $\mathcal{H}^{\otimes k}$ onto $\mathcal{H}^{\wedge k}$.

Proof: Let P be the defined linear operator. First we show that $P^2 = P$:

$$P^2(x_1 \otimes \cdots \otimes x_k) = \frac{1}{(k!)^{3/2}} \sum_{\pi} (-1)^{\sigma(\pi)} x_{\pi(1)} \wedge \cdots \wedge x_{\pi(k)}$$

$$= \frac{1}{(k!)^{3/2}} \sum_{\pi} (-1)^{\sigma(\pi)+\sigma(\pi)} x_1 \wedge \cdots \wedge x_k$$

$$= \frac{1}{\sqrt{k!}} x_1 \wedge \cdots \wedge x_k = P(x_1 \otimes \cdots \otimes x_k).$$

Moreover, $P = P^*$:

$$\langle P(x_1 \otimes \cdots \otimes x_k), y_1 \otimes \cdots \otimes y_k \rangle = \frac{1}{k!} \sum_{\pi} (-1)^{\sigma(\pi)} \prod_{i=1}^{k} \langle x_{\pi(i)}, y_i \rangle$$

$$= \frac{1}{k!} \sum_{\pi} (-1)^{\sigma(\pi^{-1})} \prod_{i=1}^{k} \langle x_i, y_{\pi^{-1}(i)} \rangle$$

$$= \langle x_1 \otimes \cdots \otimes x_k, P(y_1 \otimes \cdots \otimes y_k) \rangle.$$

So P is an orthogonal projection.   □

Example 1.59 A transposition is a permutation of $1, 2, \ldots, n$ which exchanges the place of two entries. For a transposition κ, there is a unitary $U_\kappa : \mathcal{H}^{\otimes k} \to \mathcal{H}^{\otimes k}$ such that

$$U_\kappa(v_1 \otimes v_2 \otimes \cdots \otimes v_n) = v_{\kappa(1)} \otimes v_{\kappa(2)} \otimes \cdots \otimes v_{\kappa(n)}.$$

Then

$$\mathcal{H}^{\wedge k} = \{x \in \mathcal{H}^{\otimes k} : U_\kappa x = -x \text{ for every } \kappa\}. \tag{1.24}$$

The terminology "antisymmetric" comes from this description.   □

If $e_1, e_2, \ldots, e_n$ is a basis in $\mathcal{H}$, then

$$\{e_{i(1)} \wedge e_{i(2)} \wedge \cdots \wedge e_{i(k)} \: : \: 1 \le i(1) < i(2) < \cdots < i(k)) \le n\}$$

is a basis in $\mathcal{H}^{\wedge k}$. It follows that the dimension of $\mathcal{H}^{\wedge k}$ is

$$\binom{n}{k} \quad \text{if} \;\; k \le n,$$

otherwise for $k > n$ the power $\mathcal{H}^{\wedge k}$ has dimension 0. Consequently, $\mathcal{H}^{\wedge n}$ has dimension 1.

If $A \in B(\mathcal{H})$, then the transformation $A^{\otimes k}$ leaves the subspace $\mathcal{H}^{\wedge k}$ invariant. Its restriction is denoted by $A^{\wedge k}$ which is equivalently defined as

$$A^{\wedge k}(v_1 \wedge v_2 \wedge \cdots \wedge v_k) = Av_1 \wedge Av_2 \wedge \cdots \wedge Av_k.$$

For any operators $A, B \in B(\mathcal{H})$, we have

$$(A^*)^{\wedge k} = (A^{\wedge k})^*, \qquad (AB)^{\wedge k} = A^{\wedge k} B^{\wedge k}$$

and

$$A^{\wedge n} = \lambda \cdot \text{identity}. \tag{1.25}$$

The constant λ is the determinant:

Theorem 1.60 *For $A \in \mathbb{M}_n$, the constant λ in (1.25) is $\det A$.*

Proof: If $e_1, e_2, \ldots, e_n$ is a basis in $\mathcal{H}$, then in the space $\mathcal{H}^{\wedge n}$ the vector $e_1 \wedge e_2 \wedge \cdots \wedge e_n$ forms a basis. We should compute $A^{\wedge k}(e_1 \wedge e_2 \wedge \cdots \wedge e_n)$.

$$(A^{\wedge k})(e_1 \wedge e_2 \wedge \cdots \wedge e_n) = (Ae_1) \wedge (Ae_2) \wedge \cdots \wedge (Ae_n)$$

$$= \Big(\sum_{i(1)=1}^{n} A_{i(1),1} e_{i(1)} \Big) \wedge \Big(\sum_{i(2)=1}^{n} A_{i(2),2} e_{i(2)} \Big) \wedge \cdots \wedge \Big(\sum_{i(n)=1}^{n} A_{i(n),n} e_{i(n)} \Big)$$

$$= \sum_{i(1),i(2),\ldots,i(n)=1}^{n} A_{i(1),1} A_{i(2),2} \cdots A_{i(n),n} e_{i(1)} \wedge \cdots \wedge e_{i(n)}$$

$$= \sum_{\pi} A_{\pi(1),1} A_{\pi(2),2} \cdots A_{\pi(n),n} e_{\pi(1)} \wedge \cdots \wedge e_{\pi(n)}$$

$$= \sum_{\pi} A_{\pi(1),1} A_{\pi(2),2} \cdots A_{\pi(n),n} (-1)^{\sigma(\pi)} e_1 \wedge \cdots \wedge e_n.$$

Here we used the fact that $e_{i(1)} \wedge \cdots \wedge e_{i(n)}$ can be non-zero if the vectors $e_{i(1)}, \ldots, e_{i(n)}$ are all different, in other words, this is a permutation of $e_1, e_2, \ldots, e_n$. $\qquad\square$

Example 1.61 Let $A \in \mathbb{M}_n$ be a self-adjoint matrix with eigenvalues $\lambda_1 \geq \lambda_2 \geq \cdots \geq \lambda_n$. The corresponding eigenvectors $v_1, v_2, \cdots, v_n$ form a good basis. The largest eigenvalue of the antisymmetric power $A^{\wedge k}$ is $\prod_{i=1}^k \lambda_i$:

$$A^{\wedge k}(v_1 \wedge v_2 \wedge \cdots \wedge v_k) = Av_1 \wedge Av_2 \wedge \cdots \wedge Av_k$$
$$= \left(\prod_{i=1}^k \lambda_i\right)(v_1 \wedge v_2 \wedge \cdots \wedge v_k).$$

All other eigenvalues can be obtained from the basis of the antisymmetric product (as in the proof of the next lemma). $\qquad\square$

The next lemma describes a relationship between singular values and antisymmetric powers.

Lemma 1.62 *For $A \in \mathbb{M}_n$ and for $k = 1, \ldots, n$, we have*

$$\prod_{i=1}^k s_i(A) = s_1(A^{\wedge k}) = \|A^{\wedge k}\|.$$

Proof: Since $|A|^{\wedge k} = |A^{\wedge k}|$, we may assume that $A \geq 0$. Then there exists an orthonormal basis $\{u_1, \cdots, u_n\}$ of $\mathcal{H}$ such that $Au_i = s_i(A)u_i$ for all i. We have

$$A^{\wedge k}(u_{i(1)} \wedge \cdots \wedge u_{i(k)}) = \left(\prod_{j=1}^k s_{i(j)}(A)\right)u_{i(1)} \wedge \cdots \wedge u_{i(k)},$$

and so $\{u_{i(1)} \wedge \cdots \wedge u_{i(k)} : 1 \leq i(1) < \cdots < i(k) \leq n\}$ is a complete set of eigenvectors of $A^{\wedge k}$. Hence the assertion follows. $\qquad\square$

The **symmetric tensor product** of the vectors $v_1, v_2, \ldots, v_k \in \mathcal{H}$ is

$$v_1 \vee v_2 \vee \cdots \vee v_k := \frac{1}{\sqrt{k!}} \sum_{\pi} v_{\pi(1)} \otimes v_{\pi(2)} \otimes \cdots \otimes v_{\pi(k)},$$

where the summation is over all permutations π of the set $\{1, 2, \ldots, k\}$ again. The linear span of the symmetric tensors is the symmetric tensor power $\mathcal{H}^{\vee k}$. Similarly to (1.24), we have

$$\mathcal{H}^{\vee k} = \{x \in \otimes^k \mathcal{H} : U_\kappa x = x \text{ for every } \kappa\}.$$

It follows immediately that $\mathcal{H}^{\vee k} \perp \mathcal{H}^{\wedge k}$ for any $k \geq 2$. Let $u \in \mathcal{H}^{\vee k}$ and $v \in \mathcal{H}^{\wedge k}$. Then

$$\langle u, v \rangle = \langle U_\kappa u, -U_\kappa v \rangle = -\langle u, v \rangle$$

and $\langle u, v \rangle = 0$.

If $e_1, e_2, \ldots, e_n$ is a basis in $\mathcal{H}$, then $\vee^k \mathcal{H}$ has the basis

$$\{ e_{i(1)} \vee e_{i(2)} \vee \cdots \vee e_{i(k)} \; : \; 1 \leq i(1) \leq i(2) \leq \cdots \leq i(k) \leq n \}.$$

Similarly to the proof of Lemma 1.57 we have

$$\langle v_1 \vee v_2 \vee \cdots \vee v_k, w_1 \vee w_2 \vee \cdots \vee w_k \rangle = \sum_\pi \langle v_1, w_{\pi(1)} \rangle \langle v_2, w_{\pi(2)} \rangle \ldots \langle v_k, w_{\pi(k)} \rangle.$$

The right-hand side is similar to a determinant, but the sign does not change.

The **permanent** is defined as

$$\mathrm{per}\, A = \sum_\pi A_{1,\pi(1)} A_{2,\pi(2)} \cdots A_{n,\pi(n)} \tag{1.26}$$

similarly to the determinant formula (1.1).

1.8 Notes and Remarks

The history of matrices goes back to ancient times. Their first appearance in applications to linear equations was in ancient China. The notion of determinants preceded the introduction and development of matrices and linear algebra. Determinants were first studied by a Japanese mathematician Takakazu Seki in 1683 and by Gottfried Leibniz (1646–1716) in 1693. In 1750 Gabriel **Cramer** (1704–1752) discovered his famous determinant-based formula of solutions to systems of linear equations. From the 18th century to the beginning of the 19th, theoretical studies of determinants were made by Vandermonde (famous for the determinant named after him), Joseph-Louis Lagrange (1736–1813) who characterized the maxima and minima of multivariate functions by his method known as the method of Lagrange multipliers, Pierre-Simon Laplace (1749–1827), and Augustin Louis Cauchy (1789–1857). Gaussian elimination to solve systems of linear equations by successively eliminating variables was developed around 1800 by Johann Carl Friedrich **Gauss** (1777–1855). However, as mentioned above, a prototype of this method appeared in important ancient Chinese texts. The method is also referred to as Gauss–Jordan elimination since it was published in 1887 in an extended form by Wilhelm Jordan.

As explained above, determinants had been more dominant than matrices in the first stage of the history of matrix theory up to the middle of the 19th century. The modern treatment of matrices emerged when Arthur **Cayley** (1821–1895) published his monumental work, Memoir on the Theory of Matrices, in 1858. Before that, in 1851 James Joseph **Sylvester** (1814–1897) introduced the term "matrix" after the Latin word for "womb". Cayley studied matrices in the modern style by making connections with linear transformations. The axiomatic definition of vector spaces was finally introduced by Giuseppe Peano (1858–1932) in 1888. The computation of determinants of concrete special matrices has a huge literature; for example, the book by Thomas Muir, *A Treatise on the Theory of Determinants*, (originally published in 1928) has more than 700 pages.

In this book matrices are mostly complex matrices, which can be studied from three different perspectives. The first aspect is algebraic. The $n \times n$ matrices form a $*$-algebra with linear operations, product AB and adjoint A^* as described in the first section. The second is the topological/analytic aspect of matrices as described in Sect. 1.2. Since a matrix corresponds to a linear transformation between finite-dimensional vector spaces, the operator norm is naturally assigned to a matrix. It is also important that the $n \times n$ matrices form a Hilbert space with the Hilbert–Schmidt inner product. In this respect, Sect. 1.2 may be regarded as a concise introduction to Hilbert spaces, though mostly restricted to the finite-dimensional case. The third aspect is the order structure of matrices described in Sect. 1.6 and in further detail in Sect. 2.2. These three structures are closely related to each other, and their interplay is an essential feature of the study of matrix analysis.

Cauchy proved in 1829 that the eigenvalues of a symmetric matrix are all real numbers (see the proof of Theorem 1.23). The cofactor expansion for the determinant in Theorem 1.31 was shown by Laplace. The famous Cayley–Hamilton theorem (Theorem 1.18) for 2×2 and 3×3 matrices was contained in Cayley's work mentioned above, and later William Rowan Hamilton proved it for 4×4 matrices. The formula for inverse matrices (Theorem 1.33) was also established by Cayley.

Spectral and polar decompositions are fundamental in operator theory. The phase operator U of the polar decomposition $A = U|A|$ cannot always be a unitary in the infinite-dimensional case. However, a unitary U can be chosen for matrices, although it is not unique. A finer decomposition for general square matrices is the Jordan canonical form introduced in Sect. 1.3, which is due to Camille Jordan (1771–1821), not the same Jordan as that of Gauss–Jordan elimination.

The useful minimax principle in Theorem 1.27 is often called the Courant–Fisher–Weyl minimax principle due to their contributions (see Section III.7 of 20 for details). A similar expression for singular values will be given in (6.5), and another expression called Ky Fan's maximum principle for the sum $\sum_{j=1}^{k} \lambda_j$ of the k largest eigenvalues of a self-adjoint matrix is also useful in matrix theory. Theorem 1.30 is the Hadamard inequality established by Jacques Hadamard in 1893. Weakly positive matrices were introduced by Eugene P. **Wigner** in 1963. He showed that if the product of two or three weakly positive matrices is self-adjoint, then it is positive definite.

Hilbert spaces and operators are essential in the mathematical formulation of quantum mechanics. John **von Neumann** (1903–1957) introduced several concepts in connection with operator/matrix theory and quantum physics in *Mathematische Grundlagen der Quantenmechanik*, 1932. Nowadays, matrix theory plays an essential role in the theory of quantum information as well, see [73]. When quantum theory appeared in the 1920s, some matrices had already appeared in the work of Werner Heisenberg. Later the physicist Paul Adrien Maurice Dirac (1902–1984) introduced the bra-ket notation, which is sometimes used in this book. But for column vectors x, y, matrix theorists prefer to write x^*y for the inner product $\langle x|y \rangle$ and xy^* for the rank one operator $|x\rangle\langle y|$.

Note that the Kronecker sum is often denoted by $A \oplus B$ in the literature, but in this book $\oplus$ is the notation for the direct sum. The antisymmetric and symmetric tensor products are used in the construction of antisymmetric (Fermion) and symmetric (Boson) Fock spaces in the study of quantum mechanics and quantum field theory. The antisymmetric tensor product is a powerful technique in matrix theory as well, as will be seen in Chap. 6.

Concerning the permanent (1.26), a famous conjecture of **Van der Waerden** made in 1926 was that if A is an $n \times n$ **doubly stochastic** matrix then

$$\text{per } A \geq \frac{n!}{n^n},$$

and equality holds if and only if $A_{ij} = 1/n$ for all $1 \leq i, j \leq n$. (The proof was given in 1981 by G. P. Egorychev and D. Falikman. It is included in the book [87].)

1.9 Exercises

1. Let $A : \mathcal{H}_2 \to \mathcal{H}_1, B : \mathcal{H}_3 \to \mathcal{H}_2$ and $C : \mathcal{H}_4 \to \mathcal{H}_3$ be linear mappings. Show that

$$\text{rank } AB + \text{rank } BC \leq \text{rank } B + \text{rank } ABC.$$

   (This is called **Frobenius' inequality**.)

2. Let $A : \mathcal{H} \to \mathcal{H}$ be a linear mapping. Show that

$$\dim \ker A^{n+1} = \dim \ker A + \sum_{k=1}^{n} \dim(\text{ran } A^k \cap \ker A).$$

3. Show that in the Schwarz inequality (1.2) equality occurs if and only if x and y are linearly dependent.

4. Show that

$$\|x - y\|^2 + \|x + y\|^2 = 2\|x\|^2 + 2\|y\|^2$$

   for the norm in a Hilbert space. (This is called the **parallelogram law**.)

5. Prove the polarization identity (1.6).

6. Show that an orthonormal family of vectors is linearly independent.
7. Show that the vectors $|x_1\rangle, |x_2,\rangle, \ldots, |x_n\rangle$ form an orthonormal basis in an n-dimensional Hilbert space if and only if

$$\sum_i |x_i\rangle\langle x_i| = I.$$

8. Show that the Gram–Schmidt procedure constructs an orthonormal basis e_1, $e_2, \ldots, e_n$. Show that e_k is a linear combination of $v_1, v_2, \ldots, v_k$ $(1 \le k \le n)$.
9. Show that the upper triangular matrices form an algebra.
10. Verify that the inverse of an upper triangular matrix is upper triangular if the inverse exists.
11. Compute the determinant of the matrix

$$\begin{bmatrix} 1 & 1 & 1 & 1 \\ 1 & 2 & 3 & 4 \\ 1 & 3 & 6 & 10 \\ 1 & 4 & 10 & 20 \end{bmatrix}.$$

Give an $n \times n$ generalization.
12. Compute the determinant of the matrix

$$\begin{bmatrix} 1 & -1 & 0 & 0 \\ x & h & -1 & 0 \\ x^2 & hx & h & -1 \\ x^3 & hx^2 & hx & h \end{bmatrix}.$$

Give an $n \times n$ generalization.
13. Let $A, B \in \mathbb{M}_n$ and

$$B_{ij} = (-1)^{i+j} A_{ij} \quad (1 \le i, j \le n).$$

Show that $\det A = \det B$.
14. Show that the determinant of the **Vandermonde matrix**

$$\begin{bmatrix} 1 & 1 & \cdots & 1 \\ a_1 & a_2 & \cdots & a_n \\ \vdots & \vdots & \ddots & \vdots \\ a_1^{n-1} & a_2^{n-1} & \cdots & a_n^{n-1} \end{bmatrix}$$

is $\prod_{i<j}(a_j - a_i)$.
15. Prove the following properties:

$$(|u\rangle\langle v|)^* = |v\rangle\langle u|, \quad (|u_1\rangle\langle v_1|)(|u_2\rangle\langle v_2|) = \langle v_1, u_2\rangle |u_1\rangle\langle v_2|,$$
$$A(|u\rangle\langle v|) = |Au\rangle\langle v|, \quad (|u\rangle\langle v|)A = |u\rangle\langle A^*v| \quad \text{for all } A \in B(\mathcal{H}).$$

16. Let $A, B \in B(\mathcal{H})$. Show that $\|AB\| \le \|A\|\,\|B\|$.
17. Let $\mathcal{H}$ be an n-dimensional Hilbert space. For $A \in B(\mathcal{H})$ let $\|A\|_2 := \sqrt{\operatorname{Tr} A^*A}$.
    Show that $\|A + B\|_2 \le \|A\|_2 + \|B\|_2$. Is it true that $\|AB\|_2 \le \|A\|_2 \times \|B\|_2$?
18. Find constants $c(n)$ and $d(n)$ such that

$$c(n)\|A\| \le \|A\|_2 \le d(n)\|A\|$$

    for every matrix $A \in \mathbb{M}_n(\mathbb{C})$.
19. Show that $\|A^*A\| = \|A\|^2$ for every $A \in B(\mathcal{H})$.
20. Let $\mathcal{H}$ be an n-dimensional Hilbert space. Show that given an operator $A \in B(\mathcal{H})$
    we can choose an orthonormal basis such that the matrix of A is upper triangular.
21. Let $A, B \in \mathbb{M}_n$ be invertible matrices. Show that $A + B$ is invertible if and only
    if $A^{-1} + B^{-1}$ is invertible, and moreover

$$(A + B)^{-1} = A^{-1} - A^{-1}(A^{-1} + B^{-1})^{-1}A^{-1}.$$

22. Let $A \in \mathbb{M}_n$ be self-adjoint. Show that

$$U = (I - iA)(I + iA)^{-1}$$

    is a unitary. (U is the **Cayley transform** of A.)
23. The self-adjoint matrix

$$0 \le \begin{bmatrix} a & b \\ b & c \end{bmatrix}$$

    has eigenvalues α and β. Show that

$$|b|^2 \le \left(\frac{\alpha - \beta}{\alpha + \beta}\right)^2 ac. \tag{1.27}$$

24. Show that

$$\begin{bmatrix} \lambda + z & x - iy \\ x + iy & \lambda - z \end{bmatrix}^{-1} = \frac{1}{\lambda^2 - x^2 - y^2 - z^2} \begin{bmatrix} \lambda - z & -x + iy \\ -x - iy & \lambda + z \end{bmatrix}$$

    for real parameters λ, x, y, z.
25. Let $m \le n$, $A \in \mathbb{M}_n$, $B \in \mathbb{M}_m$, $Y \in \mathbb{M}_{n \times m}$ and $Z \in \mathbb{M}_{m \times n}$. Assume that A and
    B are invertible. Show that $A + YBZ$ is invertible if and only if $B^{-1} + ZA^{-1}Y$ is
    invertible. Moreover,

$$(A + YBZ)^{-1} = A^{-1} - A^{-1}Y(B^{-1} + ZA^{-1}Y)^{-1}ZA^{-1}.$$

26. Let $\lambda_1, \lambda_2, \ldots, \lambda_n$ be the eigenvalues of a matrix $A \in \mathbb{M}_n(\mathbb{C})$. Show that A is normal if and only if

$$\sum_{i=1}^{n} |\lambda_i|^2 = \sum_{i,j=1}^{n} |A_{ij}|^2.$$

27. Show that $A \in \mathbb{M}_n$ is normal if and only if $A^* = AU$ for a unitary $U \in \mathbb{M}_n$.
28. Give an example such that $A^2 = A$, but A is not an orthogonal projection.
29. $A \in \mathbb{M}_n$ is called idempotent if $A^2 = A$. Show that each eigenvalue of an idempotent matrix is either 0 or 1.
30. Compute the eigenvalues and eigenvectors of the Pauli matrices:

$$\sigma_1 = \begin{bmatrix} 0 & 1 \\ 1 & 0 \end{bmatrix}, \qquad \sigma_2 = \begin{bmatrix} 0 & -i \\ i & 0 \end{bmatrix}, \qquad \sigma_3 = \begin{bmatrix} 1 & 0 \\ 0 & -1 \end{bmatrix}. \qquad (1.28)$$

31. Show that the Pauli matrices (1.28) are orthogonal to each other (with respect to the Hilbert–Schmidt inner product). What are the matrices which are orthogonal to all Pauli matrices?
32. The $n \times n$ **Pascal matrix** is defined as

$$P_{ij} = \binom{i+j-2}{i-1} \qquad (1 \le i, j \le n).$$

What is the determinant? (Hint: Generalize the particular relation

$$\begin{bmatrix} 1 & 1 & 1 & 1 \\ 1 & 2 & 3 & 4 \\ 1 & 3 & 6 & 10 \\ 1 & 4 & 10 & 20 \end{bmatrix} = \begin{bmatrix} 1 & 0 & 0 & 0 \\ 1 & 1 & 0 & 0 \\ 1 & 2 & 1 & 0 \\ 1 & 3 & 3 & 1 \end{bmatrix} \times \begin{bmatrix} 1 & 1 & 1 & 1 \\ 0 & 1 & 2 & 3 \\ 0 & 0 & 1 & 3 \\ 0 & 0 & 0 & 1 \end{bmatrix}$$

to $n \times n$ matrices.)

33. Let λ be an eigenvalue of a unitary operator. Show that $|\lambda| = 1$.
34. Let A be an $n \times n$ matrix and let $k \ge 1$ be an integer. Assume that $A_{ij} = 0$ if $j \ge i + k$. Show that A^{n-k} is the 0 matrix.
35. Show that $|\det U| = 1$ for a unitary U.
36. Let $U \in \mathbb{M}_n$ and $u_1, \ldots, u_n$ be n column vectors of U, i.e., $U = [u_1 \, u_2 \, \ldots \, u_n]$. Prove that U is a unitary matrix if and only if $\{u_1, \ldots, u_n\}$ is an orthonormal basis of $\mathbb{C}^n$.
37. Let a matrix $U = [u_1 \, u_2 \, \ldots \, u_n] \in \mathbb{M}_n$ be described by column vectors. Assume that $\{u_1, \ldots, u_k\}$ are given and orthonormal in $\mathbb{C}^n$. Show that $u_{k+1}, \ldots, u_n$ can be chosen in such a way that U will be a unitary matrix.
38. Compute $\det(\lambda I - A)$ when A is the tridiagonal matrix (1.9).

39. Let $U \in B(\mathcal{H})$ be a unitary. Show that

$$\lim_{n \to \infty} \frac{1}{n} \sum_{i=1}^{n} U^n x$$

exists for every vector $x \in \mathcal{H}$. (Hint: Consider the subspaces $\{x \in \mathcal{H} : Ux = x\}$ and $\{Ux - x : x \in \mathcal{H}\}$.) What is the limit

$$\lim_{n \to \infty} \frac{1}{n} \sum_{i=1}^{n} U^n ?$$

(This is the **ergodic theorem**.)

40. Let

$$|\beta_0\rangle = \frac{1}{\sqrt{2}}(|00\rangle + |11\rangle) \in \mathbb{C}^2 \otimes \mathbb{C}^2$$

and

$$|\beta_i\rangle = (\sigma_i \otimes I_2)|\beta_0\rangle \qquad (i = 1, 2, 3)$$

where the σ_i are the Pauli matrices. Show that $\{|\beta_i\rangle : 0 \le i \le 3\}$ is the Bell basis.

41. Show that the vectors of the Bell basis are eigenvectors of the matrices $\sigma_i \otimes \sigma_i$, $1 \le i \le 3$.

42. Prove the identity

$$|\psi\rangle \otimes |\beta_0\rangle = \frac{1}{2} \sum_{k=0}^{3} |\beta_k\rangle \otimes \sigma_k|\psi\rangle$$

in $\mathbb{C}^2 \otimes \mathbb{C}^2 \otimes \mathbb{C}^2$, where $|\psi\rangle \in \mathbb{C}^2$ and $|\beta_i\rangle \in \mathbb{C}^2 \otimes \mathbb{C}^2$ is defined above.

43. Write the so-called **Dirac matrices** in the form of elementary tensors (of two 2×2 matrices):

$$\gamma_1 = \begin{bmatrix} 0 & 0 & 0 & -i \\ 0 & 0 & -i & 0 \\ 0 & -i & 0 & 0 \\ -i & 0 & 0 & 0 \end{bmatrix}, \qquad \gamma_2 = \begin{bmatrix} 0 & 0 & 0 & -1 \\ 0 & 0 & 1 & 0 \\ 0 & 1 & 0 & 0 \\ -1 & 0 & 0 & 0 \end{bmatrix},$$

$$\gamma_3 = \begin{bmatrix} 0 & 0 & -i & 0 \\ 0 & 0 & 0 & i \\ i & 0 & 0 & 0 \\ 0 & -i & 0 & 0 \end{bmatrix}, \qquad \gamma_4 = \begin{bmatrix} 1 & 0 & 0 & 0 \\ 0 & 1 & 0 & 0 \\ 0 & 0 & -1 & 0 \\ 0 & 0 & 0 & -1 \end{bmatrix}.$$

44. Give the dimension of $\mathcal{H}^{\vee k}$ if $\dim(\mathcal{H}) = n$.
45. Let $A \in B(\mathcal{K})$ and $B \in B(\mathcal{H})$ be operators on the finite-dimensional spaces $\mathcal{H}$ and $\mathcal{K}$. Show that

$$\det(A \otimes B) = (\det A)^m (\det B)^n,$$

where $n = \dim \mathcal{H}$ and $m = \dim \mathcal{K}$. (Hint: The determinant is the product of the eigenvalues.)
46. Show that $\|A \otimes B\| = \|A\| \cdot \|B\|$.
47. Use Theorem 1.60 to prove that $\det(AB) = \det A \times \det B$. (Hint: Show that $(AB)^{\wedge k} = (A^{\wedge k})(B^{\wedge k})$.)
48. Let $x^n + c_1 x^{n-1} + \cdots + c_n$ be the characteristic polynomial of $A \in \mathbb{M}_n$. Show that $c_k = \operatorname{Tr} A^{\wedge k}$.
49. Show that

$$\mathcal{H} \otimes \mathcal{H} = (\mathcal{H} \vee \mathcal{H}) \oplus (\mathcal{H} \wedge \mathcal{H})$$

for a Hilbert space $\mathcal{H}$.
50. Give an example of an $A \in \mathbb{M}_n(\mathbb{C})$ such that the spectrum of A is in $\mathbb{R}^+$ and A is not positive.
51. Let $A \in \mathbb{M}_n(\mathbb{C})$. Show that A is positive if and only if X^*AX is positive for every $X \in \mathbb{M}_n(\mathbb{C})$.
52. Let $A \in B(\mathcal{H})$. Prove the equivalence of the following assertions: (i) $\|A\| \leq 1$, (ii) $A^*A \leq I$, and (iii) $AA^* \leq I$.
53. Let $A \in \mathbb{M}_n(\mathbb{C})$. Show that A is positive if and only if $\operatorname{Tr} XA$ is positive for every positive $X \in \mathbb{M}_n(\mathbb{C})$.
54. Let $\|A\| \leq 1$. Show that there are unitaries U and V such that

$$A = \frac{1}{2}(U + V).$$

(Hint: Use Example 1.39.)
55. Show that a matrix is weakly positive if and only if it is the product of two positive definite matrices.
56. Let $V : \mathbb{C}^n \to \mathbb{C}^n \otimes \mathbb{C}^n$ be defined as $V e_i = e_i \otimes e_i$. Show that

$$V^*(A \otimes B)V = A \circ B$$

for $A, B \in \mathbb{M}_n(\mathbb{C})$. Conclude **Schur's theorem**.
57. Show that

$$|\operatorname{per}(AB)|^2 \leq \operatorname{per}(AA^*)\operatorname{per}(B^*B).$$

58. Let $A \in \mathbb{M}_n$ and $B \in \mathbb{M}_m$. Show that

$$\text{Tr}\,(I_m \otimes A + B \otimes I_n) = m\text{Tr}\,A + n\text{Tr}\,B.$$

59. For a vector $f \in \mathcal{H}$ the linear operator $a^+(f) : \vee^k\mathcal{H} \to \vee^{k+1}\mathcal{H}$ is defined as

$$a^+(f)\, v_1 \vee v_2 \vee \cdots \vee v_k = f \vee v_1 \vee v_2 \vee \cdots \vee v_k.$$

Compute the adjoint of $a^+(f)$, which is denoted by $a(f)$.

60. For $A \in B(\mathcal{H})$ let $\mathcal{F}(A) : \vee^k\mathcal{H} \to \vee^k\mathcal{H}$ be defined as

$$\mathcal{F}(A)\, v_1 \vee v_2 \vee \cdots \vee v_k = \sum_{i=1}^{k} v_1 \vee v_2 \vee \cdots \vee v_{i-1} \vee Av_i \vee v_{i+1} \vee \cdots \vee v_k.$$

Show that

$$\mathcal{F}(|f\rangle\langle g|) = a^+(f)a(g)$$

for $f, g \in \mathcal{H}$. (Recall that a and a^+ are defined in the previous exercise.)

61. The group

$$\mathcal{G} = \left\{ \begin{bmatrix} a & b \\ 0 & c \end{bmatrix} : a, b, c \in \mathbb{R}, a \neq 0, c \neq 0 \right\}$$

is locally compact. Show that the left invariant Haar measure μ can be defined as

$$\mu(H) = \int_H p(A)\, dA,$$

where

$$A = \begin{bmatrix} x & y \\ 0 & z \end{bmatrix}, \quad p(A) = \frac{1}{x^2|z|}, \quad dA = dx\, dy\, dz.$$

Show that the right invariant Haar measure is similar, but

$$p(A) = \frac{1}{|x|z^2}.$$

Chapter 2
Mappings and Algebras

Most of the statements and definitions in this chapter are formulated in the Hilbert space setting. The Hilbert space is always assumed to be finite-dimensional, so instead of operators one can consider matrices. The idea of block matrices provides quite a useful tool in matrix theory. Some basic facts on block matrices are given in Sect. 2.1. Matrices have two primary structures; one is of course their algebraic structure with addition, multiplication, adjoint, etc., and another is the order structure coming from the partial order of positive semidefiniteness, as explained in Sect. 2.2. Based on this order one can consider several notions of positivity for linear maps between matrix algebras, which are discussed in Sect. 2.6.

2.1 Block Matrices

If $\mathcal{H}_1$ and $\mathcal{H}_2$ are Hilbert spaces, then $\mathcal{H}_1 \oplus \mathcal{H}_2$ consists of all the pairs (f_1, f_2), where $f_1 \in \mathcal{H}_1$ and $f_2 \in \mathcal{H}_2$. The linear combinations of the pairs are computed entrywise and the inner product is defined as

$$\langle (f_1, f_2), (g_1, g_2) \rangle := \langle f_1, g_1 \rangle + \langle f_2, g_2 \rangle.$$

It follows that the subspaces $\{(f_1, 0) : f_1 \in \mathcal{H}_1\}$ and $\{(0, f_2) : f_2 \in \mathcal{H}_2\}$ are orthogonal and span the direct sum $\mathcal{H}_1 \oplus \mathcal{H}_2$.

Assume that $\mathcal{H} = \mathcal{H}_1 \oplus \mathcal{H}_2$, $\mathcal{K} = \mathcal{K}_1 \oplus \mathcal{K}_2$ and $A : \mathcal{H} \to \mathcal{K}$ is a linear operator. A general element of $\mathcal{H}$ has the form $(f_1, f_2) = (f_1, 0) + (0, f_2)$. We have $A(f_1, 0) = (g_1, g_2)$ and $A(0, f_2) = (g_1', g_2')$ for some $g_1, g_1' \in \mathcal{K}_1$ and $g_2, g_2' \in \mathcal{K}_2$. The linear mapping A is determined uniquely by the following four linear mappings:

$$A_{i1} : f_1 \mapsto g_i, \quad A_{i1} : \mathcal{H}_1 \to \mathcal{K}_i \quad (1 \le i \le 2)$$

F. Hiai and D. Petz, *Introduction to Matrix Analysis and Applications*, Universitext, DOI: 10.1007/978-3-319-04150-6_2, © Hindustan Book Agency 2014

and
$$A_{i2} : f_2 \mapsto g_i', \quad A_{i2} : \mathcal{H}_2 \to \mathcal{K}_i \quad (1 \le i \le 2).$$

We write A in the form
$$\begin{bmatrix} A_{11} & A_{12} \\ A_{21} & A_{22} \end{bmatrix}.$$

The advantage of this notation is the formula
$$\begin{bmatrix} A_{11} & A_{12} \\ A_{21} & A_{22} \end{bmatrix} \begin{bmatrix} f_1 \\ f_2 \end{bmatrix} = \begin{bmatrix} A_{11} f_1 + A_{12} f_2 \\ A_{21} f_1 + A_{22} f_2 \end{bmatrix}.$$

(The right-hand side is $A(f_1, f_2)$ written in the form of a column vector.)

Assume that $e_1^i, e_2^i, \ldots, e_{m(i)}^i$ is a basis in $\mathcal{H}_i$ and $f_1^j, f_2^j, \ldots, f_{n(j)}^j$ is a basis in $\mathcal{K}_j, 1 \le i, j \le 2$. The linear operators $A_{ij} : \mathcal{H}_j \to \mathcal{K}_i$ have a matrix $[A_{ij}]$ with respect to these bases. Since
$$\{(e_t^1, 0) : 1 \le t \le m(1)\} \cup \{(0, e_u^2) : 1 \le u \le m(2)\}$$

is a basis in $\mathcal{H}$ and similarly
$$\{(f_t^1, 0) : 1 \le t \le n(1)\} \cup \{(0, f_u^2) : 1 \le u \le n(2)\}$$

is a basis in $\mathcal{K}$, the operator A has an $(n(1) + n(2)) \times (m(1) + m(2))$ matrix which is expressed by the $n(i) \times m(j)$ matrices $[A_{ij}]$ as
$$[A] = \begin{bmatrix} [A_{11}] & [A_{12}] \\ [A_{21}] & [A_{22}] \end{bmatrix}.$$

This is a 2×2 matrix with matrix entries and it is called a **block matrix**.
Computation with block matrices is similar to that of ordinary matrices:
$$\begin{bmatrix} [A_{11}] & [A_{12}] \\ [A_{21}] & [A_{22}] \end{bmatrix}^* = \begin{bmatrix} [A_{11}]^* & [A_{21}]^* \\ [A_{12}]^* & [A_{22}]^* \end{bmatrix},$$

$$\begin{bmatrix} [A_{11}] & [A_{12}] \\ [A_{21}] & [A_{22}] \end{bmatrix} + \begin{bmatrix} [B_{11}] & [B_{12}] \\ [B_{21}] & [B_{22}] \end{bmatrix} = \begin{bmatrix} [A_{11}] + [B_{11}] & [A_{12}] + [B_{12}] \\ [A_{21}] + [B_{21}] & [A_{22}] + [B_{22}] \end{bmatrix}$$

and

$$\begin{bmatrix} [A_{11}] & [A_{12}] \\ [A_{21}] & [A_{22}] \end{bmatrix} \cdot \begin{bmatrix} [B_{11}] & [B_{12}] \\ [B_{21}] & [B_{22}] \end{bmatrix}$$
$$= \begin{bmatrix} [A_{11}] \cdot [B_{11}] + [A_{12}] \cdot [B_{21}] & [A_{11}] \cdot [B_{12}] + [A_{12}] \cdot [B_{22}] \\ [A_{21}] \cdot [B_{11}] + [A_{22}] \cdot [B_{21}] & [A_{21}] \cdot [B_{12}] + [A_{22}] \cdot [B_{22}] \end{bmatrix}.$$

In several cases we do not emphasize the entries of a block matrix

$$\begin{bmatrix} A & B \\ C & D \end{bmatrix}.$$

However, if this matrix is self-adjoint we assume that $A = A^*$, $B^* = C$ and $D = D^*$. (These conditions include that A and D are square matrices, $A \in \mathbb{M}_n$ and $B \in \mathbb{M}_m$.)

The block matrix is used for the definition of **reducible matrices**. $A \in \mathbb{M}_n$ is reducible if there is a permutation matrix $P \in \mathbb{M}_n$ such that

$$P^t A P = \begin{bmatrix} B & C \\ 0 & D \end{bmatrix}.$$

A matrix $A \in \mathbb{M}_n$ is **irreducible** if it is not reducible.

For a 2×2 matrix, it is very easy to check the positivity:

$$\begin{bmatrix} a & b \\ \bar{b} & c \end{bmatrix} \geq 0 \quad \text{if and only if} \quad a \geq 0 \quad \text{and} \quad b\bar{b} \leq ac.$$

If the entries are matrices, then the condition for positivity is similar but it is a bit more complicated. It is obvious that a diagonal block matrix

$$\begin{bmatrix} A & 0 \\ 0 & D \end{bmatrix}$$

is positive if and only if the diagonal entries A and D are positive.

Theorem 2.1 *Assume that A is invertible. The self-adjoint block matrix*

$$\begin{bmatrix} A & B \\ B^* & C \end{bmatrix} \tag{2.1}$$

is positive if and only if A is positive and

$$B^* A^{-1} B \leq C.$$

Proof: First assume that $A = I$. The positivity of

$$\begin{bmatrix} I & B \\ B^* & C \end{bmatrix}$$

is equivalent to the condition

$$\langle (f_1, f_2), \begin{bmatrix} I & B \\ B^* & C \end{bmatrix} (f_1, f_2) \rangle \geq 0$$

for all vectors f_1 and f_2. A computation gives that this condition is

$$\langle f_1, f_1 \rangle + \langle f_2, C f_2 \rangle \geq -2\mathrm{Re}\,\langle B f_2, f_1 \rangle.$$

If we replace f_1 by $e^{i\varphi} f_1$ with real φ, then the left-hand side does not change, while the right-hand side becomes $2|\langle B f_2, f_1 \rangle|$ for an appropriate φ. Choosing $f_1 = B f_2$, we obtain the condition

$$\langle f_2, C f_2 \rangle \geq \langle f_2, B^* B f_2 \rangle$$

for every f_2. This means that positivity implies the condition $C \geq B^* B$. The converse is also true, since the right-hand side of the equation

$$\begin{bmatrix} I & B \\ B^* & C \end{bmatrix} = \begin{bmatrix} I & 0 \\ B^* & 0 \end{bmatrix} \begin{bmatrix} I & B \\ 0 & 0 \end{bmatrix} + \begin{bmatrix} 0 & 0 \\ 0 & C - B^* B \end{bmatrix}$$

is the sum of two positive block matrices.

For a general positive invertible A, the positivity of (2.1) is equivalent to the positivity of the block matrix

$$\begin{bmatrix} A^{-1/2} & 0 \\ 0 & I \end{bmatrix} \begin{bmatrix} A & B \\ B^* & C \end{bmatrix} \begin{bmatrix} A^{-1/2} & 0 \\ 0 & I \end{bmatrix} = \begin{bmatrix} I & A^{-1/2} B \\ B^* A^{-1/2} & C \end{bmatrix}.$$

This gives the condition $C \geq B^* A^{-1} B$.   □

Another important characterization of the positivity of (2.1) is the condition that $A, C \geq 0$ and $B = A^{1/2} W C^{1/2}$ with a contraction W. (Here the invertibility of A or C is not necessary.)

Theorem 2.1 has applications in different areas, see for example the Cramér–Rao inequality, Sect. 7.5.

Theorem 2.2 *For an invertible A, we have the so-called* **Schur factorization**

$$\begin{bmatrix} A & B \\ C & D \end{bmatrix} = \begin{bmatrix} I & 0 \\ C A^{-1} & I \end{bmatrix} \cdot \begin{bmatrix} A & 0 \\ 0 & D - C A^{-1} B \end{bmatrix} \cdot \begin{bmatrix} I & A^{-1} B \\ 0 & I \end{bmatrix}. \tag{2.2}$$

The proof is simply the computation of the product on the right-hand side. Since

$$\begin{bmatrix} I & 0 \\ C A^{-1} & I \end{bmatrix}^{-1} = \begin{bmatrix} I & 0 \\ -C A^{-1} & I \end{bmatrix}$$

is invertible, the positivity of the left-hand side of (2.2) with $C = B^*$ is equivalent to the positivity of the middle factor of the right-hand side. This fact gives the second proof of Theorem 2.1.

In the Schur factorization the first factor is lower triangular, the second factor is block diagonal and the third one is upper triangular. This structure allows an easy computation of the determinant and the inverse.

Theorem 2.3 *The determinant can be computed as follows.*

$$\det \begin{bmatrix} A & B \\ C & D \end{bmatrix} = \det A \cdot \det (D - CA^{-1}B).$$

If

$$M = \begin{bmatrix} A & B \\ C & D \end{bmatrix},$$

then $D - CA^{-1}B$ is called the **Schur complement** of A in M, and is denoted by M/A. Hence the determinant formula becomes $\det M = \det A \cdot \det (M/A)$.

Theorem 2.4 *Let*

$$M = \begin{bmatrix} A & B \\ B^* & C \end{bmatrix}$$

be a positive invertible matrix. Then

$$M/C = A - BC^{-1}B^* = \sup \left\{ X \geq 0 : \begin{bmatrix} X & 0 \\ 0 & 0 \end{bmatrix} \leq \begin{bmatrix} A & B \\ B^* & C \end{bmatrix} \right\}.$$

Proof: The condition

$$\begin{bmatrix} A - X & B \\ B^* & C \end{bmatrix} \geq 0$$

is equivalent to

$$A - X \geq BC^{-1}B^*,$$

and this gives the result. □

Theorem 2.5 *For a block matrix*

$$0 \leq \begin{bmatrix} A & X \\ X^* & B \end{bmatrix} \in \mathbb{M}_n,$$

we have

$$\begin{bmatrix} A & X \\ X^* & B \end{bmatrix} = U \begin{bmatrix} A & 0 \\ 0 & 0 \end{bmatrix} U^* + V \begin{bmatrix} 0 & 0 \\ 0 & B \end{bmatrix} V^*$$

for some unitaries $U, V \in \mathbb{M}_n$.

Proof: We can take

$$0 \le \begin{bmatrix} C & Y \\ Y^* & D \end{bmatrix} \in \mathbb{M}_n$$

such that

$$\begin{bmatrix} A & X \\ X^* & B \end{bmatrix} = \begin{bmatrix} C & Y \\ Y^* & D \end{bmatrix} \begin{bmatrix} C & Y \\ Y^* & D \end{bmatrix} = \begin{bmatrix} C^2 + YY^* & CV + YD \\ Y^*C + DY^* & Y^*Y + D^2 \end{bmatrix}.$$

It follows that

$$\begin{bmatrix} A & X \\ X^* & B \end{bmatrix} = \begin{bmatrix} C & 0 \\ Y^* & 0 \end{bmatrix} \begin{bmatrix} C & Y \\ 0 & 0 \end{bmatrix} + \begin{bmatrix} 0 & Y \\ 0 & D \end{bmatrix} \begin{bmatrix} 0 & 0 \\ Y^* & D \end{bmatrix} = T^*T + S^*S,$$

where

$$T = \begin{bmatrix} C & Y \\ 0 & 0 \end{bmatrix} \quad \text{and} \quad S = \begin{bmatrix} 0 & 0 \\ Y^* & D \end{bmatrix}.$$

When $T = U|T|$ and $S = V|S|$ for unitaries $U, V \in \mathbb{M}_n$, then

$$T^*T = U(TT^*)U^* \quad \text{and} \quad S^*S = V(SS^*)V^*.$$

From the formulas

$$TT^* = \begin{bmatrix} C^2 + YY^* & 0 \\ 0 & 0 \end{bmatrix} = \begin{bmatrix} A & 0 \\ 0 & 0 \end{bmatrix}, \quad SS^* = \begin{bmatrix} 0 & 0 \\ 0 & Y^*Y + D^2 \end{bmatrix} = \begin{bmatrix} 0 & 0 \\ 0 & B \end{bmatrix},$$

we have the result. $\square$

Example 2.6 Similarly to the previous theorem we take a block matrix

$$0 \le \begin{bmatrix} A & X \\ X^* & B \end{bmatrix} \in \mathbb{M}_n.$$

For a unitary

$$W := \frac{1}{\sqrt{2}} \begin{bmatrix} iI & -I \\ iI & I \end{bmatrix}$$

we notice that

$$W \begin{bmatrix} A & X \\ X^* & B \end{bmatrix} W^* = \begin{bmatrix} \frac{A+B}{2} + \mathrm{Im}\, X & \frac{A-B}{2} + i\mathrm{Re}\, X \\ \frac{A-B}{2} - i\mathrm{Re}\, X & \frac{A+B}{2} - \mathrm{Im}\, X \end{bmatrix}.$$

So Theorem 2.5 gives

$$\begin{bmatrix} A & X \\ X^* & B \end{bmatrix} = U \begin{bmatrix} \frac{A+B}{2} + \mathrm{Im}\, X & 0 \\ 0 & 0 \end{bmatrix} U^* + V \begin{bmatrix} 0 & 0 \\ 0 & \frac{A+B}{2} - \mathrm{Im}\, X \end{bmatrix} V^*$$

for some unitaries $U, V \in \mathbb{M}_n$. $\qquad\qquad\qquad\qquad\qquad\qquad\qquad$ □

We have two remarks. If C is not invertible, then the supremum in Theorem 2.4 is $A - BC^\dagger B^*$, where $C^\dagger$ is the Moore–Penrose generalized inverse. The supremum of that theorem can be formulated without the block matrix formalism. Assume that P is an ortho-projection (see Sect. 2.3). Then

$$[P]M := \sup\{N : 0 \le N \le M, \quad PN = N\}. \qquad (2.3)$$

If

$$P = \begin{bmatrix} I & 0 \\ 0 & 0 \end{bmatrix} \quad \text{and} \quad M = \begin{bmatrix} A & B \\ B^* & C \end{bmatrix},$$

then $[P]M = M/C$. The formula (2.3) makes clear that if Q is another ortho-projection such that $P \le Q$, then $[P]M \le [P]QMQ$.

It follows from the factorization that for an invertible block matrix

$$\begin{bmatrix} A & B \\ C & D \end{bmatrix},$$

both A and $D - CA^{-1}B$ must be invertible. This implies that

$$\begin{bmatrix} A & B \\ C & D \end{bmatrix}^{-1} = \begin{bmatrix} I & -A^{-1}B \\ 0 & I \end{bmatrix} \cdot \begin{bmatrix} A^{-1} & 0 \\ 0 & (D - CA^{-1}B)^{-1} \end{bmatrix} \cdot \begin{bmatrix} I & 0 \\ -CA^{-1} & I \end{bmatrix}.$$

After multiplication on the right-hand side, we have the following:

$$\begin{bmatrix} A & B \\ C & D \end{bmatrix}^{-1} = \begin{bmatrix} A^{-1} + A^{-1}BW^{-1}CA^{-1} & -A^{-1}BW^{-1} \\ -W^{-1}CA^{-1} & W^{-1} \end{bmatrix}$$

$$= \begin{bmatrix} V^{-1} & -V^{-1}BD^{-1} \\ -D^{-1}CV^{-1} & D^{-1} + D^{-1}CV^{-1}BD^{-1} \end{bmatrix}, \qquad (2.4)$$

where $W = M/A := D - CA^{-1}B$ and $V = M/D := A - BD^{-1}C$.

Example 2.7 Let $X_1, X_2, \ldots, X_{m+k}$ be real random variables with (Gaussian) joint probability distribution

$$f_M(\mathbf{z}) := \sqrt{\frac{\det M}{(2\pi)^{m+k}}} \exp\left(-\tfrac{1}{2}\langle \mathbf{z}, M\mathbf{z}\rangle\right),$$

where $\mathbf{z} = (z_1, z_2, \ldots, z_{m+k})$ and M is a positive definite real $(m + k) \times (m + k)$ matrix, see Example 1.43. We want to compute the distribution of the random variables $X_1, X_2, \ldots, X_m$.

Let

$$M = \begin{bmatrix} A & B \\ B^* & D \end{bmatrix}$$

be written in the form of a block matrix, where A is $m \times m$ and D is $k \times k$. Let $z = (x_1, x_2)$, where $x_1 \in \mathbb{R}^m$ and $x_2 \in \mathbb{R}^k$. Then the marginal of the Gaussian probability distribution

$$f_M(x_1, x_2) = \sqrt{\frac{\det M}{(2\pi)^{m+k}}} \exp\left(-\tfrac{1}{2}\langle (x_1, x_2), M(x_1, x_2)\rangle\right)$$

on $\mathbb{R}^m$ is the distribution

$$f_1(x_1) = \sqrt{\frac{\det M}{(2\pi)^m \det D}} \exp\left(-\tfrac{1}{2}\langle x_1, (A - BD^{-1}B^*)x_1\rangle\right). \qquad (2.5)$$

We have

$$\langle (x_1, x_2), M(x_1, x_2)\rangle = \langle Ax_1 + Bx_2, x_1\rangle + \langle B^*x_1 + Dx_2, x_2\rangle$$
$$= \langle Ax_1, x_1\rangle + \langle Bx_2, x_1\rangle + \langle B^*x_1, x_2\rangle + \langle Dx_2, x_2\rangle$$
$$= \langle Ax_1, x_1\rangle + 2\langle B^*x_1, x_2\rangle + \langle Dx_2, x_2\rangle$$
$$= \langle Ax_1, x_1\rangle + \langle D(x_2 + Wx_1), (x_2 + Wx_1)\rangle - \langle DWx_1, Wx_1\rangle,$$

where $W = D^{-1}B^*$. We integrate on $\mathbb{R}^k$ as

$$\int \exp\left(-\tfrac{1}{2}(x_1, x_2)M(x_1, x_2)^t\right) dx_2$$
$$= \exp\left(-\tfrac{1}{2}(\langle Ax_1, x_1\rangle - \langle DWx_1, Wx_1\rangle)\right)$$
$$\times \int \exp\left(-\tfrac{1}{2}\langle D(x_2 + Wx_1), (x_2 + Wx_1)\rangle\right) dx_2$$
$$= \exp\left(-\frac{1}{2}\langle (A - BD^{-1}B^*)x_1, x_1\rangle\right)\sqrt{\frac{(2\pi)^k}{\det D}}$$

and obtain (2.5).

This computation gives a proof of Theorem 2.3 (for a real positive definite matrix) as well. If we know that $f_1(x_1)$ is Gaussian, then its quadratic matrix can be obtained from formula (2.4). The covariance of $X_1, X_2, \ldots, X_{m+k}$ is M^{-1}. Therefore, the covariance of $X_1, X_2, \ldots, X_m$ is $(A - BD^{-1}B^*)^{-1}$. It follows that the quadratic matrix is the inverse: $A - BD^{-1}B^* \equiv M/D$. $\qquad \square$

Theorem 2.8 *Let A be a positive $n \times n$ block matrix with $k \times k$ entries. Then A is the sum of block matrices B of the form $[B]_{ij} = X_i^* X_j$ for some $k \times k$ matrices $X_1, X_2, \ldots, X_n$.*

Proof: A can be written as $C^* C$ for some

$$C = \begin{bmatrix} C_{11} & C_{12} & \ldots & C_{1n} \\ C_{21} & C_{22} & \ldots & C_{2n} \\ \vdots & \vdots & \ddots & \vdots \\ C_{n1} & C_{n2} & \ldots & C_{nn} \end{bmatrix}.$$

Let B_i be the block matrix such that its ith row is the same as in C and all other elements are 0. Then $C = B_1 + B_2 + \cdots + B_n$ and for $t \neq i$ we have $B_t^* B_i = 0$. Therefore,

$$A = (B_1 + B_2 + \cdots + B_n)^* (B_1 + B_2 + \cdots + B_n) = B_1^* B_1 + B_2^* B_2 + \cdots + B_n^* B_n.$$

The (i, j) entry of $B_t^* B_t$ is $C_{ti}^* C_{tj}$; hence this matrix is of the required form. □

Example 2.9 Let $\mathcal{H}$ be an n-dimensional Hilbert space and $A \in B(\mathcal{H})$ be a positive operator with eigenvalues $\lambda_1 \geq \lambda_2 \geq \cdots \geq \lambda_n$. If $x, y \in \mathcal{H}$ are orthogonal vectors, then

$$|\langle x, Ay \rangle|^2 \leq \left(\frac{\lambda_1 - \lambda_n}{\lambda_1 + \lambda_n} \right)^2 \langle x, Ax \rangle \langle y, Ay \rangle,$$

which is called the **Wielandt inequality**. (It also appeared in Theorem 1.45.) The argument presented here includes a block matrix.

We can assume that x and y are unit vectors and we extend them to a basis. Let

$$M = \begin{bmatrix} \langle x, Ax \rangle & \langle x, Ay \rangle \\ \langle y, Ax \rangle & \langle y, Ay \rangle \end{bmatrix},$$

where A has a block matrix

$$\begin{bmatrix} M & B \\ B^* & C \end{bmatrix}. \tag{2.6}$$

We can see that $M \geq 0$ and its determinant is positive:

$$|\langle x, Ay \rangle|^2 \leq \langle x, Ax \rangle \langle y, Ay \rangle.$$

If $\lambda_n = 0$, then the proof is complete. Now we assume that $\lambda_n > 0$. Let α and β be the eigenvalues of M. Formula (1.27) tells us that

$$|\langle x, Ay \rangle|^2 \leq \left(\frac{\alpha - \beta}{\alpha + \beta} \right)^2 \langle x, Ax \rangle \langle y, Ay \rangle.$$

We need the inequality

$$\frac{\alpha - \beta}{\alpha + \beta} \leq \frac{\lambda_1 - \lambda_n}{\lambda_1 + \lambda_n}$$

when $\alpha \geq \beta$. This is true, since $\lambda_1 \geq \alpha \geq \beta \geq \lambda_n$. $\square$

As an application of the block matrix technique, we consider the following result, called the **UL-factorization** (or the Cholesky factorization).

Theorem 2.10 *Let X be an $n \times n$ invertible positive matrix. Then there is a unique upper triangular matrix T with positive diagonal such that $X = TT^*$.*

Proof: The proof is by mathematical induction on n. For $n = 1$ the statement is clear. We assume that the factorization is true for $(n - 1) \times (n - 1)$ matrices and write X in the form

$$\begin{bmatrix} A & B \\ B^* & C \end{bmatrix},$$ (2.7)

where A is an (invertible) $(n - 1) \times (n - 1)$ matrix and C is a number. If

$$T = \begin{bmatrix} T_{11} & T_{12} \\ 0 & T_{22} \end{bmatrix}$$

is written in a similar form, then

$$TT^* = \begin{bmatrix} T_{11}T_{11}^* + T_{12}T_{12}^* & T_{12}T_{22}^* \\ T_{22}T_{12}^* & T_{22}T_{22}^* \end{bmatrix}.$$

The condition $X = TT^*$ leads to the equations

$$T_{11}T_{11}^* + T_{12}T_{12}^* = A,$$
$$T_{12}T_{22}^* = B,$$
$$T_{22}T_{22}^* = C.$$

If $C = 0$, then the positivity of (2.7) forces $B = 0$ so that we can apply the induction hypothesis to A. So we may assume that $C > 0$. If the number T_{22} is positive, then $T_{22} = \sqrt{C}$ is the unique solution and moreover

$$T_{12} = BC^{-1/2}, \qquad T_{11}T_{11}^* = A - BC^{-1}B^*.$$

From the positivity of (2.7), we have $A - BC^{-1}B^* \geq 0$ by Theorem 2.1. The induction hypothesis gives that the latter can be written in the form $T_{11}T_{11}^*$ where T_{11} is upper triangular. Therefore T is upper triangular, too. □

If $0 \leq A \in \mathbb{M}_n$ and $0 \leq B \in \mathbb{M}_m$, then $0 \leq A \otimes B$. More generally, if $0 \leq A_i \in \mathbb{M}_n$ and $0 \leq B_i \in \mathbb{M}_m$, then

$$\sum_{i=1}^{k} A_i \otimes B_i$$

is positive. These matrices in $\mathbb{M}_n \otimes \mathbb{M}_m$ are called **separable positive matrices**. Is it true that every positive matrix in $\mathbb{M}_n \otimes \mathbb{M}_m$ is separable? A counterexample follows.

Example 2.11 Let $\mathbb{M}_4 = \mathbb{M}_2 \otimes \mathbb{M}_2$ and

$$D := \frac{1}{2} \begin{bmatrix} 0 & 0 & 0 & 0 \\ 0 & 1 & 1 & 0 \\ 0 & 1 & 1 & 0 \\ 0 & 0 & 0 & 0 \end{bmatrix}.$$

D is a rank 1 positive operator, it is a projection. If $D = \sum_i D_i$, then $D_i = \lambda_i D$. If D is separable, then it is a tensor product. If D is a tensor product, then up to a constant factor it is equal to $(\mathrm{Tr}_2 D) \otimes (\mathrm{Tr}_1 D)$ (as noted in Example 1.56). We have

$$\mathrm{Tr}_1 D = \mathrm{Tr}_2 D = \frac{1}{2} \begin{bmatrix} 1 & 0 \\ 0 & 1 \end{bmatrix}.$$

Their tensor product has rank 4 and it cannot be λD. It follows that this D is not separable. □

In quantum theory the non-separable positive operators are said to be **entangled**. The positive operator D is **maximally entangled** if it has minimal rank (meaning rank 1) and the partial traces have maximal rank. The matrix D in the previous example is maximally entangled.

It is interesting that there is no effective procedure to decide if a positive operator in a tensor product space is separable or entangled.

2.2 Partial Ordering

Let $A, B \in B(\mathcal{H})$ be self-adjoint operators. The **partial ordering** $A \leq B$ holds if $B - A$ is positive, or equivalently

$$\langle x, Ax \rangle \leq \langle x, Bx \rangle$$

for all vectors x. From this formulation one can easily see that $A \leq B$ implies $XAX^* \leq XBX^*$ for every operator X.

Example 2.12 Assume that for the orthogonal projections P and Q the inequality $P \leq Q$ holds. If $Px = x$ for a unit vector x, then $\langle x, Px \rangle \leq \langle x, Qx \rangle \leq 1$ shows that $\langle x, Qx \rangle = 1$. Therefore the relation

$$\|x - Qx\|^2 = \langle x - Qx, x - Qx \rangle = \langle x, x \rangle - \langle x, Qx \rangle = 0$$

gives that $Qx = x$. The range of Q includes the range of P.   □

Let A_n be a sequence of operators on a finite-dimensional Hilbert space. Fix a basis and let $[A_n]$ be the matrix of A_n. Similarly, the matrix of the operator A is $[A]$. Let the Hilbert space be m-dimensional, so the matrices are $m \times m$. Recall that the following conditions are equivalent:

(1) $\|A - A_n\| \to 0$.
(2) $A_n x \to Ax$ for every vector x.
(3) $\langle x, A_n y \rangle \to \langle x, Ay \rangle$ for all vectors x and y.
(4) $\langle x, A_n x \rangle \to \langle x, Ax \rangle$ for every vector x.
(5) $\mathrm{Tr}\, (A - A_n)^*(A - A_n) \to 0$.
(6) $[A_n]_{ij} \to [A]_{ij}$ for every $1 \leq i, j \leq m$.

These conditions describe in several ways the **convergence** of a sequence of operators or matrices.

Theorem 2.13 *Let A_n be an increasing sequence of operators with an upper bound: $A_1 \leq A_2 \leq \cdots \leq B$. Then there is an operator $A \leq B$ such that $A_n \to A$.*

Proof: Let $\phi_n(x, y) := \langle x, A_n y \rangle$ be a sequence of complex bilinear functionals. Then $\phi_n(x, x)$ is a bounded increasing real sequence and it is convergent. By the polarization identity, $\phi_n(x, y)$ is convergent as well and the limit gives a complex bilinear functional ϕ. If the corresponding operator is denoted by A, then

$$\langle x, A_n y \rangle \to \langle x, Ay \rangle$$

for all vectors x and y. This is the convergence $A_n \to A$. The condition $\langle x, Ax \rangle \leq \langle x, Bx \rangle$ means $A \leq B$.   □

Example 2.14 Assume that $0 \leq A \leq I$ for an operator A. Define a sequence T_n of operators by recursion. Let $T_1 = 0$ and

$$T_{n+1} = T_n + \frac{1}{2}(A - T_n^2) \quad (n \in \mathbb{N}).$$

T_n is a polynomial in A with real coefficients. Thus these operators commute with each other. Since

$$I - T_{n+1} = \frac{1}{2}(I - T_n)^2 + \frac{1}{2}(I - A),$$

induction shows that $T_n \leq I$.

We show that $T_1 \leq T_2 \leq T_3 \leq \cdots$ by mathematical induction again. In the recursion

$$T_{n+1} - T_n = \frac{1}{2}\left((I - T_{n-1})(T_n - T_{n-1}) + (I - T_n)(T_n - T_{n-1})\right),$$

$I - T_{n-1} \geq 0$ and $T_n - T_{n-1} \geq 0$ by the assumption. Since they commute their product is positive. Similarly $(I - T_n)(T_n - T_{n-1}) \geq 0$. It follows that the right-hand side is positive.

Theorem 2.13 tells us that T_n converges to an operator B. The limit of the recursion formula yields

$$B = B + \frac{1}{2}(A - B^2).$$

Therefore $A = B^2$. This example is a constructive proof of Theorem 1.38. □

Theorem 2.15 *Assume that $0 < A, B \in \mathbb{M}_n$ are invertible matrices and $A \leq B$. Then $B^{-1} \leq A^{-1}$.*

Proof: The condition $A \leq B$ is equivalent to $B^{-1/2}AB^{-1/2} \leq I$ and the statement $B^{-1} \leq A^{-1}$ is equivalent to $I \leq B^{1/2}A^{-1}B^{1/2}$. If $X = B^{-1/2}AB^{-1/2}$, then we have to show that $X \leq I$ implies $X^{-1} \geq I$. The condition $X \leq I$ means that all eigenvalues of X are in the interval $(0, 1]$. This implies that all eigenvalues of X^{-1} are in $[1, \infty)$. □

Assume that $A \leq B$. It follows from (1.13) that the largest eigenvalue of A is smaller than the largest eigenvalue of B. Let $\lambda(A) = (\lambda_1(A), \ldots, \lambda_n(A))$ denote the vector of the eigenvalues of A in decreasing order (counting multiplicities).

The next result is called **Weyl's monotonicity theorem**.

Theorem 2.16 *If $A \leq B$, then $\lambda_k(A) \leq \lambda_k(B)$ for all k.*

This is a consequence of the minimax principle, Theorem 1.27.

Corollary 2.17 *Let $A, B \in B(\mathcal{H})$ be self-adjoint operators.*

(1) *If $A \leq B$, then $\mathrm{Tr}\, A \leq \mathrm{Tr}\, B$.*
(2) *If $0 \leq A \leq B$, then $\det A \leq \det B$.*

Theorem 2.18 (Schur's theorem) *Let A and B be positive $n \times n$ matrices. Then*

$$C_{ij} = A_{ij}B_{ij} \qquad (1 \leq i, j \leq n)$$

determines a positive matrix.

Proof: If $A_{ij} = \overline{\lambda}_i \lambda_j$ and $B_{ij} = \overline{\mu}_i \mu_j$, then $C_{ij} = \overline{\lambda_i \mu_i} \lambda_j \mu_j$ and C is positive by Example 1.40. The general case reduces to this one.   □

The matrix C of the previous theorem is called the **Hadamard** (or Schur) **product** of the matrices A and B and is denoted by $C = A \circ B$.

Corollary 2.19 *Assume that* $0 \leq A \leq B$ *and* $0 \leq C \leq D$. *Then* $A \circ C \leq B \circ D$.

Proof: The equation

$$B \circ D - A \circ C = (B - A) \circ D + A \circ (D - C)$$

implies the statement.   □

Theorem 2.20 (Oppenheim's inequality) *If* $0 \leq A, B \in \mathbb{M}_n$, *then*

$$\det(A \circ B) \geq \left(\prod_{i=1}^{n} A_{ii} \right) \det B.$$

Proof: For $n = 1$ the statement is obvious. The argument will be by induction on n. We take the Schur complementation and the block matrix formalism

$$A = \begin{bmatrix} a & A_1 \\ A_2 & A_3 \end{bmatrix} \quad \text{and} \quad B = \begin{bmatrix} b & B_1 \\ B_2 & B_3 \end{bmatrix},$$

where $a, b \in [0, \infty)$. We may assume that $a, b > 0$. From the inductive assumption we have

$$\det(A_3 \circ (B/b)) \geq A_{2,2} A_{3,3} \dots A_{n,n} \det(B/b). \tag{2.8}$$

From Theorem 2.3 we have $\det(A \circ B) = ab \det(A \circ B/ab)$ and

$$A \circ B/ab = A_3 \circ B_3 - (A_2 \circ B_2)a^{-1}b^{-1}(A_1 \circ B_1)$$
$$= A_3 \circ (B/b) + (A/a) \circ (B_2 B_1 b^{-1}).$$

The matrices A/a and B/b are positive, see Theorem 2.4. So the matrices

$$A_3 \circ (B/b) \quad \text{and} \quad (A/a) \circ (B_2 B_1 b^{-1})$$

are positive as well. Thus

$$\det(A \circ B) \geq ab \det(A_3 \circ (B/b)).$$

Finally the inequality (2.8) gives

$$\det(A \circ B) \geq \left(\prod_{i=1}^{n} A_{ii} \right) b \det(B/b).$$

Since $\det B = b \det(B/b)$, the proof is complete. $\qquad\square$

A linear mapping $\alpha : \mathbb{M}_n \to \mathbb{M}_n$ is called **completely positive** if it has the form

$$\alpha(B) = \sum_{i=1}^{k} V_i^* B V_i$$

for some matrices V_i. The sum of completely positive mappings is completely positive. (More details concerning completely positive mappings can be found in Theorem 2.49.)

Example 2.21 Let $A \in \mathbb{M}_n$ be a positive matrix. The mapping $S_A : B \mapsto A \circ B$ sends positive matrices to positive matrices. Therefore it is a positive mapping.

We want to show that S_A is completely positive. Since S_A is additive in A, it is enough to prove the case $A_{ij} = \overline{\lambda}_i \lambda_j$. Then

$$S_A(B) = \mathrm{Diag}(\overline{\lambda}_1, \overline{\lambda}_2, \dots, \overline{\lambda}_n) \, B \, \mathrm{Diag}(\lambda_1, \lambda_2, \dots, \lambda_n)$$

and S_A is completely positive. $\qquad\square$

2.3 Projections

Let $\mathcal{K}$ be a closed subspace of a Hilbert space $\mathcal{H}$. Any vector $x \in \mathcal{H}$ can be written in the form $x_0 + x_1$, where $x_0 \in \mathcal{K}$ and $x_1 \perp \mathcal{K}$, see Theorem 1.11. The linear mapping $P : x \mapsto x_0$ is called the (orthogonal) **projection** onto $\mathcal{K}$. The orthogonal projection P has the properties $P = P^2 = P^*$. If an operator $P \in B(\mathcal{H})$ satisfies $P = P^2 = P^*$, then it is an (orthogonal) projection (onto its range). Instead of orthogonal projection the terminology **ortho-projection** is also used.

The partial ordering is very simple for projections, see Example 2.12. If P and Q are projections, then the relation $P \leq Q$ means that the range of P is included in the range of Q. An equivalent algebraic formulation is $PQ = P$. The largest projection in $\mathbb{M}_n$ is the identity I and the smallest one is 0. Therefore $0 \leq P \leq I$ for any projection $P \in \mathbb{M}_n$.

Example 2.22 In $\mathbb{M}_2$ the non-trivial ortho-projections have rank 1 and they have the form

$$P = \frac{1}{2} \begin{bmatrix} 1 + a_3 & a_1 - ia_2 \\ a_1 + ia_2 & 1 - a_3 \end{bmatrix},$$

where $a_1, a_2, a_3 \in \mathbb{R}$ and $a_1^2 + a_2^2 + a_3^2 = 1$. In terms of the **Pauli matrices**

$$\sigma_0 = \begin{bmatrix} 1 & 0 \\ 0 & 1 \end{bmatrix}, \quad \sigma_1 = \begin{bmatrix} 0 & 1 \\ 1 & 0 \end{bmatrix}, \quad \sigma_2 = \begin{bmatrix} 0 & -i \\ i & 0 \end{bmatrix}, \quad \sigma_3 = \begin{bmatrix} 1 & 0 \\ 0 & -1 \end{bmatrix} \tag{2.9}$$

we have

$$P = \frac{1}{2}\left(\sigma_0 + \sum_{i=1}^{3} a_i \sigma_i \right).$$

An equivalent formulation is $P = |x\rangle\langle x|$, where $x \in \mathbb{C}^2$ is a unit vector. This can be extended to an arbitrary ortho-projection $Q \in \mathbb{M}_n(\mathbb{C})$:

$$Q = \sum_{i=1}^{k} |x_i\rangle\langle x_i|,$$

where the set $\{x_i : 1 \le i \le k\}$ is a family of orthogonal unit vectors in $\mathbb{C}^n$. (k is the rank of the image of Q, or Tr Q.)   □

If P is a projection, then $I - P$ is a projection as well and it is often denoted by $P^\perp$, since the range of $I - P$ is the orthogonal complement of the range of P.

Example 2.23 Let P and Q be projections. The relation $P \perp Q$ means that the range of P is orthogonal to the range of Q. An equivalent algebraic formulation is $PQ = 0$. Since the orthogonality relation is symmetric, $PQ = 0$ if and only if $QP = 0$. (We can also arrive at this statement by taking the adjoint.)

We show that $P \perp Q$ if and only if $P + Q$ is a projection as well. $P + Q$ is self-adjoint and it is a projection if

$$(P + Q)^2 = P^2 + PQ + QP + Q^2 = P + Q + PQ + QP = P + Q$$

or equivalently

$$PQ + QP = 0.$$

This is true if $P \perp Q$. On the other hand, the condition $PQ + QP = 0$ implies that $PQP + QP^2 = PQP + QP = 0$ and QP must be self-adjoint. We conclude that $PQ = 0$, which is the orthogonality.   □

Assume that P and Q are projections on the same Hilbert space. Among the projections which are smaller than P and Q there is a maximal projection, denoted by $P \wedge Q$, which is the orthogonal projection onto the intersection of the ranges of P and Q.

Theorem 2.24 *Assume that P and Q are ortho-projections. Then*

$$P \wedge Q = \lim_{n \to \infty} (PQP)^n = \lim_{n \to \infty} (QPQ)^n.$$

Proof: The operator $A := PQP$ is a positive contraction. Therefore the sequence A^n is monotone decreasing and Theorem 2.13 implies that A^n has the limit R. The operator R is self-adjoint. Since $(A^n)^2 \to R^2$ we have $R = R^2$; in other words, R is an ortho-projection. If $Px = x$ and $Qx = x$ for a vector x, then $Ax = x$ and it follows that $Rx = x$. This means that $R \geq P \wedge Q$.

From the inequality $PQP \leq P$, $R \leq P$ follows. Taking the limit of $(PQP)^n Q$ $(PQP)^n = (PQP)^{2n+1}$, we have $RQR = R$. From this we have $R(I - Q)R = 0$ and $(I - Q)R = 0$. This gives $R \leq Q$.

It has been proved that $R \leq P$, Q and $R \geq P \wedge Q$. So $R = P \wedge Q$ is the only possibility.   □

Corollary 2.25 *Assume that P and Q are ortho-projections and $0 \leq H \leq P, Q$. Then $H \leq P \wedge Q$.*

Proof: Since $(I - P)H(I - P) = 0$ implies $H^{1/2}(I - P) = 0$, we have $H^{1/2}P = H^{1/2}$ so that $PHP = H$, and similarly $QHQ = H$. These imply $(PQP)^n H(PQP)^n = H$ and the limit $n \to \infty$ gives $RHR = H$, where $R = P \wedge Q$. Hence $H \leq R$.   □

Let P and Q be ortho-projections. If the ortho-projection R has the property $R \geq P, Q$, then the image of R includes the images of P and Q. The smallest such R projects to the linear subspace generated by the images of P and Q. This ortho-projection is denoted by $P \vee Q$. The set of ortho-projections becomes a lattice with the operations $\wedge$ and $\vee$. However, the so-called distributivity

$$A \vee (B \wedge C) = (A \vee B) \wedge (A \vee C)$$

does not hold.

Example 2.26 We show that any operator $X \in \mathbb{M}_n(\mathbb{C})$ is a linear combination of ortho-projections. We write

$$X = \frac{1}{2}(X + X^*) + \frac{1}{2i}(iX - iX^*),$$

where $X + X^*$ and $iX - iX^*$ are self-adjoint operators. Therefore, it is enough to find linear combinations of ortho-projections for self-adjoint operators. This is essentially the spectral decomposition (1.11).

Assume that φ_0 is defined on projections of $\mathbb{M}_n(\mathbb{C})$ and it has the properties

$$\varphi_0(0) = 0, \quad \varphi_0(I) = 1, \quad \varphi_0(P + Q) = \varphi_0(P) + \varphi_0(Q) \quad \text{if} \quad P \perp Q.$$

It is a famous theorem of **Gleason** that in the case $n > 2$ the mapping φ_0 has a linear extension $\varphi : \mathbb{M}_n(\mathbb{C}) \to \mathbb{C}$. The linearity implies φ is of the form

$$\varphi(X) = \operatorname{Tr} \rho X \quad (X \in \mathbb{M}_n(\mathbb{C}))$$

for some matrix $\rho \in \mathbb{M}_n(\mathbb{C})$. However, from the properties of φ_0 we have $\rho \geq 0$ and $\text{Tr}\,\rho = 1$. Such a ρ is usually called a density matrix in the quantum applications. It is clear that if ρ has rank 1, then it is a projection. $\qquad\qquad\qquad\qquad\square$

In quantum information theory the traditional **variance** is

$$\text{Var}_\rho(A) = \text{Tr}\,\rho A^2 - (\text{Tr}\,\rho A)^2 \qquad\qquad (2.10)$$

where ρ is a density matrix and $A \in \mathbb{M}_n(\mathbb{C})$ is a self-adjoint operator. This is a straightforward analogy of the variance in probability theory; a standard notation is $\langle A^2 \rangle - \langle A \rangle^2$ in both formalisms. We note that for two self-adjoint operators the corresponding notion is **covariance**:

$$\text{Cov}_\rho(A, B) = \text{Tr}\,\rho AB - (\text{Tr}\,\rho A)(\text{Tr}\,\rho B).$$

It is rather different from probability theory that the variance (2.10) can be strictly positive even in the case where ρ has rank 1. If ρ has rank 1, then it is an ortho-projection of rank 1, also known as a pure state.

It is easy to show that

$$\text{Var}_\rho(A + \lambda I) = \text{Var}_\rho(A) \quad \text{for} \quad \lambda \in \mathbb{R}$$

and the concavity of the variance functional $\rho \mapsto \text{Var}_\rho(A)$:

$$\text{Var}_\rho(A) \geq \sum_i \lambda_i \text{Var}_{\rho_i}(A) \quad \text{if} \quad \rho = \sum_i \lambda_i \rho_i.$$

(Here $\lambda_i \geq 0$ and $\sum_i \lambda_i = 1$.)

The formulation is easier if ρ is diagonal. We can change the basis of the n-dimensional space so that $\rho = \text{Diag}(p_1, p_2, \ldots, p_n)$; then we have

$$\text{Var}_\rho(A) = \sum_{i,j} \frac{p_i + p_j}{2} |A_{ij}|^2 - \left(\sum_i p_i A_{ii}\right)^2. \qquad\qquad (2.11)$$

In the projection example $P = \text{Diag}(1, 0, \ldots, 0)$, formula (2.11) gives

$$\text{Var}_P(A) = \sum_{i \neq 1} |A_{1i}|^2$$

and this can be strictly positive.

Theorem 2.27 *Let ρ be a density matrix. Take all the decompositions such that*

$$\rho = \sum_i q_i Q_i,$$ (2.12)

where Q_i are pure states and (q_i) is a probability distribution. Then

$$\text{Var}_\rho(A) = \sup\left(\sum_i q_i \left(\text{Tr } Q_i A^2 - (\text{Tr } Q_i A)^2\right)\right),$$ (2.13)

where the supremum is over all decompositions (2.12).

The proof will be an application of matrix theory. The first lemma contains a trivial computation on block matrices.

Lemma 2.28 *Assume that*

$$\rho = \begin{bmatrix} \rho^\wedge & 0 \\ 0 & 0 \end{bmatrix}, \qquad \rho_i = \begin{bmatrix} \rho_i^\wedge & 0 \\ 0 & 0 \end{bmatrix}, \qquad A = \begin{bmatrix} A^\wedge & B \\ B^* & C \end{bmatrix}$$

and

$$\rho = \sum_i \lambda_i \rho_i, \qquad \rho^\wedge = \sum_i \lambda_i \rho_i^\wedge.$$

Then

$$\left(\text{Tr } \rho^\wedge (A^\wedge)^2 - (\text{Tr } \rho^\wedge A^\wedge)^2\right) - \sum_i \lambda_i \left(\text{Tr } \rho_i^\wedge (A^\wedge)^2 - (\text{Tr } \rho_i^\wedge A^\wedge)^2\right)$$

$$= (\text{Tr } \rho A^2 - (\text{Tr } \rho A)^2) - \sum_i \lambda_i \left(\text{Tr } \rho_i A^2 - (\text{Tr } \rho_i A)^2\right).$$

This lemma shows that if $\rho \in \mathbb{M}_n(\mathbb{C})$ has a rank $k < n$, then the computation of a variance $\text{Var}_\rho(A)$ can be reduced to $k \times k$ matrices. The equality in (2.13) is rather obvious for a rank 2 density matrix and, by the previous lemma, the computations will be with 2×2 matrices.

Lemma 2.29 *For a rank 2 matrix ρ, equality holds in (2.13).*

Proof: By Lemma 2.28 we can make a computation with 2×2 matrices. We can assume that

$$\rho = \begin{bmatrix} p & 0 \\ 0 & 1-p \end{bmatrix}, \qquad A = \begin{bmatrix} a_1 & b \\ \bar{b} & a_2 \end{bmatrix}.$$

Then

$$\text{Tr } \rho A^2 = p(a_1^2 + |b|^2) + (1-p)(a_2^2 + |b|^2).$$

We can assume that

$$\mathrm{Tr}\,\rho A = pa_1 + (1-p)a_2 = 0.$$

Let

$$Q_1 = \begin{bmatrix} p & ce^{-i\varphi} \\ ce^{i\varphi} & 1-p \end{bmatrix},$$

where $c = \sqrt{p(1-p)}$. This is a projection and

$$\mathrm{Tr}\,Q_1 A = a_1 p + a_2(1-p) + bc\,e^{-i\varphi} + \bar{b}c\,e^{i\varphi} = 2c\,\mathrm{Re}\,b\,e^{-i\varphi}.$$

We choose φ such that $\mathrm{Re}\,b\,e^{-i\varphi} = 0$. Then $\mathrm{Tr}\,Q_1 A = 0$ and

$$\mathrm{Tr}\,Q_1 A^2 = p(a_1^2 + |b|^2) + (1-p)(a_2^2 + |b|^2) = \mathrm{Tr}\,\rho A^2.$$

Let

$$Q_2 = \begin{bmatrix} p & -ce^{-i\varphi} \\ -ce^{i\varphi} & 1-p \end{bmatrix}.$$

Then

$$\rho = \frac{1}{2}Q_1 + \frac{1}{2}Q_2$$

and we have

$$\frac{1}{2}(\mathrm{Tr}\,Q_1 A^2 + \mathrm{Tr}\,Q_2 A^2) = p(a_1^2 + |b|^2) + (1-p)(a_2^2 + |b|^2) = \mathrm{Tr}\,\rho A^2.$$

Therefore we have an equality.   □

We denote by $r(\rho)$ the rank of an operator ρ. The idea of the proof is to reduce the rank and the block diagonal formalism will be used.

Lemma 2.30 *Let ρ be a density matrix and $A = A^*$ be in $\mathbb{M}_n(\mathbb{C})$. Assume the block matrix forms*

$$\rho = \begin{bmatrix} \rho_1 & 0 \\ 0 & \rho_2 \end{bmatrix}, \quad A = \begin{bmatrix} A_1 & A_2 \\ A_2^* & A_3 \end{bmatrix}$$

and $r(\rho_1), r(\rho_2) > 1$. We construct

$$\rho' := \begin{bmatrix} \rho_1 & X^* \\ X & \rho_2 \end{bmatrix}$$

such that

$$\mathrm{Tr}\,\rho A = \mathrm{Tr}\,\rho' A, \quad \rho' \geq 0, \quad r(\rho') < r(\rho).$$

Proof: The condition $\operatorname{Tr} \rho A = \operatorname{Tr} \rho' A$ is equivalent to $\operatorname{Tr} X A_2 + \operatorname{Tr} X^* A_2^* = 0$ and this holds if and only if $\operatorname{Re} \operatorname{Tr} X A_2 = 0$.

There exist unitaries U and W such that $U \rho_1 U^*$ and $W \rho_2 W^*$ are diagonal:

$$U \rho_1 U^* = \operatorname{Diag}(0, \ldots, 0, a_1, \ldots, a_k), \qquad W \rho_2 W^* = \operatorname{Diag}(b_1, \ldots, b_l, 0, \ldots, 0)$$

where $a_i, b_j > 0$. Then ρ has the same rank, $k + l$, as the matrix

$$\begin{bmatrix} U & 0 \\ 0 & W \end{bmatrix} \rho \begin{bmatrix} U^* & 0 \\ 0 & W^* \end{bmatrix} = \begin{bmatrix} U \rho_1 U^* & 0 \\ 0 & W \rho_2 W^* \end{bmatrix}.$$

A possible modification of this matrix is $Y :=$

$$\begin{bmatrix} \operatorname{Diag}(0, \ldots, 0, a_1, \ldots, a_{k-1}) & 0 & 0 & 0 \\ 0 & a_k & \sqrt{a_k b_1} & 0 \\ 0 & \sqrt{a_k b_1} & b_1 & 0 \\ 0 & 0 & 0 & \operatorname{Diag}(b_2, \ldots, b_l, 0, \ldots, 0) \end{bmatrix}$$

$$= \begin{bmatrix} U \rho_1 U^* & M \\ M & W \rho_2 W^* \end{bmatrix}$$

and $r(Y) = k + l - 1$. So Y has a smaller rank than ρ. Next we take

$$\begin{bmatrix} U^* & 0 \\ 0 & W^* \end{bmatrix} Y \begin{bmatrix} U & 0 \\ 0 & W \end{bmatrix} = \begin{bmatrix} \rho_1 & U^* M W \\ W^* M U & \rho_2 \end{bmatrix}$$

which has the same rank as Y. If $X_1 := W^* M U$ is multiplied by $e^{i\alpha}$ ($\alpha > 0$), then the positivity condition and the rank remain. On the other hand, we can choose $\alpha > 0$ such that $\operatorname{Re} \operatorname{Tr} e^{i\alpha} X_1 A_2 = 0$. Then $X := e^{i\alpha} X_1$ is the matrix we wanted. $\qquad \square$

Lemma 2.31 *Let ρ be a density matrix of rank $m > 0$ and $A = A^*$ be in $\mathbb{M}_n(\mathbb{C})$. We claim the existence of a decomposition*

$$\rho = p\rho_- + (1 - p)\rho_+$$

such that $r(\rho_-) < m$, $r(\rho_+) < m$, and

$$\operatorname{Tr} A \rho_+ = \operatorname{Tr} A \rho_- = \operatorname{Tr} \rho A.$$

Proof: By unitary transformation we can obtain the setup of the previous lemma:

$$\rho = \begin{bmatrix} \rho_1 & 0 \\ 0 & \rho_2 \end{bmatrix}, \qquad A = \begin{bmatrix} A_1 & A_2 \\ A_2^* & A_3 \end{bmatrix}.$$

With ρ' as in the previous lemma we choose

$$\rho_+ = \rho' = \begin{bmatrix} \rho_1 & X^* \\ X & \rho_2 \end{bmatrix}, \qquad \rho_- = \begin{bmatrix} \rho_1 & -X^* \\ -X & \rho_2 \end{bmatrix}.$$

Then

$$\rho = \frac{1}{2}\rho_- + \frac{1}{2}\rho_+$$

and the requirements $\operatorname{Tr} A\rho_+ = \operatorname{Tr} A\rho_- = \operatorname{Tr} \rho A$ also hold.   □

Proof of Theorem 2.27: For rank 2 states, the theorem is true by Lemma 2.29. Any state with a rank larger than 2 can be decomposed into a mixture of lower rank states, according to Lemma 2.31, that have the same expectation value for A as the original ρ has. The lower rank states can then be decomposed into a mixture of states with an even lower rank, until we reach states of rank ≤ 2. Thus, any state ρ can be decomposed into a mixture of pure states

$$\rho = \sum p_k Q_k$$

such that $\operatorname{Tr} A Q_k = \operatorname{Tr} A\rho$. Hence the statement of the theorem follows.   □

2.4 Subalgebras

A unital **∗-subalgebra** of $\mathbb{M}_n(\mathbb{C})$ is a subspace $\mathcal{A}$ that contains the identity I and is closed under matrix multiplication and adjoint. That is, if $A, B \in \mathcal{A}$, then so are AB and A^*. In what follows, to simplify the notation, we shall use the term subalgebra for all ∗-subalgebras.

Example 2.32 A simple subalgebra is

$$\mathcal{A} = \left\{ \begin{bmatrix} z & w \\ w & z \end{bmatrix} : z, w \in \mathbb{C} \right\} \subset \mathbb{M}_2(\mathbb{C}).$$

Since $A, B \in \mathcal{A}$ implies $AB = BA$, this is a commutative subalgebra. In terms of the Pauli matrices (2.9) we have

$$\mathcal{A} = \{ z\sigma_0 + w\sigma_1 : z, w \in \mathbb{C} \}.$$

This example will be generalized.   □

Assume that $P_1, P_2, \ldots, P_n$ are projections of rank 1 in $\mathbb{M}_n(\mathbb{C})$ such that $P_i P_j = 0$ for $i \neq j$ and $\sum_i P_i = I$. Then

$$\mathcal{A} = \left\{ \sum_{i=1}^{n} \alpha_i P_i : \alpha_i \in \mathbb{C} \right\}$$

is a maximal commutative $*$-subalgebra of $\mathbb{M}_n(\mathbb{C})$. The usual name is **MASA**, which is an acronym for Maximal Abelian Sub-Algebra.

Let $\mathcal{A}$ be any subset of $\mathbb{M}_n(\mathbb{C})$. Then $\mathcal{A}'$, the **commutant** of $\mathcal{A}$, is given by

$$\mathcal{A}' = \{ B \in \mathbb{M}_n(\mathbb{C}) : BA = AB \text{ for all } A \in \mathcal{A} \}.$$

It is easy to see that for any set $\mathcal{A} \subset \mathbb{M}_n(\mathbb{C})$, $\mathcal{A}'$ is a subalgebra. If $\mathcal{A}$ is a MASA, then $\mathcal{A}'' = \mathcal{A}$.

Theorem 2.33 *If $\mathcal{A} \subset \mathbb{M}_n(\mathbb{C})$ is a unital $*$-subalgebra, then $\mathcal{A}'' = \mathcal{A}$.*

Proof: We first show that for any $*$-subalgebra $\mathcal{A}$, $B \in \mathcal{A}''$ and any $v \in \mathbb{C}^n$, there exists an $A \in \mathcal{A}$ such that $Av = Bv$. Let $\mathcal{K}$ be the subspace of $\mathbb{C}^n$ given by

$$\mathcal{K} = \{ Av : A \in \mathcal{A} \}.$$

Let P be the orthogonal projection onto $\mathcal{K}$ in $\mathbb{C}^n$. Since, by construction, $\mathcal{K}$ is invariant under the action of $\mathcal{A}$, $PAP = AP$ for all $A \in \mathcal{A}$. Taking the adjoint, $PA^*P = PA^*$ for all $A \in \mathcal{A}$. Since $\mathcal{A}$ is a $*$-algebra, this implies $PA = AP$ for all $A \in \mathcal{A}$. That is, $P \in \mathcal{A}'$. Thus, for any $B \in \mathcal{A}''$, $BP = PB$ and so $\mathcal{K}$ is invariant under the action of $\mathcal{A}''$. In particular, $Bv \in \mathcal{K}$ and hence, by the definition of $\mathcal{K}$, $Bv = Av$ for some $A \in \mathcal{A}$.

We apply the previous statement to the $*$-subalgebra

$$\mathcal{M} = \{ A \otimes I_n : A \in \mathcal{A} \} \subset \mathbb{M}_n(\mathbb{C}) \otimes \mathbb{M}_n(\mathbb{C}) = \mathbb{M}_{n^2}(\mathbb{C}).$$

It is easy to see that

$$\mathcal{M}'' = \{ B \otimes I_n : B \in \mathcal{A}'' \} \subset \mathbb{M}_n(\mathbb{C}) \otimes \mathbb{M}_n(\mathbb{C}).$$

Now let $\{v_1, \ldots, v_n\}$ be any basis of $\mathbb{C}^n$ and form the vector

$$v = \begin{bmatrix} v_1 \\ v_2 \\ \vdots \\ v_n \end{bmatrix} \in \mathbb{C}^{n^2}.$$

Then

$$(A \otimes I_n)v = (B \otimes I_n)v$$

and $Av_j = Bv_j$ for every $1 \leq j \leq n$. Since $\{v_1, \ldots, v_n\}$ is a basis of $\mathbb{C}^n$, this means $B = A \in \mathcal{A}$. Since B was an arbitrary element of $\mathcal{A}''$, this shows that $\mathcal{A}'' \subset \mathcal{A}$. Since $\mathcal{A} \subset \mathcal{A}''$ is an automatic consequence of the definitions, this proves that $\mathcal{A}'' = \mathcal{A}$. $\square$

Next we study subalgebras $\mathcal{A} \subset \mathcal{B} \subset \mathbb{M}_n(\mathbb{C})$. A **conditional expectation** $\mathcal{E}$: $\mathcal{B} \to \mathcal{A}$ is a unital positive mapping which has the property

$$\mathcal{E}(AB) = A\mathcal{E}(B) \quad \text{for every} \quad A \in \mathcal{A} \quad \text{and} \quad B \in \mathcal{B}.$$

Choosing $B = I$, we obtain that $\mathcal{E}$ acts identically on $\mathcal{A}$. It follows from the positivity of $\mathcal{E}$ that $\mathcal{E}(C^*) = \mathcal{E}(C)^*$. Therefore, $\mathcal{E}(BA) = \mathcal{E}(B)A$ for all $A \in \mathcal{A}$ and $B \in \mathcal{B}$. Another standard notation for a conditional expectation $\mathcal{B} \to \mathcal{A}$ is $\mathcal{E}_{\mathcal{A}}^{\mathcal{B}}$.

Theorem 2.34 *Assume that* $\mathcal{A} \subset \mathcal{B} \subset \mathbb{M}_n(\mathbb{C})$. *If* $\alpha : \mathcal{A} \to \mathcal{B}$ *is the embedding, then the dual* $\mathcal{E} : \mathcal{B} \to \mathcal{A}$ *of* α *with respect to the Hilbert–Schmidt inner product is a conditional expectation.*

Proof: From the definition

$$\text{Tr}\, \alpha(A)B = \text{Tr}\, A\mathcal{E}(B) \quad (A \in \mathcal{A},\ B \in \mathcal{B})$$

of the dual, we see that $\mathcal{E} : \mathcal{B} \to \mathcal{A}$ is a positive unital mapping and $\mathcal{E}(A) = A$ for every $A \in \mathcal{A}$. For every $A, A_1 \in \mathcal{A}$ and $B \in \mathcal{B}$ we further have

$$\text{Tr}\, A\mathcal{E}(A_1 B) = \text{Tr}\, \alpha(A)A_1 B = \text{Tr}\, \alpha(AA_1)B = \text{Tr}\, AA_1\mathcal{E}(B),$$

which implies that $\mathcal{E}(A_1 B) = A_1\mathcal{E}(B)$. $\square$

Note that a conditional expectation $\mathcal{E} : \mathcal{B} \to \mathcal{A}$ has norm 1, that is, $\|\mathcal{E}(B)\| \leq \|B\|$ for every $B \in \mathcal{B}$. This follows from Corollary 2.45.

The subalgebras $\mathcal{A}_1, \mathcal{A}_2 \subset \mathbb{M}_n(\mathbb{C})$ cannot be orthogonal since I is in $\mathcal{A}_1$ and in $\mathcal{A}_2$. They are called **complementary** or **quasi-orthogonal** if $A_i \in \mathcal{A}_i$ and $\text{Tr}\, A_i = 0$ for $i = 1, 2$ imply that $\text{Tr}\, A_1 A_2 = 0$.

Example 2.35 In $\mathbb{M}_2(\mathbb{C})$ the subalgebras

$$\mathcal{A}_i := \{a\sigma_0 + b\sigma_i : a, b \in \mathbb{C}\} \quad (1 \leq i \leq 3)$$

are commutative and quasi-orthogonal. This follows from the facts that $\text{Tr}\, \sigma_i = 0$ for $1 \leq i \leq 3$ and

$$\sigma_1\sigma_2 = i\sigma_3, \quad \sigma_2\sigma_3 = i\sigma_1 \quad \sigma_3\sigma_1 = i\sigma_2.$$

So $\mathbb{M}_2(\mathbb{C})$ has 3 quasi-orthogonal MASAs.

In $\mathbb{M}_4(\mathbb{C}) = \mathbb{M}_2(\mathbb{C}) \otimes \mathbb{M}_2(\mathbb{C})$ we can give five quasi-orthogonal MASAs. Each of them is the linear span of four operators in one of the following lines:

$$\sigma_0 \otimes \sigma_0, \quad \sigma_0 \otimes \sigma_1, \quad \sigma_1 \otimes \sigma_0, \quad \sigma_1 \otimes \sigma_1,$$
$$\sigma_0 \otimes \sigma_0, \quad \sigma_0 \otimes \sigma_2, \quad \sigma_2 \otimes \sigma_0, \quad \sigma_2 \otimes \sigma_2,$$
$$\sigma_0 \otimes \sigma_0, \quad \sigma_0 \otimes \sigma_3, \quad \sigma_3 \otimes \sigma_0, \quad \sigma_3 \otimes \sigma_3,$$
$$\sigma_0 \otimes \sigma_0, \quad \sigma_1 \otimes \sigma_2, \quad \sigma_2 \otimes \sigma_3, \quad \sigma_3 \otimes \sigma_1,$$
$$\sigma_0 \otimes \sigma_0, \quad \sigma_1 \otimes \sigma_3, \quad \sigma_2 \otimes \sigma_1, \quad \sigma_3 \otimes \sigma_2.$$

$\square$

Theorem 2.36 *Assume that $\{A_i : 1 \leq i \leq k\}$ is a set of quasi-orthogonal MASAs in $\mathbb{M}_n(\mathbb{C})$. Then $k \leq n + 1$.*

Proof: The argument is rather simple. The traceless part of $\mathbb{M}_n(\mathbb{C})$ has dimension $n^2 - 1$ and the traceless part of a MASA has dimension $n - 1$. Therefore $k \leq (n^2 - 1)/(n - 1) = n + 1$. $\square$

Determining the maximal number of quasi-orthogonal MASAs is a hard problem. For example, if $n = 2^m$, then $n + 1$ MASAs is possible, but for an arbitrary n there is no definite result.

The next theorem gives a characterization of complementarity.

Theorem 2.37 *Let A_1 and A_2 be subalgebras of $\mathbb{M}_n(\mathbb{C})$ and denote Tr $/n$ by τ. The following conditions are equivalent:*

(i) *If $P \in A_1$ and $Q \in A_2$ are minimal projections, then $\tau(PQ) = \tau(P)\tau(Q)$.*
(ii) *The subalgebras A_1 and A_2 are quasi-orthogonal in $\mathbb{M}_n(\mathbb{C})$.*
(iii) *$\tau(A_1 A_2) = \tau(A_1)\tau(A_2)$ if $A_1 \in A_1, A_2 \in A_2$.*
(iv) *If $\mathcal{E}_1 : \mathbb{M}_n(\mathbb{C}) \to A_1$ is the trace-preserving conditional expectation, then $\mathcal{E}_1$ restricted to A_2 is a linear functional (times I).*

Proof: Note that $\tau((A_1 - \tau(A_1)I)(A_2 - \tau(A_2)I)) = 0$ and $\tau(A_1 A_2) = \tau(A_1)\tau(A_2)$ are equivalent. If they hold for minimal projections, they hold for arbitrary operators as well. Moreover, (iv) is equivalent to the property $\tau(A_1 \mathcal{E}_1(A_2)) = \tau(A_1(\tau(A_2)I))$ for every $A_1 \in A_1$ and $A_2 \in A_2$, and note that $\tau(A_1 \mathcal{E}_1(A_2)) = \tau(A_1 A_2)$. $\square$

Example 2.38 A simple example of quasi-orthogonal subalgebras can be formulated with tensor products. If $A = \mathbb{M}_n(\mathbb{C}) \otimes \mathbb{M}_n(\mathbb{C})$, $A_1 = \mathbb{M}_n(\mathbb{C}) \otimes \mathbb{C}I_n \subset A$ and $A_2 = \mathbb{C}I_n \otimes \mathbb{M}_n(\mathbb{C}) \subset A$, then A_1 and A_2 are quasi-orthogonal subalgebras of A. This comes from the property Tr $(A \otimes B) = \text{Tr } A \cdot \text{Tr } B$.

For $n = 2$ we give another example using the Pauli matrices. The 4-dimensional subalgebra $A_1 = \mathbb{M}_2(\mathbb{C}) \otimes \mathbb{C}I_2$ is the linear span of the set

$$\{\sigma_0 \otimes \sigma_0, \sigma_1 \otimes \sigma_0, \sigma_2 \otimes \sigma_0, \sigma_3 \otimes \sigma_0\}.$$

Together with the identity, each of the following triplets linearly spans a subalgebra A_j isomorphic to $M_2(\mathbb{C})$ ($2 \leq j \leq 4$):

$$\{\sigma_3 \otimes \sigma_1,\ \sigma_3 \otimes \sigma_2,\ \sigma_0 \otimes \sigma_3\},$$
$$\{\sigma_2 \otimes \sigma_3,\ \sigma_2 \otimes \sigma_1,\ \sigma_0 \otimes \sigma_2\},$$
$$\{\sigma_1 \otimes \sigma_2,\ \sigma_1 \otimes \sigma_3,\ \sigma_0 \otimes \sigma_1\}.$$

It is easy to check that the subalgebras $\mathcal{A}_1, \dots, \mathcal{A}_4$ are complementary.

The orthogonal complement of the four subalgebras is spanned by $\{\sigma_0 \otimes \sigma_3, \sigma_3 \otimes \sigma_0, \sigma_3 \otimes \sigma_3\}$. The linear span of this together with $\sigma_0 \otimes \sigma_0$ is a commutative subalgebra.   □

The previous example describes the general situation for $M_4(\mathbb{C})$. This will be the content of the next theorem. It is easy to calculate that the number of complementary subalgebras isomorphic to $\mathbb{M}_2(\mathbb{C})$ is at most $(16 - 1)/3 = 5$. However, the next theorem says that 5 is not possible.

If $x = (x_1, x_2, x_3) \in \mathbb{R}^3$, then the notation

$$x \cdot \sigma = x_1\sigma_1 + x_2\sigma_2 + x_3\sigma_3$$

will be used and shall be called a Pauli triplet.

Theorem 2.39 *Assume that $\{\mathcal{A}_i : 0 \le i \le 3\}$ is a family of pairwise quasi-orthogonal subalgebras of $M_4(\mathbb{C})$ which are isomorphic to $M_2(\mathbb{C})$. For every $0 \le i \le 3$, there exists a Pauli triplet $A(i, j)$ $(j \neq i)$ such that $\mathcal{A}'_i \cap \mathcal{A}_j$ is the linear span of I and $A(i, j)$. Moreover, the subspace linearly spanned by*

$$I \quad and \quad \left(\bigcup_{i=0}^{3} \mathcal{A}_i\right)^{\perp}$$

is a maximal Abelian subalgebra.

Proof: Since the intersection $\mathcal{A}'_0 \cap \mathcal{A}_j$ is a 2-dimensional commutative subalgebra, we can find a self-adjoint unitary $A(0, j)$ such that $\mathcal{A}'_0 \cap \mathcal{A}_j$ is spanned by I and $A(0, j) = x(0, j) \cdot \sigma \otimes I$, where $x(0, j) \in \mathbb{R}^3$. Due to the quasi-orthogonality of $\mathcal{A}_1, \mathcal{A}_2$ and $\mathcal{A}_3$, the unit vectors $x(0, j)$ are pairwise orthogonal (see (2.18)). The matrices $A(0, j)$ anti-commute:

$$A(0, i)A(0, j) = i(x(0, i) \times x(0, j)) \cdot \sigma \otimes I$$
$$= -i(x(0, j) \times x(0, i)) \cdot \sigma \otimes I = -A(0, j)A(0, i)$$

for $i \neq j$. Moreover,

$$A(0, 1)A(0, 2) = i(x(0, 1) \times x(0, 2)) \cdot \sigma$$

and $x(0, 1) \times x(0, 2) = \pm x(0, 3)$ because $x(0, 1) \times x(0, 2)$ is orthogonal to both $x(0, 1)$ and $x(0, 2)$. If necessary, we can change the sign of $x(0, 3)$ so that $A(0, 1)A(0, 2) = iA(0, 3)$ holds.

Starting with the subalgebras $\mathcal{A}'_1$, $\mathcal{A}'_2$, $\mathcal{A}'_3$ we can similarly construct the other Pauli triplets. In this way, we arrive at the four Pauli triplets, the rows of the following table:

$$\begin{matrix}
\star & A(0, 1) & A(0, 2) & A(0, 3) \\
A(1, 0) & \star & A(1, 2) & A(1, 3) \\
A(2, 0) & A(2, 1) & \star & A(2, 3) \\
A(3, 0) & A(3, 1) & A(3, 2) & \star
\end{matrix} \tag{2.14}$$

When $\{\mathcal{A}_i : 1 \le i \le 3\}$ is a family of pairwise quasi-orthogonal subalgebras, then the commutants $\{\mathcal{A}'_i : 1 \le i \le 3\}$ are pairwise quasi-orthogonal as well. $\mathcal{A}''_j = \mathcal{A}_j$ and $\mathcal{A}'_i$ have nontrivial intersection for $i \ne j$, actually the previously defined $A(i, j)$ is in the intersection. For a fixed j the three unitaries $A(i, j)$ $(i \ne j)$ form a Pauli triplet up to a sign. (It follows that changing sign we can always reach the situation where the first three columns of table (2.14) form Pauli triplets. $A(0, 3)$ and $A(1, 3)$ anti-commute, but it may happen that $A(0, 3)A(1, 3) = -iA(2, 3)$.)

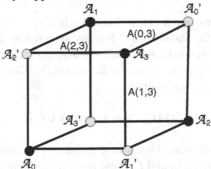

This picture shows a family $\{\mathcal{A}_i : 0 \le i \le 3\}$ of pairwise quasi-orthogonal subalgebras of $\mathbb{M}_4(\mathbb{C})$ which are isomorphic to $\mathbb{M}_2(\mathbb{C})$. The edges between two vertices represent the one-dimensional traceless intersection of the two subalgebras corresponding to two vertices. The three edges starting from a vertex represent a Pauli triplet.

Let $C_0 := \{\pm A(i, j)A(j, i) : i \ne j\} \cup \{\pm I\}$ and $C := C_0 \cup iC_0$. We want to show that C is a commutative group (with respect to the multiplication of unitaries).

Note that the products in C_0 have factors in symmetric position in (2.14) with respect to the main diagonal indicated by stars. Moreover, $A(i, j) \in \mathcal{A}(j)$ and $A(j, k) \in \mathcal{A}(j)'$, and these operators commute.

We have two cases for a product in C. Taking the product of $A(i, j)A(j, i)$ and $A(u, v)A(v, u)$, we have

$$(A(i, j)A(j, i))(A(i, j)A(j, i)) = I$$

in the simplest case, since $A(i, j)$ and $A(j, i)$ are commuting self-adjoint unitaries. The situation is slightly more complicated if the cardinality of the set $\{i, j, u, v\}$ is 3 or 4. First,

$$
\begin{aligned}
(A(1,0)A(0,1))(A(3,0)A(0,3)) &= A(0,1)(A(1,0)A(3,0))A(0,3) \\
&= \pm i(A(0,1)A(2,0))A(0,3) \\
&= \pm iA(2,0)(A(0,1)A(0,3)) \\
&= \pm A(2,0)A(0,2),
\end{aligned}
$$

and secondly,

$$
\begin{aligned}
(A(1,0)A(0,1))(A(3,2)A(2,3)) &= \pm iA(1,0)A(0,2)(A(0,3)A(3,2))A(2,3) \\
&= \pm iA(1,0)A(0,2)A(3,2)(A(0,3)A(2,3)) \\
&= \pm A(1,0)(A(0,2)A(3,2))A(1,3) \\
&= \pm iA(1,0)(A(1,2)A(1,3)) \\
&= \pm A(1,0)A(1,0) = \pm I. \qquad (2.15)
\end{aligned}
$$

So the product of any two operators in C is again in C.

Now we show that the subalgebra C linearly spanned by the unitaries $\{A(i, j) A(j, i) : i \neq j\} \cup \{I\}$ is a maximal Abelian subalgebra. Since we know the commutativity of this algebra, we estimate the dimension. It follows from (2.15) and the self-adjointness of $A(i, j)A(j, i)$ that

$$
A(i, j)A(j, i) = \pm A(k, \ell)A(\ell, k)
$$

when i, j, k and ℓ are different. Therefore C is linearly spanned by $A(0, 1)A(1, 0)$, $A(0, 2)A(2, 0)$, $A(0, 3)A(3, 0)$ and I. These are four different self-adjoint unitaries.

Finally, we check that the subalgebra C is quasi-orthogonal to $\mathcal{A}(i)$. If the cardinality of the set $\{i, j, k, \ell\}$ is 4, then we have

$$
\mathrm{Tr}\, A(i, j)(A(i, j)A(j, i)) = \mathrm{Tr}\, A(j, i) = 0
$$

and

$$
\mathrm{Tr}\, A(k, \ell)A(i, j)A(j, i) = \pm\mathrm{Tr}\, A(k, \ell)A(k, l)A(\ell, k) = \pm\mathrm{Tr}\, A(\ell, k) = 0.
$$

Moreover, because $\mathcal{A}(k)$ is quasi-orthogonal to $\mathcal{A}(i)$, we also have $A(i, k) \perp A(j, i)$, so

$$
\mathrm{Tr}\, A(i, \ell)(A(i, j)A(j, i)) = \pm i\,\mathrm{Tr}\, A(i, k)A(j, i) = 0.
$$

From this we can conclude that

$$
A(k, \ell) \perp A(i, j)A(j, i)
$$

for all $k \neq \ell$ and $i \neq j$. $\square$

2.5 Kernel Functions

Let $\mathcal{X}$ be a non-empty set. A function $\psi : \mathcal{X} \times \mathcal{X} \to \mathbb{C}$ is often called a **kernel**. A kernel $\psi : \mathcal{X} \times \mathcal{X} \to \mathbb{C}$ is called **positive definite** if

$$\sum_{j,k=1}^{n} c_j \overline{c_k} \psi(x_j, x_k) \geq 0$$

for all finite sets $\{c_1, c_2, \ldots, c_n\} \subset \mathbb{C}$ and $\{x_1, x_2, \ldots, x_n\} \subset \mathcal{X}$.

Example 2.40 It follows from Schur's theorem that the product of positive definite kernels is a positive definite kernel as well.

If $\psi : \mathcal{X} \times \mathcal{X} \to \mathbb{C}$ is positive definite, then

$$e^{\psi} = \sum_{n=0}^{\infty} \frac{1}{n!} \psi^m$$

and $\tilde{\psi}(x, y) = f(x)\psi(x, y)\overline{f(y)}$ are positive definite for any function $f : \mathcal{X} \to \mathbb{C}$. $\square$

The function $\psi : \mathcal{X} \times \mathcal{X} \to \mathbb{C}$ is called a **conditionally negative definite** kernel if $\psi(x, y) = \overline{\psi(y, x)}$ and

$$\sum_{j,k=1}^{n} c_j \overline{c_k} \psi(x_j, x_k) \leq 0$$

for all finite sets $\{c_1, c_2, \ldots, c_n\} \subset \mathbb{C}$ and $\{x_1, x_2, \ldots, x_n\} \subset \mathcal{X}$ when $\sum_{j=1}^{n} c_j = 0$.

The above properties of a kernel depend on the matrices

$$\begin{bmatrix} \psi(x_1, x_1) & \psi(x_1, x_2) & \ldots & \psi(x_1, x_n) \\ \psi(x_2, x_1) & \psi(x_2, x_2) & \ldots & \psi(x_2, x_n) \\ \vdots & \vdots & \ddots & \vdots \\ \psi(x_n, x_1) & \psi(x_n, x_2) & \ldots & \psi(x_n, x_n) \end{bmatrix}.$$

If a kernel is positive definite, then $-f$ is conditionally negative definite, but the converse is not true.

Lemma 2.41 *Assume that the function $\psi : \mathcal{X} \times \mathcal{X} \to \mathbb{C}$ has the property $\psi(x, y) = \overline{\psi(y, x)}$ and fix $x_0 \in \mathcal{X}$. Then*

$$\varphi(x, y) := -\psi(x, y) + \psi(x, x_0) + \psi(x_0, y) - \psi(x_0, x_0)$$

is positive definite if and only if ψ is conditionally negative definite.

The proof is rather straightforward, but an interesting particular case is the following:

Example 2.42 Assume that $f : \mathbb{R}^+ \to \mathbb{R}$ is a C^1-function with the property $f(0) = f'(0) = 0$. Let $\psi : \mathbb{R}^+ \times \mathbb{R}^+ \to \mathbb{R}$ be defined as

$$\psi(x, y) = \begin{cases} \dfrac{f(x) - f(y)}{x - y} & \text{if } x \neq y, \\[3mm] f'(x) & \text{if } x = y. \end{cases}$$

(This is the so-called kernel of divided difference.) Assume that this is conditionally negative definite. Now we apply the lemma with $x_0 = \varepsilon$:

$$-\frac{f(x) - f(y)}{x - y} + \frac{f(x) - f(\varepsilon)}{x - \varepsilon} + \frac{f(\varepsilon) - f(y)}{\varepsilon - y} - f'(\varepsilon)$$

is positive definite and from the limit $\varepsilon \to 0$, we have the positive definite kernel

$$-\frac{f(x) - f(y)}{x - y} + \frac{f(x)}{x} + \frac{f(y)}{y} = -\frac{f(x)y^2 - f(y)x^2}{x(x - y)y}.$$

Assume that $f(x) > 0$ for all $x > 0$. Multiplication by $xy/(f(x)f(y))$ gives a positive definite kernel

$$\frac{\dfrac{x^2}{f(x)} - \dfrac{y^2}{f(y)}}{x - y},$$

which is a divided difference of the function $g(x) := x^2/f(x)$ on $(0, \infty)$. $\square$

Theorem 2.43 (Schoenberg's theorem) *Let $\mathcal{X}$ be a non-empty set and let $\psi : \mathcal{X} \times \mathcal{X} \to \mathbb{C}$ be a kernel. Then ψ is conditionally negative definite if and only if $\exp(-t\psi)$ is positive definite for every $t > 0$.*

Proof: If $\exp(-t\psi)$ is positive definite, then $1 - \exp(-t\psi)$ is conditionally negative definite and so is

$$\psi = \lim_{t \to 0} \frac{1}{t}(1 - \exp(-t\psi)).$$

Assume now that ψ is conditionally negative definite. Take $x_0 \in \mathcal{X}$ and set

$$\varphi(x, y) := -\psi(x, y) + \psi(x, x_0) + \psi(x_0, y) - \psi(x_0, x_0),$$

which is positive definite due to the previous lemma. Then

$$e^{-\psi(x,y)} = e^{\varphi(x,y)} e^{-\psi(x,x_0)} \overline{e^{-\psi(y,x_0)}} e^{\psi(x_0,x_0)}$$

is positive definite. This proves the case $t = 1$, and the argument is similar for general $t > 0$. $\square$

The kernel functions are a kind of generalization of matrices. If $A \in \mathbb{M}_n$, then the corresponding kernel function is given by $\mathcal{X} := \{1, 2, \ldots, n\}$ and

$$\psi_A(i, j) = A_{ij} \qquad (1 \leq i, j \leq n).$$

Therefore the results of this section have matrix consequences.

2.6 Positivity-Preserving Mappings

Let $\alpha : \mathbb{M}_n \to \mathbb{M}_k$ be a linear mapping. It is called **positive** (or positivity-preserving) if it sends positive (semidefinite) matrices to positive (semidefinite) matrices. α is **unital** if $\alpha(I_n) = I_k$.

The **dual** $\alpha^* : \mathbb{M}_k \to \mathbb{M}_n$ of α is defined by the equation

$$\mathrm{Tr}\, \alpha(A)B = \mathrm{Tr}\, A\alpha^*(B) \qquad (A \in \mathbb{M}_n, B \in \mathbb{M}_k).$$

It is easy to see that α is positive if and only if α^* is positive and α is trace-preserving if and only if α^* is unital.

The inequality
$$\alpha(AA^*) \geq \alpha(A)\alpha(A)^*$$

is called the **Schwarz inequality**. If the Schwarz inequality holds for a linear mapping α, then α is positivity-preserving. If α is a positive mapping, then this inequality holds for normal matrices. This result is called the **Kadison inequality**.

Theorem 2.44 *Let $\alpha : \mathbb{M}_n(\mathbb{C}) \to \mathbb{M}_k(\mathbb{C})$ be a positive unital mapping.*

(1) *If $A \in \mathbb{M}_n$ is a normal operator, then*

$$\alpha(AA^*) \geq \alpha(A)\alpha(A)^*.$$

(2) *If $A \in \mathbb{M}_n$ is positive such that A and $\alpha(A)$ are invertible, then*

$$\alpha(A^{-1}) \geq \alpha(A)^{-1}.$$

Proof: A has a spectral decomposition $\sum_i \lambda_i P_i$, where the P_i's are pairwise orthogonal projections. We have $A^*A = \sum_i |\lambda_i|^2 P_i$ and

$$\begin{bmatrix} I & \alpha(A) \\ \alpha(A)^* & \alpha(A^*A) \end{bmatrix} = \sum_i \begin{bmatrix} 1 & \lambda_i \\ \overline{\lambda_i} & |\lambda_i|^2 \end{bmatrix} \otimes \alpha(P_i).$$

Since $\alpha(P_i)$ is positive, the left-hand side is positive as well. Reference to Theorem 2.1 gives the first inequality.

To prove the second inequality, use the identity

$$\begin{bmatrix} \alpha(A) & I \\ I & \alpha(A^{-1}) \end{bmatrix} = \sum_i \begin{bmatrix} \lambda_i & 1 \\ 1 & \lambda_i^{-1} \end{bmatrix} \otimes \alpha(P_i)$$

to conclude that the left-hand side is a positive block matrix. The positivity implies our statement.   □

Corollary 2.45 *A positive unital mapping* $\alpha : \mathbb{M}_n(\mathbb{C}) \to \mathbb{M}_k(\mathbb{C})$ *has norm 1, i.e.,* $\|\alpha(A)\| \le \|A\|$ *for every* $A \in \mathbb{M}_n(\mathbb{C})$.

Proof: Let $A \in \mathbb{M}_n(\mathbb{C})$ be such that $\|A\| \le 1$, and take the polar decomposition $A = U|A|$ with a unitary U. By Example 1.39 there is a unitary V such that $|A| = (V + V^*)/2$ and so $A = (UV + UV^*)/2$. Hence it suffices to show that $\|\alpha(U)\| \le 1$ for every unitary U. This follows from the Kadison inequality in (1) of the previous theorem as

$$\|\alpha(U)\|^2 = \|\alpha(U)^*\alpha(U)\| \le \|\alpha(U^*U)\| = \|\alpha(I)\| = 1.$$

□

The linear mapping $\alpha : \mathbb{M}_n \to \mathbb{M}_k$ is called **2-positive** if

$$\begin{bmatrix} A & B \\ B^* & C \end{bmatrix} \ge 0 \quad \text{implies} \quad \begin{bmatrix} \alpha(A) & \alpha(B) \\ \alpha(B^*) & \alpha(C) \end{bmatrix} \ge 0$$

when $A, B, C \in \mathbb{M}_n$.

Lemma 2.46 *Let* $\alpha : \mathbb{M}_n(\mathbb{C}) \to \mathbb{M}_k(\mathbb{C})$ *be a 2-positive mapping. If* $A, \alpha(A) > 0$, *then*

$$\alpha(B)^*\alpha(A)^{-1}\alpha(B) \le \alpha(B^*A^{-1}B)$$

for every $B \in \mathbb{M}_n$. *Hence, a 2-positive unital mapping satisfies the Schwarz inequality.*

Proof: Since

$$\begin{bmatrix} A & B \\ B^* & B^*A^{-1}B \end{bmatrix} \geq 0,$$

the 2-positivity implies

$$\begin{bmatrix} \alpha(A) & \alpha(B) \\ \alpha(B^*) & \alpha(B^*A^{-1}B) \end{bmatrix} \geq 0.$$

So Theorem 2.1 implies the statement.   □

If $B = B^*$, then the 2-positivity condition is not necessary in the previous lemma, positivity is enough.

Lemma 2.47 *Let $\alpha : \mathbb{M}_n \to \mathbb{M}_k$ be a 2-positive unital mapping. Then*

$$\mathcal{N}_\alpha := \{A \in \mathbb{M}_n : \alpha(A^*A) = \alpha(A)^*\alpha(A) \text{ and } \alpha(AA^*) = \alpha(A)\alpha(A)^*\}$$

is a subalgebra of $\mathbb{M}_n$ and

$$\alpha(AB) = \alpha(A)\alpha(B) \quad and \quad \alpha(BA) = \alpha(B)\alpha(A)$$

holds for all $A \in \mathcal{N}_\alpha$ and $B \in \mathbb{M}_n$.

Proof: The proof is based only on the Schwarz inequality. Assume that $\alpha(AA^*) = \alpha(A)\alpha(A)^*$. Then

$$\begin{aligned}
t&\big(\alpha(A)\alpha(B) + \alpha(B)^*\alpha(A)^*\big) \\
&= \alpha(tA^* + B)^*\alpha(tA^* + B) - t^2\alpha(A)\alpha(A)^* - \alpha(B)^*\alpha(B) \\
&\leq \alpha\big((tA^* + B)^*(tA^* + B)\big) - t^2\alpha(AA^*) - \alpha(B)^*\alpha(B) \\
&= t\alpha(AB + B^*A^*) + \alpha(B^*B) - \alpha(B)^*\alpha(B)
\end{aligned}$$

for a real t. Divide the inequality by t and let $t \to \pm\infty$. Then

$$\alpha(A)\alpha(B) + \alpha(B)^*\alpha(A)^* = \alpha(AB + B^*A^*)$$

and similarly

$$\alpha(A)\alpha(B) - \alpha(B)^*\alpha(A)^* = \alpha(AB - B^*A^*).$$

Adding these two equalities we have

$$\alpha(AB) = \alpha(A)\alpha(B).$$

The other identity is proven similarly.   □

It follows from the previous lemma that if α is a 2-positive unital mapping and its inverse is 2-positive as well, then α is multiplicative. Indeed, the assumption implies $\alpha(A^*A) = \alpha(A)^*\alpha(A)$ for every A.

A linear mapping $\mathcal{E} : \mathbb{M}_n \to \mathbb{M}_k$ is called **completely positive** if

$$\mathcal{E} \otimes \mathrm{id}_n : \mathbb{M}_n \otimes \mathbb{M}_n \to \mathbb{M}_k \otimes \mathbb{M}_n$$

is a positive mapping, where $\mathrm{id}_n : \mathbb{M}_n \to \mathbb{M}_n$ is the identity mapping and $\mathcal{E} \otimes \mathrm{id}_n$ is defined by

$$(\mathcal{E} \otimes \mathrm{id}_n)\big([X_{ij}]_{i,j=1}^n\big) := [\mathcal{E}(X_{ij})]_{i,j=1}^n.$$

(Here, $B(\mathcal{H}) \otimes \mathbb{M}_n$ is identified with the $n \times n$ block matrices whose entries are operators in $B(\mathcal{H})$.) Note that if a linear mapping $\mathcal{E} : \mathbb{M}_n \to \mathbb{M}_k$ is completely positive in the above sense, then $\mathcal{E} \otimes \mathrm{id}_m : \mathbb{M}_n \otimes \mathbb{M}_m \to \mathbb{M}_k \otimes \mathbb{M}_m$ is positive for every $m \in \mathbb{N}$.

Example 2.48 Consider the transpose mapping $\mathcal{E} : A \mapsto A^t$ on 2×2 matrices:

$$\begin{bmatrix} x & y \\ z & w \end{bmatrix} \mapsto \begin{bmatrix} x & z \\ y & w \end{bmatrix}.$$

$\mathcal{E}$ is obviously positive. The matrix

$$\begin{bmatrix} 2 & 0 & 0 & 2 \\ 0 & 1 & 1 & 0 \\ 0 & 1 & 1 & 0 \\ 2 & 0 & 0 & 2 \end{bmatrix}$$

is positive. The extension of $\mathcal{E}$ maps this to

$$\begin{bmatrix} 2 & 0 & 0 & 1 \\ 0 & 1 & 2 & 0 \\ 0 & 2 & 1 & 0 \\ 1 & 0 & 0 & 2 \end{bmatrix}.$$

This is not positive, so $\mathcal{E}$ is not completely positive.   □

Theorem 2.49 *Let $\mathcal{E} : \mathbb{M}_n \to \mathbb{M}_k$ be a linear mapping. Then the following conditions are equivalent:*

(1) *$\mathcal{E}$ is completely positive.*
(2) *The block matrix X defined by*

$$X_{ij} = \mathcal{E}(E(ij)) \qquad (1 \le i, j \le n) \tag{2.16}$$

    is positive, where $E(ij)$ are the matrix units of $\mathbb{M}_n$.
(3) *There are operators $V_t : \mathbb{C}^n \to \mathbb{C}^k$ $(1 \le t \le k^2)$ such that*

$$\mathcal{E}(A) = \sum_t V_t A V_t^* .$$

(2.17)

(4) *For finite families* $A_i \in \mathbb{M}_n(\mathbb{C})$ *and* $B_i \in \mathbb{M}_k(\mathbb{C})$ $(1 \le i \le n)$, *the inequality*

$$\sum_{i,j} B_i^* \mathcal{E}(A_i^* A_j) B_j \ge 0$$

holds.

Proof: (1) implies (2): The matrix

$$\sum_{i,j} E(ij) \otimes E(ij) = \frac{1}{n} \left(\sum_{i,j} E(ij) \otimes E(ij) \right)^2$$

is positive. Therefore,

$$(\mathrm{id}_n \otimes \mathcal{E}) \left(\sum_{i,j} E(ij) \otimes E(ij) \right) = \sum_{i,j} E(ij) \otimes \mathcal{E}(E(ij)) = X$$

is positive as well.

(2) implies (3): Assume that the block matrix X is positive. There are orthogonal projections P_i $(1 \le i \le n)$ on $\mathbb{C}^{nk}$ such that they are pairwise orthogonal and

$$P_i X P_j = \mathcal{E}(E(ij)).$$

We have a decomposition

$$X = \sum_{t=1}^{nk} |f_t\rangle\langle f_t|,$$

where $|f_t\rangle$ are appropriately normalized eigenvectors of X. Since P_i is a partition of unity, we have

$$|f_t\rangle = \sum_{i=1}^{n} P_i |f_t\rangle$$

and we define $V_t : \mathbb{C}^n \to \mathbb{C}^k$ by

$$V_t |i\rangle = P_i |f_t\rangle.$$

($|i\rangle$ are the canonical basis vectors.) In this notation,

$$X = \sum_t \sum_{i,j} P_i |f_t\rangle\langle f_t| P_j = \sum_{i,j} P_i \left(\sum_t V_t |i\rangle\langle j| V_t^* \right) P_j$$

and hence

$$\mathcal{E}(E(ij)) = P_i X P_j = \sum_t V_t E(ij) V_t^*.$$

Since this holds for all matrix units $E(ij)$, we obtain

$$\mathcal{E}(A) = \sum_t V_t A V_t^*.$$

(3) implies (4): Assume that $\mathcal{E}$ is of the form (2.17). Then

$$\sum_{i,j} B_i^* \mathcal{E}(A_i^* A_j) B_j = \sum_t \sum_{i,j} B_i^* V_t (A_i^* A_j) V_t^* B_j$$

$$= \sum_t \left(\sum_i A_i V_t^* B_i \right)^* \left(\sum_j A_j V_t^* B_j \right) \geq 0$$

follows.

(4) implies (1): We consider

$$\mathcal{E} \otimes \mathrm{id}_n : \mathbb{M}_n \otimes \mathbb{M}_n \to \mathbb{M}_k \otimes \mathbb{M}_n.$$

Since any positive operator in $\mathbb{M}_n \otimes \mathbb{M}_n$ is the sum of operators in the form $\sum_{i,j} A_i^* A_j \otimes E(ij)$ (Theorem 2.8), it is enough to show that

$$Y := \mathcal{E} \otimes \mathrm{id}_n \left(\sum_{i,j} A_i^* A_j \otimes E(ij) \right) = \sum_{i,j} \mathcal{E}(A_i^* A_j) \otimes E(ij)$$

is positive. On the other hand, $Y = [Y_{ij}]_{i,j=1}^n \in \mathbb{M}_k \otimes \mathbb{M}_n$ is positive if and only if

$$\sum_{i,j} B_i^* Y_{ij} B_j = \sum_{i,j} B_i^* \mathcal{E}(A_i^* A_j) B_j \geq 0.$$

The positivity of this operator is assumed in (4). Hence (1) follows. $\qquad\square$

The representation (2.17) is called the **Kraus representation**. The block matrix X defined by (2.16) is called the **representing block matrix** (or the **Choi matrix**).

Example 2.50 We take $\mathcal{A} \subset \mathcal{B} \subset \mathbb{M}_n(\mathbb{C})$ and a conditional expectation $\mathcal{E} : \mathcal{B} \to \mathcal{A}$. Using condition (4) of the previous theorem we can argue that $\mathcal{E}$ is completely positive. For $A_i \in \mathcal{A}$ and $B_i \in \mathcal{B}$ we have

$$\sum_{i,j} A_i^* \mathcal{E}(B_i^* B_j) A_j = \mathcal{E}\left(\left(\sum_i B_i A_i\right)^* \left(\sum_j B_j A_j\right)\right) \geq 0$$

and this is enough. □

The next example is slightly different.

Example 2.51 Let $\mathcal{H}$ and $\mathcal{K}$ be Hilbert spaces and (f_i) be a basis in $\mathcal{K}$. For each i define the linear operator $V_i : \mathcal{H} \to \mathcal{H} \otimes \mathcal{K}$ by $V_i e = e \otimes f_i$ ($e \in \mathcal{H}$). These operators are isometries with pairwise orthogonal ranges and the adjoints act as $V_i^*(e \otimes f) = \langle f_i, f \rangle e$.

The **partial trace** $\mathrm{Tr}_2 : B(\mathcal{H} \otimes \mathcal{K}) \to B(\mathcal{H})$ introduced in Sect. 1.7 can be written as

$$\mathrm{Tr}_2(A) = \sum_i V_i^* A V_i \quad (A \in B(\mathcal{H} \otimes \mathcal{K})).$$

The reason for the terminology is the formula $\mathrm{Tr}_2(X \otimes Y) = X \mathrm{Tr}\, Y$. The above expression implies that Tr_2 is completely positive. It is actually a conditional expectation up to a constant factor. □

Example 2.52 The trace $\mathrm{Tr} : \mathbb{M}_k(\mathbb{C}) \to \mathbb{C}$ is completely positive if $\mathrm{Tr} \otimes \mathrm{id}_n : \mathbb{M}_k(\mathbb{C}) \otimes \mathbb{M}_n(\mathbb{C}) \to \mathbb{M}_n(\mathbb{C})$ is a positive mapping. However, this is a partial trace which is known to be positive (even completely positive).

It follows that any positive linear functional $\psi : \mathbb{M}_k(\mathbb{C}) \to \mathbb{C}$ is completely positive. Since $\psi(A) = \mathrm{Tr}\, DA$ for some positive D, ψ is the composition of the completely positive mappings $A \mapsto D^{1/2} A D^{1/2}$ and Tr. □

Example 2.53 Let $\mathcal{E} : \mathbb{M}_n \to \mathbb{M}_k$ be a positive linear mapping such that $\mathcal{E}(A)$ and $\mathcal{E}(B)$ commute for any $A, B \in \mathbb{M}_n$. We want to show that $\mathcal{E}$ is completely positive.

Any two self-adjoint matrices in the range of $\mathcal{E}$ commute, so we can change the basis so that all of them become diagonal. It follows that $\mathcal{E}$ has the form

$$\mathcal{E}(A) = \sum_i \psi_i(A) E_{ii},$$

where E_{ii} are the diagonal matrix units and ψ_i are positive linear functionals. Since the sum of completely positive mappings is completely positive, it is enough to show that $A \mapsto \psi(A) F$ is completely positive for a positive functional ψ and for a positive matrix F. The complete positivity of this mapping means that for an $m \times m$ block matrix X with entries $X_{ij} \in \mathbb{M}_n$, if $X \geq 0$ then the block matrix $[\psi(X_{ij}) F]_{i,j=1}^n$ should be positive. This is true, since the matrix $[\psi(X_{ij})]_{i,j=1}^n$ is positive (due to the complete positivity of ψ). □

Example 2.54 A linear mapping $\mathcal{E} : \mathbb{M}_2 \to \mathbb{M}_2$ is defined by the formula

$$\mathcal{E} : \begin{bmatrix} 1+z & x-iy \\ x+iy & 1-z \end{bmatrix} \mapsto \begin{bmatrix} 1+\gamma z & \alpha x - i\beta y \\ \alpha x + i\beta y & 1-\gamma z \end{bmatrix}$$

where α, β, γ are real parameters.

The condition for positivity is

$$-1 \le \alpha, \beta, \gamma \le 1.$$

It is not difficult to compute the representing block matrix as follows:

$$X = \frac{1}{2} \begin{bmatrix} 1+\gamma & 0 & 0 & \alpha+\beta \\ 0 & 1-\gamma & \alpha-\beta & 0 \\ 0 & \alpha-\beta & 1-\gamma & 0 \\ \alpha+\beta & 0 & 0 & 1+\gamma \end{bmatrix}.$$

This matrix is positive if and only if

$$|1 \pm \gamma| \ge |\alpha \pm \beta|.$$

In quantum information theory this mapping $\mathcal{E}$ is called the **Pauli channel**. □

Example 2.55 Fix a positive definite matrix $A \in \mathbb{M}_n$ and set

$$T_A(K) = \int_0^\infty (t+A)^{-1} K (t+A)^{-1} \, dt \qquad (K \in \mathbb{M}_n).$$

This mapping $T_A : \mathbb{M}_n \to \mathbb{M}_n$ is obviously positivity-preserving and approximation of the integral by a finite sum also shows the complete positivity.

If $A = \mathrm{Diag}(\lambda_1, \lambda_2, \ldots, \lambda_n)$, then we see from integration that the entries of $T_A(K)$ are

$$T_A(K)_{ij} = \frac{\log \lambda_i - \log \lambda_j}{\lambda_i - \lambda_j} K_{ij}.$$

Another integration gives that the mapping

$$\alpha : L \mapsto \int_0^1 A^t L A^{1-t} \, dt$$

acts as

$$(\alpha(L))_{ij} = \frac{\lambda_i - \lambda_j}{\log \lambda_i - \log \lambda_j} L_{ij}.$$

This shows that

$$T_A^{-1}(L) = \int_0^1 A^t L A^{1-t} \, dt.$$

To show that T_A^{-1} is not positive, we take $n = 2$ and consider

$$T_A^{-1} \begin{bmatrix} 1 & 1 \\ 1 & 1 \end{bmatrix} = \begin{bmatrix} \lambda_1 & \dfrac{\lambda_1 - \lambda_2}{\log \lambda_1 - \log \lambda_2} \\ \dfrac{\lambda_1 - \lambda_2}{\log \lambda_1 - \log \lambda_2} & \lambda_2 \end{bmatrix}.$$

The positivity of this matrix is equivalent to the inequality

$$\sqrt{\lambda_1 \lambda_2} \geq \frac{\lambda_1 - \lambda_2}{\log \lambda_1 - \log \lambda_2}$$

between the geometric and logarithmic means. The opposite inequality holds, see Example 5.22, and therefore T_A^{-1} is not positive. $\qquad\square$

The next result tells us that the **Kraus representation** of a completely positive mapping is unique up to a unitary matrix.

Theorem 2.56 *Let* $\mathcal{E} : \mathbb{M}_n(\mathbb{C}) \to \mathbb{M}_m(\mathbb{C})$ *be a linear mapping which is represented as*

$$\mathcal{E}(A) = \sum_{t=1}^{k} V_t A V_t^* \quad \text{and} \quad \mathcal{E}(A) = \sum_{t=1}^{k} W_t A W_t^*$$

with operators $V_t, W_t : \mathbb{C}^n \to \mathbb{C}^m$. *Then there exists a* $k \times k$ *unitary matrix* $[c_{tu}]$ *such that*

$$W_t = \sum_u c_{tu} V_u \quad (1 \leq t \leq k).$$

Proof: Without loss of generality we may assume that $m \geq n$. Indeed, we can embed $\mathbb{M}_m = B(\mathbb{C}^m)$ into a bigger $\mathbb{M}_{m'} = B(\mathbb{C}^{m'})$ and consider $\mathcal{E}$ as a mapping $\mathbb{M}_n \to \mathbb{M}_{m'}$. Let x_i be a basis in $\mathbb{C}^m$ and y_j be a basis in $\mathbb{C}^n$. Consider the vectors

$$v_t := \sum_{j=1}^{n} x_j \otimes V_t y_j \quad \text{and} \quad w_t := \sum_{j=1}^{n} x_j \otimes W_t y_j .$$

We have

$$|v_t\rangle\langle v_t| = \sum_{j,j'} |x_j\rangle\langle x_{j'}| \otimes V_t |y_j\rangle\langle y_{j'}| V_t^*$$

and

$$|w_t\rangle\langle w_t| = \sum_{j,j'} |x_j\rangle\langle x_{i'}| \otimes W_t |y_j\rangle\langle y_{j'}| W_t^*.$$

Our hypothesis implies that

$$\sum_t |v_t\rangle\langle v_t| = \sum_t |w_t\rangle\langle w_t|.$$

Lemma 1.24 tells us that there is a unitary matrix $[c_{tu}]$ such that

$$w_t = \sum_u c_{tu} v_u.$$

This implies that

$$W_t y_j = \sum_u c_{tu} V_u y_j \qquad (1 \le j \le n).$$

Hence we conclude the statement of the theorem. $\square$

2.7 Notes and Remarks

Theorem 2.5 is from the paper J.-C. Bourin and E.-Y. Lee, Unitary orbits of Hermitian operators with convex or concave functions, Bull. London Math. Soc. **44**(2012), 1085–1102.

The **Wielandt inequality** has an extension to matrices. Let A be an $n \times n$ positive matrix with eigenvalues $\lambda_1 \ge \lambda_2 \ge \cdots \ge \lambda_n$. Let X and Y be $n \times p$ and $n \times q$ matrices such that $X^*Y = 0$. The generalized inequality is

$$X^*AY(Y^*AY)^- Y^*AX \le \left(\frac{\lambda_1 - \lambda_n}{\lambda_1 + \lambda_n}\right)^2 X^*AX,$$

where a generalized inverse $(Y^*AY)^-$ is included: $BB^-B = B$. See Song-Gui Wang and Wai-Cheung Ip, A matrix version of the Wielandt inequality and its applications to statistics, Linear Algebra Appl. **296**(1999), 171–181.

The lattice of ortho-projections has applications in quantum theory. The cited **Gleason theorem** was obtained by A. M. Gleason in 1957, see also R. Cooke, M. Keane and W. Moran, An elementary proof of Gleason's theorem, Math. Proc. Cambridge Philos. Soc. **98**(1985), 117–128.

Theorem 2.27 is from the paper D. Petz and G. Tóth, Matrix variances with projections, Acta Sci. Math. (Szeged), **78**(2012), 683–688. An extension of this result is in the paper Z. Léka and D. Petz, Some decompositions of matrix variances, to be published.

Theorem 2.33 is the double commutant theorem of **von Neumann** from 1929; the original proof was for operators on an infinite-dimensional Hilbert space. (There is a relevant difference between finite and infinite dimensions; in a finite-dimensional space all subspaces are closed.) The conditional expectation in Theorem 2.34 was first introduced in the paper H. **Umegaki**, Conditional expectation in an operator algebra,

Tôhoku Math. J. **6**(1954), 177–181, and it is related to the so-called **Tomiyama theorem**.

The maximum number of complementary MASAs in $\mathbb{M}_n(\mathbb{C})$ is a popular subject. If n is a prime power, then $n+1$ MASAs can be constructed, but $n = 6$ is an unknown problematic case. (The expected number of complementary MASAs is 3 here.) It is interesting that n MASAs cannot exist in $\mathbb{M}_n(\mathbb{C})$ for any $n > 1$, see the paper [83] of M. Weiner.

Theorem 2.39 is from the paper H. Ohno, D. **Petz** and A. Szántó, Quasi-orthogonal subalgebras of 4×4 matrices, Linear Algebra Appl. **425**(2007), 109–118. It was conjectured that in the case $n = 2^k$ the algebra $\mathbb{M}_n(\mathbb{C})$ cannot have $N_k := (4^k - 1)/3$ complementary subalgebras isomorphic to $\mathbb{M}_2$, but it was proved that there are $N_k - 1$ copies. 2 is not a typical prime number in this situation. If $p > 2$ is a prime number, then in the case $n = p^k$ the algebra $\mathbb{M}_n(\mathbb{C})$ has $N_k := (p^{2k} - 1)/(p^2 - 1)$ complementary subalgebras isomorphic to $\mathbb{M}_p$, see the paper H. **Ohno**, Quasi-orthogonal subalgebras of matrix algebras, Linear Algebra Appl. **429**(2008), 2146–2158.

Positive and conditionally negative definite kernel functions are well discussed in the book C. Berg, J. P. R. Christensen and P. Ressel, *Harmonic Analysis on Semigroups. Theory of Positive Definite and Related Functions*, Graduate Texts in Mathematics, vol. 100. Springer, New York, 1984. (It is noteworthy that conditionally negative definite is called there 'negative definite'.)

2.8 Exercises

1. Show that
$$\begin{bmatrix} A & B \\ B^* & C \end{bmatrix} \geq 0$$
   if and only if $B = A^{1/2} Z C^{1/2}$ for a matrix Z with $\|Z\| \leq 1$.

2. Let $X, U, V \in \mathbb{M}_n$ and assume that U and V are unitaries. Prove that
$$\begin{bmatrix} I & U & X \\ U^* & I & V \\ X^* & V^* & I \end{bmatrix} \geq 0$$
   if and only if $X = UV$.

3. Show that for $A, B \in \mathbb{M}_n$ the formula
$$\begin{bmatrix} I & A \\ 0 & I \end{bmatrix}^{-1} \begin{bmatrix} AB & 0 \\ B & 0 \end{bmatrix} \begin{bmatrix} I & A \\ 0 & I \end{bmatrix} = \begin{bmatrix} 0 & 0 \\ B & BA \end{bmatrix}$$
   holds. Conclude that AB and BA have the same eigenvectors.

4. Assume that $0 < A \in \mathbb{M}_n$. Show that $A + A^{-1} \geq 2I$.

5. Assume that

$$A = \begin{bmatrix} A_1 & B \\ B^* & A_2 \end{bmatrix} > 0.$$

Show that $\det A \le \det A_1 \times \det A_2$.

6. Assume that the eigenvalues of the self-adjoint matrix

$$\begin{bmatrix} A & B \\ B^* & C \end{bmatrix}$$

are $\lambda_1 \le \lambda_2 \le \ldots \lambda_n$ and the eigenvalues of A are $\beta_1 \le \beta_2 \le \cdots \le \beta_m$. Show that

$$\lambda_i \le \beta_i \le \lambda_{i+n-m}.$$

7. Show that a matrix $A \in \mathbb{M}_n$ is irreducible if and only if for every $1 \le i, j \le n$ there is a power k such that $(A^k)_{ij} \ne 0$.

8. Let $A, B, C, D \in \mathbb{M}_n$ and $AC = CA$. Show that

$$\det \begin{bmatrix} A & B \\ C & D \end{bmatrix} = \det(AD - CB).$$

9. Let $A, B, C \in \mathbb{M}_n$ and

$$\begin{bmatrix} A & B \\ B^* & C \end{bmatrix} \ge 0.$$

Show that $B^* \circ B \le A \circ C$.

10. Let $A, B \in \mathbb{M}_n$. Show that $A \circ B$ is a submatrix of $A \otimes B$.

11. Assume that P and Q are projections. Show that $P \le Q$ is equivalent to $PQ = P$.

12. Assume that $P_1, P_2, \ldots, P_n$ are projections and $P_1 + P_2 + \cdots + P_n = I$. Show that the projections are pairwise orthogonal.

13. Let $A_1, A_2, \cdots, A_k \in \mathbb{M}_n^{sa}$ and $A_1 + A_2 + \ldots + A_k = I$. Show that the following statements are equivalent:

(1) All operators A_i are projections.
(2) For all $i \ne j$ the product $A_i A_j = 0$ holds.
(3) rank (A_1) + rank (A_2) + $\cdots$ + rank $(A_k) = n$.

14. Let $U|A|$ be the polar decomposition of $A \in \mathbb{M}_n$. Show that A is normal if and only if $U|A| = |A|U$.

15. The matrix $M \in \mathbb{M}_n(\mathbb{C})$ is defined as

$$M_{ij} = \min\{i, j\}.$$

Show that M is positive.

16. Let $A \in \mathbb{M}_n$ and define the mapping $S_A : \mathbb{M}_n \to \mathbb{M}_n$ by $S_A : B \mapsto A \circ B$. Show that the following statements are equivalent.

   (1) A is positive.
   (2) $S_A : \mathbb{M}_n \to \mathbb{M}_n$ is positive.
   (3) $S_A : \mathbb{M}_n \to \mathbb{M}_n$ is completely positive.

17. Let A, B, C be operators on a Hilbert space $\mathcal{H}$ and $A, C \geq 0$. Show that

$$\begin{bmatrix} A & B \\ B^* & C \end{bmatrix} \geq 0$$

   if and only if $|\langle Bx, y \rangle| \leq \langle Ay, y \rangle \cdot \langle Cx, x \rangle$ for every $x, y \in \mathcal{H}$.

18. Let $P \in \mathbb{M}_n$ be idempotent, i.e. $P^2 = P$. Show that P is an ortho-projection if and only if $\|P\| \leq 1$.

19. Let $P \in \mathbb{M}_n$ be an ortho-projection and $0 < A \in \mathbb{M}_n$. Prove the following formulas:

$$[P](A^2) \leq ([P]A)^2, \qquad ([P]A)^{1/2} \leq [P](A^{1/2}), \qquad [P](A^{-1}) \leq ([P]A)^{\dagger}.$$

20. Show that the kernels

$$\psi(x, y) = \cos(x - y), \quad \cos(x^2 - y^2), \quad (1 + |x - y|)^{-1}$$

   are positive semidefinite on $\mathbb{R} \times \mathbb{R}$.

21. Show that the equality

$$A \vee (B \wedge C) = (A \vee B) \wedge (A \vee C)$$

   is not true for ortho-projections.

22. Assume that the kernel $\psi : \mathcal{X} \times \mathcal{X} \to \mathbb{C}$ is positive definite and $\psi(x, x) > 0$ for every $x \in \mathcal{X}$. Show that

$$\bar{\psi}(x, y) = \frac{\psi(x, y)}{\psi(x, x)\psi(y, y)}$$

   is a positive definite kernel.

23. Assume that the kernel $\psi : \mathcal{X} \times \mathcal{X} \to \mathbb{C}$ is negative definite and $\psi(x, x) \geq 0$ for every $x \in \mathcal{X}$. Show that

$$\log(1 + \psi(x, y))$$

   is a negative definite kernel.

24. Show that the kernel $\psi(x, y) = (\sin(x - y))^2$ is negative semidefinite on $\mathbb{R} \times \mathbb{R}$.

25. Show that the linear mapping $\mathcal{E}_{p,n} : \mathbb{M}_n \to \mathbb{M}_n$ defined as

$$\mathcal{E}_{p,n}(A) = pA + (1 - p)\frac{I}{n}\mathrm{Tr}\,A$$

is completely positive if and only if

$$-\frac{1}{n^2 - 1} \le p \le 1.$$

26. Show that the linear mapping $\mathcal{E} : \mathbb{M}_n \to \mathbb{M}_n$ defined as

$$\mathcal{E}(D) = \frac{1}{n - 1}(\mathrm{Tr}\,(D)I - D^t)$$

is a completely positive unital mapping. (Here D^t denotes the transpose of D.)
Show that $\mathcal{E}$ has a negative eigenvalue. (This mapping is called the **Holevo–Werner channel**.)

27. Define $\mathcal{E} : \mathbb{M}_n \to \mathbb{M}_n$ by

$$\mathcal{E}(A) = \frac{1}{n - 1}(I\,\mathrm{Tr}\,A - A).$$

Show that $\mathcal{E}$ is positive but not completely positive.

28. Let p be a real number. Show that the mapping $\mathcal{E}_{p,2} : \mathbb{M}_2 \to \mathbb{M}_2$ defined as

$$\mathcal{E}_{p,2}(A) = pA + (1 - p)\frac{I}{2}\mathrm{Tr}\,A$$

is positive if and only if $-1 \le p \le 1$. Show that $\mathcal{E}_{p,2}$ is completely positive if and only if $-1/3 \le p \le 1$.

29. Show that $\|(f_1, f_2)\|^2 = \|f_1\|^2 + \|f_2\|^2$.

30. Give the analogue of Theorem 2.1 when C is assumed to be invertible.

31. Let $0 \le A \le I$. Find the matrices B and C such that

$$\begin{bmatrix} A & B \\ B^* & C \end{bmatrix}$$

is a projection.

32. Let $\dim \mathcal{H} = 2$ and $0 \le A, B \in B(\mathcal{H})$. Show that there is an orthogonal basis such that

$$A = \begin{bmatrix} a & 0 \\ 0 & b \end{bmatrix}, \qquad B = \begin{bmatrix} c & d \\ d & e \end{bmatrix}$$

with positive numbers $a, b, c, d, e \ge 0$.

33. Let

$$M = \begin{bmatrix} A & B \\ B & A \end{bmatrix}$$

and assume that A and B are self-adjoint. Show that M is positive if and only if
$-A \leq B \leq A$.

34. Determine the inverses of the matrices

$$
A = \begin{bmatrix} a & -b \\ b & a \end{bmatrix} \quad \text{and} \quad B = \begin{bmatrix} a & b & c & d \\ -b & a & -d & c \\ -c & d & a & b \\ -d & c & -b & a \end{bmatrix}.
$$

35. Give the analogue of the factorization (2.2) when D is assumed to be invertible.

36. Show that the self-adjoint invertible matrix

$$
\begin{bmatrix} A & B & C \\ B^* & D & 0 \\ C^* & 0 & E \end{bmatrix}
$$

has inverse in the form

$$
\begin{bmatrix} Q^{-1} & -P & -R \\ -P^* & D^{-1}(I + B^*P) & D^{-1}B^*R \\ -R^* & R^*BD^{-1} & E^{-1}(I + C^*R) \end{bmatrix},
$$

where

$$
Q = A - BD^{-1}B^* - CE^{-1}C^*, \qquad P = Q^{-1}BD^{-1}, \qquad R = Q^{-1}CE^{-1}.
$$

37. Find the determinant and the inverse of the block matrix

$$
\begin{bmatrix} A & 0 \\ a & 1 \end{bmatrix}.
$$

38. Let $A \in \mathbb{M}_n$ be an invertible matrix and $d \in \mathbb{C}$. Show that

$$
\det \begin{bmatrix} A & b \\ c & d \end{bmatrix} = (d - cA^{-1}b)\det A
$$

where $c = [c_1, \ldots, c_n]$ and $b = [b_1, \ldots, n_n]^t$.

39. Prove the concavity of the variance functional $\rho \mapsto \mathrm{Var}_\rho(A)$ defined in (2.10). The concavity is

$$
\mathrm{Var}_\rho(A) \geq \sum_i \lambda_i \mathrm{Var}_{\rho_i}(A) \quad \text{if} \quad \rho = \sum_i \lambda_i \rho_i
$$

when $\lambda_i \geq 0$ and $\sum_i \lambda_i = 1$.

40. For $x, y \in \mathbb{R}^3$ and

$$x \cdot \sigma := \sum_{i=1}^{3} x_i \sigma_i, \qquad y \cdot \sigma := \sum_{i=1}^{3} y_i \sigma_i$$

show that

$$(x \cdot \sigma)(y \cdot \sigma) = \langle x, y \rangle \sigma_0 + \mathrm{i}(x \times y) \cdot \sigma, \qquad (2.18)$$

where $x \times y$ is the vectorial product in $\mathbb{R}^3$.

Chapter 3
Functional Calculus and Derivation

Let $A \in \mathbb{M}_n(\mathbb{C})$ and $p(x) := \sum_i c_i x^i$ be a polynomial. It is quite obvious that by $p(A)$ we mean the matrix $\sum_i c_i A^i$. So the functional calculus is trivial for polynomials. Slightly more generally, let f be a holomorphic function with the Taylor expansion $f(z) = \sum_{k=0}^{\infty} c_k (z-a)^k$. Then for every $A \in \mathbb{M}_n(\mathbb{C})$ such that the operator norm $\|A - aI\|$ is less than the radius of convergence of f, one can define the analytic functional calculus $f(A) := \sum_{k=0}^{\infty} c_k (A - aI)^k$. This analytic functional calculus can be generalized via the Cauchy integral:

$$f(A) := \frac{1}{2\pi i} \int_\Gamma f(z)(zI - A)^{-1} \, dz$$

if f is holomorphic in a domain G containing the eigenvalues of A, where Γ is a simple closed contour in G surrounding the eigenvalues of A. On the other hand, when $A \in \mathbb{M}_n(\mathbb{C})$ is self-adjoint and f is a general function defined on an interval containing the eigenvalues of A, the functional calculus $f(A)$ is defined via the spectral decomposition of A or the diagonalization of A, that is,

$$f(A) = \sum_{i=1}^{k} f(\alpha_i) P_i = U \mathrm{Diag}(f(\lambda_1), \ldots, f(\lambda_n)) U^*$$

for the spectral decomposition $A = \sum_{i=1}^{k} \alpha_i P_i$ and the diagonalization $A = U \mathrm{Diag}(\lambda_1, \ldots, \lambda_n) U^*$. In this way, one has some types of functional calculus for matrices (and also operators). When different types of functional calculus can be defined for one $A \in \mathbb{M}_n(\mathbb{C})$, they yield the same result. The second half of this chapter contains several formulas for derivatives

$$\frac{d}{dt} f(A + tT)$$

and Fréchet derivatives of functional calculus.

F. Hiai and D. Petz, *Introduction to Matrix Analysis and Applications*,
Universitext, DOI: 10.1007/978-3-319-04150-6_3,
© Hindustan Book Agency 2014

3.1 The Exponential Function

The exponential function is well-defined for all complex numbers. It has a convenient Taylor expansion and it appears in some differential equations. It is also important for matrices.

The Taylor expansion can be used to define e^A for a matrix $A \in \mathbb{M}_n(\mathbb{C})$:

$$e^A := \sum_{n=0}^{\infty} \frac{A^n}{n!}. \tag{3.1}$$

Here the right-hand side is an absolutely convergent series:

$$\sum_{n=0}^{\infty} \left\| \frac{A^n}{n!} \right\| \leq \sum_{n=0}^{\infty} \frac{\|A\|^n}{n!} = e^{\|A\|}.$$

The first example is in connection with the Jordan canonical form.

Example 3.1 We take

$$A = \begin{bmatrix} a & 1 & 0 & 0 \\ 0 & a & 1 & 0 \\ 0 & 0 & a & 1 \\ 0 & 0 & 0 & a \end{bmatrix} = aI + J.$$

Since I and J commute and $J^m = 0$ for $m > 3$, we have

$$A^n = a^n I + na^{n-1}J + \frac{n(n-1)}{2} a^{n-2}J^2 + \frac{n(n-1)(n-2)}{2 \cdot 3} a^{n-3}J^3$$

and

$$\sum_{n=0}^{\infty} \frac{A^n}{n!} = \sum_{n=0}^{\infty} \frac{a^n}{n!}I + \sum_{n=1}^{\infty} \frac{a^{n-1}}{(n-1)!}J + \frac{1}{2}\sum_{n=2}^{\infty} \frac{a^{n-2}}{(n-2)!}J^2 + \frac{1}{6}\sum_{n=3}^{\infty} \frac{a^{n-3}}{(n-3)!}J^3$$

$$= e^a I + e^a J + \frac{1}{2}e^a J^2 + \frac{1}{6}e^a J^3.$$

$$\tag{3.2}$$

So we have

$$e^A = e^a \begin{bmatrix} 1 & 1 & 1/2 & 1/6 \\ 0 & 1 & 1 & 1/2 \\ 0 & 0 & 1 & 1 \\ 0 & 0 & 0 & 1 \end{bmatrix}.$$

Note that (3.2) shows that e^A is a linear combination of I, A, A^2, A^3. (This is contained in Theorem 3.6, the coefficients are specified by differential equations.) If $B = SAS^{-1}$, then $e^B = Se^A S^{-1}$. ☐

Example 3.2 It is a basic fact in analysis that

$$e^a = \lim_{n \to \infty} \left(1 + \frac{a}{n}\right)^n$$

for a complex number a, but we also have for matrices:

$$e^A = \lim_{n \to \infty} \left(I + \frac{A}{n}\right)^n. \tag{3.3}$$

This can be checked in a similar way to the previous example:

$$e^{aI+J} = \lim_{n \to \infty} \left(I\left(1 + \frac{a}{n}\right) + \frac{1}{n}J\right)^n.$$

From the point of view of numerical computation (3.1) is a better formula, but (3.3) will be extended in the next theorem. (An extension of the exponential function will appear later in (6.45).) ☐

Theorem 3.3 *Let*

$$T_{m,n}(A) = \left[\sum_{k=0}^{m} \frac{1}{k!}\left(\frac{A}{n}\right)^k\right]^n \qquad (m, n \in \mathbb{N}).$$

Then

$$\lim_{m \to \infty} T_{m,n}(A) = \lim_{n \to \infty} T_{m,n}(A) = e^A.$$

Proof: The matrices $B = e^{\frac{A}{n}}$ and

$$T = \sum_{k=0}^{m} \frac{1}{k!}\left(\frac{A}{n}\right)^k$$

commute. Hence

$$e^A - T_{m,n}(A) = B^n - T^n = (B - T)(B^{n-1} + B^{n-2}T + \cdots + T^{n-1}).$$

We can estimate:

$$\|e^A - T_{m,n}(A)\| \le \|B - T\| n \times \max\{\|B\|^i \|T\|^{n-i-1} : 0 \le i \le n - 1\}.$$

Since $\|T\| \le e^{\frac{\|A\|}{n}}$ and $\|B\| \le e^{\frac{\|A\|}{n}}$, we have

$$\|e^A - T_{m,n}(A)\| \le n\|e^{\frac{A}{n}} - T\|e^{\frac{n-1}{n}\|A\|}.$$

By bounding the tail of the Taylor series,

$$\|e^A - T_{m,n}(A)\| \le \frac{n}{(m+1)!}\left(\frac{\|A\|}{n}\right)^{m+1} e^{\frac{\|A\|}{n}} e^{\frac{n-1}{n}\|A\|}$$

converges to 0 in the two cases $m \to \infty$ and $n \to \infty$. $\qquad\square$

Theorem 3.4 *If $AB = BA$, then*

$$e^{t(A+B)} = e^{tA}e^{tB} \quad (t \in \mathbb{R}). \tag{3.4}$$

Conversely, if this equality holds, then $AB = BA$.

Proof: First we assume that $AB = BA$ and compute the product $e^A e^B$ by multiplying term by term the series:

$$e^A e^B = \sum_{m,n=0}^{\infty} \frac{1}{m!n!} A^m B^n.$$

Therefore,

$$e^A e^B = \sum_{k=0}^{\infty} \frac{1}{k!} C_k,$$

where

$$C_k := \sum_{m+n=k} \frac{k!}{m!n!} A^m B^n.$$

By the commutation relation the binomial formula holds and $C_k = (A+B)^k$. We conclude

$$e^A e^B = \sum_{k=0}^{\infty} \frac{1}{k!}(A+B)^k$$

which is the statement.

Another proof can be obtained by differentiation. It follows from the expansion (3.1) that the derivative of the matrix-valued function $t \mapsto e^{tA}$ defined on $\mathbb{R}$ is $e^{tA}A$:

$$\frac{d}{dt} e^{tA} = e^{tA} A = A e^{tA}. \tag{3.5}$$

Therefore, when $AC = CA$,

$$\frac{d}{dt} e^{tA} e^{C-tA} = e^{tA} A e^{C-tA} - e^{tA} A e^{C-tA} = 0.$$

It follows that the function $t \mapsto e^{tA} e^{C-tA}$ is constant. In particular,

$$e^A e^{C-A} = e^C.$$

Put $A + B$ in place of C to obtain the statement (3.4).

The first derivative of (3.4) is

$$e^{t(A+B)}(A + B) = e^{tA} A e^{tB} + e^{tA} e^{tB} B$$

and the second derivative is

$$e^{t(A+B)}(A + B)^2 = e^{tA} A^2 e^{tB} + e^{tA} A e^{tB} B + e^{tA} A e^{tB} B + e^{tA} e^{tB} B^2.$$

For $t = 0$ this is $BA = AB$.   □

Example 3.5 The matrix exponential function can be used to formulate the solution of a linear first-order differential equation. Let

$$x(t) = \begin{bmatrix} x_1(t) \\ x_2(t) \\ \vdots \\ x_n(t) \end{bmatrix} \quad \text{and} \quad x_0 = \begin{bmatrix} x_1 \\ x_2 \\ \vdots \\ x_n \end{bmatrix}.$$

The solution of the differential equation

$$x'(t) = Ax(t), \qquad x(0) = x_0$$

is $x(t) = e^{tA} x_0$, by formula (3.5).   □

Theorem 3.6 *Let $A \in \mathbb{M}_n$ with characteristic polynomial*

$$p(\lambda) = \det(\lambda I - A) = \lambda^n + c_{n-1}\lambda^{n-1} + \cdots + c_1\lambda + c_0.$$

Then

$$e^{tA} = x_0(t)I + x_1(t)A + \cdots + x_{n-1}(t)A^{n-1},$$

where the vector

$$x(t) = (x_0(t), x_1(t), \ldots, x_{n-1}(t))$$

satisfies the nth order differential equation

$$x^{(n)}(t) + c_{n-1}x^{(n-1)}(t) + \cdots + c_1 x'(t) + c_0 x(t) = 0$$

with the initial condition

$$x^{(k)}(0) = (\overset{1}{\breve{0}}, \ldots, 0, \overset{k}{\breve{1}}, 0, \ldots, 0)$$

for $0 \le k \le n - 1$.

Proof: We can check that the matrix-valued functions

$$F_1(t) = x_0(t)I + x_1(t)A + \cdots + x_{n-1}(t)A^{n-1}$$

and $F_2(t) = e^{tA}$ satisfy the conditions

$$F^{(n)}(t) + c_{n-1}F^{(n-1)}(t) + \cdots + c_1 F'(t) + c_0 F(t) = 0$$

and

$$F(0) = I, \; F'(0) = A, \ldots, F^{(n-1)}(0) = A^{n-1}.$$

Therefore $F_1 = F_2$. □

Example 3.7 In the case of 2×2 matrices, the use of the **Pauli matrices**

$$\sigma_1 = \begin{bmatrix} 0 & 1 \\ 1 & 0 \end{bmatrix}, \qquad \sigma_2 = \begin{bmatrix} 0 & -i \\ i & 0 \end{bmatrix}, \qquad \sigma_3 = \begin{bmatrix} 1 & 0 \\ 0 & -1 \end{bmatrix}$$

is efficient, together with I they form an orthogonal system with respect to the Hilbert–Schmidt inner product.

Let $A \in \mathbb{M}_2^{sa}$ be such that

$$A = c_1 \sigma_1 + c_2 \sigma_2 + c_3 \sigma_3, \qquad c_1^2 + c_2^2 + c_3^2 = 1$$

in the representation with Pauli matrices. It is easy to check that $A^2 = I$. Therefore, for even powers $A^{2n} = I$, but for odd powers $A^{2n+1} = A$. Choose $c \in \mathbb{R}$ and combine these two facts with the knowledge of the relation of the exponential to sine and cosine:

$$e^{icA} = \sum_{n=0}^{\infty} \frac{i^n c^n A^n}{n!} = \sum_{n=0}^{\infty} \frac{(-1)^n c^{2n} A^{2n}}{(2n)!} + i \sum_{n=0}^{\infty} \frac{(-1)^n c^{2n+1} A^{2n+1}}{(2n+1)!}$$

$$= (\cos c)I + i(\sin c)A.$$

A general matrix has the form $C = c_0 I + cA$ and

$$e^{iC} = e^{ic_0}(\cos c)I + ie^{ic_0}(\sin c)A.$$

(e^C is similar, see Exercise 13.)   □

The next theorem gives the so-called **Lie–Trotter formula**. (A generalization is Theorem 5.17.)

Theorem 3.8 *Let $A, B \in \mathbb{M}_n(\mathbb{C})$. Then*

$$e^{A+B} = \lim_{m \to \infty} \left(e^{A/m} e^{B/m}\right)^n.$$

Proof: First we observe that the identity

$$X^n - Y^n = \sum_{j=0}^{n-1} X^{n-1-j}(X - Y)Y^j$$

implies the norm estimate

$$\|X^n - Y^n\| \le nt^{n-1}\|X - Y\|$$

for the submultiplicative operator norm when the constant t is chosen such that $\|X\|, \|Y\| \le t$.

Now we choose $X_n := \exp((A + B)/n)$ and $Y_n := \exp(A/n)\exp(B/n)$. From the above estimate we have

$$\|X_n^n - Y_n^n\| \le nu\|X_n - Y_n\|, \tag{3.6}$$

if we can find a constant u such that $\|X_n\|^{n-1}, \|Y_n\|^{n-1} \le u$. Since

$$\|X_n\|^{n-1} \le \left(\exp((\|A\| + \|B\|)/n)\right)^{n-1} \le \exp(\|A\| + \|B\|)$$

and

$$\|Y_n\|^{n-1} \le \left(\exp(\|A\|/n)\right)^{n-1}\left(\exp(\|B\|/n)\right)^{n-1} \le \exp\|A\| \cdot \exp\|B\|,$$

$u = \exp(\|A\| + \|B\|)$ can be chosen to have the estimate (3.6).

The theorem follows from (3.6) if we can show that $n\|X_n - Y_n\| \to 0$. The power series expansion of the exponential function yields

$$X_n = I + \frac{A+B}{n} + \frac{1}{2}\left(\frac{A+B}{n}\right)^2 + \cdots$$

and

$$Y_n = \left(I + \frac{A}{n} + \frac{1}{2}\left(\frac{A}{n}\right)^2 + \cdots\right)\left(I + \frac{B}{n} + \frac{1}{2}\left(\frac{B}{n}\right)^2 + \cdots\right).$$

If $X_n - Y_n$ is computed by multiplying the two series in Y_n, one can observe that all constant terms and all terms containing $1/n$ cancel. Therefore

$$\|X_n - Y_n\| \le \frac{c}{n^2}$$

for some positive constant c.   □

If A and B are self-adjoint matrices, then it can be better to reach e^{A+B} as the limit of self-adjoint matrices.

Corollary 3.9

$$e^{A+B} = \lim_{n\to\infty}\left(e^{\frac{A}{2n}}\, e^{\frac{B}{n}}\, e^{\frac{A}{2n}}\right)^n.$$

Proof: We have

$$\left(e^{\frac{A}{2n}}\, e^{\frac{B}{n}}\, e^{\frac{A}{2n}}\right)^n = e^{-\frac{A}{2n}}\left(e^{A/n}e^{B/n}\right)^n e^{\frac{A}{2n}}$$

and the limit $n \to \infty$ gives the result.   □

The Lie–Trotter formula can be extended to more matrices:

$$\left\|e^{A_1+A_2+\cdots+A_k} - (e^{A_1/n}e^{A_2/n}\cdots e^{A_k/n})^n\right\|$$

$$\le \frac{2}{n}\left(\sum_{j=1}^{k}\|A_j\|\right)\exp\left(\frac{n+2}{n}\sum_{j=1}^{k}\|A_j\|\right). \tag{3.7}$$

Theorem 3.10 *For matrices $A, B \in \mathbb{M}_n$ the Taylor expansion of the function $\mathbb{R} \ni t \mapsto e^{A+tB}$ is*

$$\sum_{k=0}^{\infty} t^k A_k(1)$$

where $A_0(s) = e^{sA}$ and

$$A_k(s) = \int_0^s dt_1 \int_0^{t_1} dt_2 \cdots \int_0^{t_{k-1}} dt_k e^{(s-t_1)A} B e^{(t_1-t_2)A} B \cdots B e^{t_k A}$$

for $s \in \mathbb{R}$.

Proof: To make differentiation easier we write

$$A_k(s) = \int_0^s e^{(s-t_1)A} B A_{k-1}(t_1)\, dt_1 = e^{sA} \int_0^s e^{-t_1 A} B A_{k-1}(t_1)\, dt_1$$

for $k \geq 1$. It follows that

$$\frac{d}{ds} A_k(s) = A e^{sA} \int_0^s e^{-t_1 A} B A_{k-1}(t_1)\, dt_1 + e^{sA} \frac{d}{ds} \int_0^s e^{-t_1 A} B A_{k-1}(t_1)\, dt_1$$

$$= A A_k(s) + B A_{k-1}(s).$$

Therefore

$$F(s) := \sum_{k=0}^{\infty} A_k(s)$$

satisfies the differential equation

$$F'(s) = (A + B)F(s), \qquad F(0) = I.$$

Therefore $F(s) = e^{s(A+B)}$. If $s = 1$ and we write tB in place of B, then we get the expansion of e^{A+tB}. $\qquad \square$

Corollary 3.11

$$\frac{\partial}{\partial t} e^{A+tB} \Big|_{t=0} = \int_0^1 e^{uA} B e^{(1-u)A}\, du.$$

Another important formula for the exponential function is the **Baker–Campbell–Hausdorff formula**:

$$e^{tA} e^{tB} = \exp\left(t(A + B) + \frac{t^2}{2}[A, B] + \frac{t^3}{12}([A, [A, B]] - [B, [A, B]]) + O(t^4) \right)$$

in which the commutator $[A, B] := AB - BA$ appears.

A function $f : \mathbb{R}^+ = [0, \infty) \to \mathbb{R}$ is **completely monotone** if the nth derivative of f has the sign $(-1)^n$ on the whole of $\mathbb{R}^+$ and for every $n \in \mathbb{N}$.

The next theorem is related to a conjecture.

Theorem 3.12 *Let $A, B \in \mathbb{M}_n^{sa}$ and let $t \in \mathbb{R}$. The following statements are equivalent:*

(i) *The polynomial $t \mapsto \mathrm{Tr}\,(A + tB)^p$ has only positive coefficients for every $A, B \geq 0$ and all $p \in \mathbb{N}$.*

(ii) *For every self-adjoint A and $B \geq 0$, the function $t \mapsto \mathrm{Tr}\,\exp\,(A - tB)$ is completely monotone on $[0, \infty)$.*

(iii) *For every $A > 0$, $B \geq 0$ and all $p \geq 0$, the function $t \mapsto \mathrm{Tr}\,(A + tB)^{-p}$ is completely monotone on $[0, \infty)$.*

Proof: (i)$\Rightarrow$(ii): We have

$$\mathrm{Tr}\,\exp\,(A - tB) = e^{-\|A\|} \sum_{k=0}^{\infty} \frac{1}{k!} \mathrm{Tr}\,(A + \|A\|I - tB)^k$$

and it follows from Bernstein's theorem and (i) that the right-hand side is the Laplace transform of a positive measure supported in $[0, \infty)$.

(ii)$\Rightarrow$(iii): By the matrix equation

$$(A + tB)^{-p} = \frac{1}{\Gamma(p)} \int_0^{\infty} \exp\,[-u(A + tB)]\,u^{p-1}du,$$

we can see the signs of the derivatives.

(iii)$\Rightarrow$(i): It suffices to assume (iii) only for $p \in \mathbb{N}$. For invertible A, by Lemma 3.31 below we observe that the rth derivative of $\mathrm{Tr}\,(A_0 + tB_0)^{-p}$ at $t = 0$ is related to the coefficient of t^r in $\mathrm{Tr}\,(A + tB)^p$ as given by (3.17), where A, A_0, B, B_0 are related as in the lemma. The left-hand side of (3.17) has the sign $(-1)^r$ because it is the derivative of a completely monotone function. Thus the right-hand side has the correct sign as stated in item (i). The case of non-invertible A follows from a continuity argument. $\square$

The **Laplace transform** of a measure μ on $\mathbb{R}^+$ is

$$f(t) = \int_0^{\infty} e^{-tx}\,d\mu(x) \qquad (t \in \mathbb{R}^+).$$

According to the **Bernstein theorem** such a measure μ exists if and only if f is a completely monotone function.

Bessis, Moussa and Villani conjectured in 1975 that the function $t \mapsto \mathrm{Tr}\,\exp(A - tB)$ is a completely monotone function if A is self-adjoint and B is positive. Theorem 3.12 due to Lieb and Seiringer gives an equivalent condition. Property (i) has a very simple formulation.

3.2 Other Functions

All reasonable functions can be approximated by polynomials and, for a polynomial $p(X)$, it is elementary to compute $p(X)$ for a matrix $X \in \mathbb{M}_n$. The canonical Jordan decomposition

$$X = S \begin{bmatrix} J_{k_1}(\lambda_1) & 0 & \cdots & 0 \\ 0 & J_{k_2}(\lambda_2) & \cdots & 0 \\ \vdots & \vdots & \ddots & \vdots \\ 0 & 0 & \cdots & J_{k_m}(\lambda_m) \end{bmatrix} S^{-1} = SJS^{-1}$$

gives that

$$p(X) = S \begin{bmatrix} p(J_{k_1}(\lambda_1)) & 0 & \cdots & 0 \\ 0 & p(J_{k_2}(\lambda_2)) & \cdots & 0 \\ \vdots & \vdots & \ddots & \vdots \\ 0 & 0 & \cdots & p(J_{k_m}(\lambda_m)) \end{bmatrix} S^{-1} = Sp(J)S^{-1}.$$

The crucial point is the computation of $(J_k(\lambda))^m$. Since $J_k(\lambda) = \lambda I_n + J_k(0) = \lambda I_n + J_k$ is the sum of commuting matrices, we can compute the mth power by using the binomial formula:

$$(J_k(\lambda))^m = \lambda^m I_n + \sum_{j=1}^{m} \binom{m}{j} \lambda^{m-j} J_k^j .$$

The powers of J_k are known, see Example 1.15. Let $m > 3$, then the example

$$J_4(\lambda)^m = \begin{bmatrix} \lambda^m & m\lambda^{m-1} & \dfrac{m(m-1)\lambda^{m-2}}{2!} & \dfrac{m(m-1)(m-2)\lambda^{m-3}}{3!} \\ 0 & \lambda^m & m\lambda^{m-1} & \dfrac{m(m-1)\lambda^{m-2}}{2!} \\ 0 & 0 & \lambda^m & m\lambda^{m-1} \\ 0 & 0 & 0 & \lambda^m \end{bmatrix}$$

demonstrates the point. In another formulation,

$$p(J_4(\lambda)) = \begin{bmatrix} p(\lambda) & p'(\lambda) & \dfrac{p''(\lambda)}{2!} & \dfrac{p^{(3)}(\lambda)}{3!} \\[2mm] 0 & p(\lambda) & p'(\lambda) & \dfrac{p''(\lambda)}{2!} \\[2mm] 0 & 0 & p(\lambda) & p'(\lambda) \\[2mm] 0 & 0 & 0 & p(\lambda) \end{bmatrix},$$

which is actually correct for all polynomials and for every smooth function. We conclude that if the canonical Jordan form is known for $X \in \mathbb{M}_n$, then $f(X)$ is computable. In particular, the above argument gives the following result.

Theorem 3.13 *For $X \in \mathbb{M}_n$ the relation*

$$\det e^X = \exp(\mathrm{Tr}\, X)$$

holds between trace and determinant.

A matrix $A \in \mathbb{M}_n$ is **diagonalizable** if

$$A = S \,\mathrm{Diag}(\lambda_1, \lambda_2, \dots, \lambda_n) S^{-1}$$

for some invertible matrix S. Observe that this condition means that in the Jordan canonical form all Jordan blocks are 1×1 and the numbers $\lambda_1, \lambda_2, \dots, \lambda_n$ are the eigenvalues of A. In this case,

$$f(A) = S \,\mathrm{Diag}(f(\lambda_1), f(\lambda_2), \dots, f(\lambda_n)) S^{-1} \tag{3.8}$$

when the complex-valued function f is defined on the set of eigenvalues of A.

If the numbers $\lambda_1, \lambda_2, \dots, \lambda_n$ are different, then we can have a polynomial $p(x)$ of order $n - 1$ such that $p(\lambda_i) = f(\lambda_i)$:

$$p(x) = \sum_{j=1}^{n} \prod_{i \neq j} \frac{x - \lambda_i}{\lambda_j - \lambda_i} f(\lambda_j).$$

(This is the so-called **Lagrange interpolation** formula.) Therefore we have

$$f(A) = p(A) = \sum_{j=1}^{n} \prod_{i \neq j} \frac{A - \lambda_i I}{\lambda_j - \lambda_i} f(\lambda_j).$$

(The relevant formulations are in Exercises 14 and 15.)

Example 3.14 Consider the self-adjoint matrix

$$X = \begin{bmatrix} 1+z & x-yi \\ x+yi & 1-z \end{bmatrix} = \begin{bmatrix} 1+z & w \\ \overline{w} & 1-z \end{bmatrix}$$

where $x, y, z \in \mathbb{R}$. From the characteristic polynomial we have the eigenvalues

$$\lambda_1 = 1+R \quad \text{and} \quad \lambda_2 = 1-R,$$

where $R = \sqrt{x^2 + y^2 + z^2}$. If $R < 1$, then X is positive and invertible. The eigenvectors are

$$u_1 = \begin{bmatrix} R+z \\ \overline{w} \end{bmatrix} \quad \text{and} \quad u_2 = \begin{bmatrix} R-z \\ -\overline{w} \end{bmatrix}.$$

Set

$$\Delta = \begin{bmatrix} 1+R & 0 \\ 0 & 1-R \end{bmatrix}, \quad S = \begin{bmatrix} R+z & R-z \\ \overline{w} & -\overline{w} \end{bmatrix}.$$

We can check that $XS = S\Delta$, hence

$$X = S\Delta S^{-1}.$$

To compute S^{-1} we use the formula

$$\begin{bmatrix} a & b \\ c & d \end{bmatrix}^{-1} = \frac{1}{ad-bc} \begin{bmatrix} d & -b \\ -c & a \end{bmatrix}.$$

Hence

$$S^{-1} = \frac{1}{2\overline{w}R} \begin{bmatrix} \overline{w} & R-z \\ \overline{w} & -R-z \end{bmatrix}.$$

It follows that

$$X^t = a_t \begin{bmatrix} b_t+z & w \\ \overline{w} & b_t-z \end{bmatrix},$$

where

$$a_t = \frac{(1+R)^t - (1-R)^t}{2R}, \quad b_t = R \frac{(1+R)^t + (1-R)^t}{(1+R)^t - (1-R)^t}.$$

The matrix $X/2$ is a density matrix and has applications in quantum theory. □

In the previous example the function $f(x) = x^t$ was used. If the eigenvalues of A are positive, then $f(A)$ is well-defined. The canonical Jordan decomposition is not the only one we might use. It is known in analysis that

$$x^p = \frac{\sin p\pi}{\pi} \int_0^\infty \frac{x\lambda^{p-1}}{\lambda + x} \, d\lambda \quad (x \in (0, \infty))$$

when $0 < p < 1$. It follows that for a positive matrix A we have

$$A^p = \frac{\sin p\pi}{\pi} \int_0^\infty \lambda^{p-1} A(\lambda I + A)^{-1} \, d\lambda.$$

For self-adjoint matrices A a simple formula for A^p is available, nevertheless the previous integral formula is still useful in some situations, for example in the context of differentiation.

Recall that self-adjoint matrices are diagonalizable and they have a spectral decomposition. Let $A = \sum_i \lambda_i P_i$ be the spectral decomposition of a self-adjoint $A \in \mathbb{M}_n(\mathbb{C})$. ($\lambda_i$ are the different eigenvalues and P_i are the corresponding eigen-projections; the rank of P_i is the multiplicity of λ_i.) Then

$$f(A) = \sum_i f(\lambda_i) P_i . \tag{3.9}$$

Usually we assume that f is continuous on an interval containing the eigenvalues of A.

Example 3.15 Consider

$$f_+(t) := \max\{t, 0\} \quad \text{and} \quad f_-(t) := \max\{-t, 0\} \quad \text{for} \quad t \in \mathbb{R}.$$

For each $A \in B(\mathcal{H})^{sa}$ define

$$A_+ := f_+(A) \quad \text{and} \quad A_- := f_-(A).$$

Since $f_+(t), f_-(t) \geq 0$, $f_+(t) - f_-(t) = t$ and $f_+(t) f_-(t) = 0$, we have

$$A_+, A_- \geq 0, \quad A = A_+ - A_-, \quad A_+ A_- = 0.$$

These A_+ and A_- are called the **positive part** and the **negative part** of A, respectively, and $A = A_+ + A_-$ is called the **Jordan decomposition** of A. □

Let f be holomorphic inside and on a positively oriented simple contour Γ in the complex plane and let A be an $n \times n$ matrix such that its eigenvalues are inside Γ. Then

$$f(A) := \frac{1}{2\pi i} \int_{\Gamma} f(z)(zI - A)^{-1} dz \qquad (3.10)$$

is defined by a contour integral. When A is self-adjoint, then (3.9) makes sense and it is an exercise to show that it gives the same result as (3.10).

Example 3.16 We can define the square root function on the set

$$G := \{re^{i\varphi} \in \mathbb{C} : r > 0, \ -\pi/2 < \varphi < \pi/2\}$$

as $\sqrt{re^{i\varphi}} := \sqrt{r}e^{i\varphi/2}$ and this is a holomorphic function on G.

When $X = S\operatorname{Diag}(\lambda_1, \lambda_2, \ldots, \lambda_n) S^{-1} \in \mathbb{M}_n$ is a weakly positive matrix, then $\lambda_1, \lambda_2, \ldots, \lambda_n > 0$ and to use (3.10) we can take a positively oriented simple contour Γ in G such that the eigenvalues are inside Γ. Then

$$\sqrt{X} = \frac{1}{2\pi i} \int_{\Gamma} \sqrt{z}(zI - X)^{-1} dz$$

$$= S \left(\frac{1}{2\pi i} \int_{\Gamma} \sqrt{z} \operatorname{Diag}(1/(z - \lambda_1), 1/(z - \lambda_2), \ldots, 1/(z - \lambda_n)) \, dz \right) S^{-1}$$

$$= S \operatorname{Diag}(\sqrt{\lambda_1}, \sqrt{\lambda_2}, \ldots, \sqrt{\lambda_n}) S^{-1}.$$

$\square$

Example 3.17 The **logarithm** is a well-defined differentiable function on positive numbers. Therefore for a strictly positive operator A formula (3.9) gives $\log A$. Since

$$\log x = \int_0^\infty \frac{1}{1+t} - \frac{1}{x+t} \, dt,$$

we can use

$$\log A = \int_0^\infty \frac{1}{1+t} I - (A + tI)^{-1} \, dt. \qquad (3.11)$$

If we have a matrix A with eigenvalues outside of $\mathbb{R}^- = (-\infty, 0]$, then we can take the domain

$$\mathcal{D} = \{re^{i\varphi} \in \mathbb{C} : r > 0, \ -\pi < \varphi < \pi\}$$

with the function $re^{i\varphi} \mapsto \log r + i\varphi$. The integral formula (3.10) can be used for the calculus. Another useful formula is

$$\log A = \int_0^1 (A - I)(t(A - I) + I)^{-1}\, dt \qquad (3.12)$$

(when A does not have eigenvalue in $\mathbb{R}^-$).

Note that $\log(ab) = \log a + \log b$ is not true for any complex numbers, so it cannot be expected to hold for (commuting) matrices. $\square$

Theorem 3.18 *If f_k and g_k are functions $(\alpha, \beta) \to \mathbb{R}$ such that for some $c_k \in \mathbb{R}$*

$$\sum_k c_k f_k(x) g_k(y) \geq 0$$

for every $x, y \in (\alpha, \beta)$, then

$$\sum_k c_k \mathrm{Tr}\, f_k(A) g_k(B) \geq 0$$

whenever A, B are self-adjoint matrices with spectra in (α, β).

Proof: Let $A = \sum_i \lambda_i P_i$ and $B = \sum_j \mu_j Q_j$ be the spectral decompositions. Then

$$\sum_k c_k \mathrm{Tr}\, f_k(A) g_k(B) = \sum_k \sum_{i,j} c_k \mathrm{Tr}\, P_i f_k(\lambda_i) g_k(\mu_j) Q_j$$

$$= \sum_{i,j} \mathrm{Tr}\, P_i Q_j \sum_k c_k f_k(\lambda_i) g_k(\mu_j) \geq 0$$

by the hypothesis. $\square$

In the theorem assume that $\sum_k c_k f_k(x) g_k(y) = 0$ if and only if $x = y$. Then we show that $\sum_k c_k \mathrm{Tr}\, f_k(A) g_k(B) = 0$ if and only if $A = B$. From the above proof it follows that $\sum_k c_k \mathrm{Tr}\, f_k(A) g_k(B) = 0$ holds if and only if $\mathrm{Tr}\, P_i Q_j > 0$ implies $\lambda_i = \mu_j$. This property yields

$$Q_j A Q_j = \sum_i \lambda_i Q_j P_i Q_j = \mu_j Q_j,$$

and similarly $Q_j A^2 Q_j = \mu_j^2 Q_j$. Hence

$$(A Q_j - \mu_j Q_j)^*(A Q_j - \mu_j Q_j) = Q_j A^2 Q_j - 2\mu_j Q_j A Q_j + \mu_j^2 Q_j = 0$$

so that $A Q_j = \mu_j Q_j = B Q_j$ for all j, which implies $A = B$. The converse is obvious.

Example 3.19 In order to exhibit an application of the previous theorem, let us assume that f is convex. Then

$$f(x) - f(y) - (x - y)f'(y) \geq 0$$

and

$$\text{Tr } f(A) \geq \text{Tr } f(B) + \text{Tr } (A - B)f'(B).$$

Replacing f by $-\eta(t) = t \log t$ we have

$$\text{Tr } A \log A \geq \text{Tr } B \log B + \text{Tr } (A - B) + \text{Tr } (A - B) \log B$$

or equivalently

$$\text{Tr } A(\log A - \log B) - \text{Tr } (A - B) \geq 0.$$

The left-hand side is the quantum **relative entropy** $S(A\|B)$ of the positive definite matrices A and B. (In fact, $S(A\|B)$ is well-defined for $A, B \geq 0$ if $\ker A \supset \ker B$; otherwise, it is defined to be $+\infty$). Moreover, since $-\eta(t)$ is strictly convex, we see that $S(A\|B) = 0$ if and only if $A = B$.

If $\text{Tr } A = \text{Tr } B$, then $S(A\|B)$ is the so-called **Umegaki relative entropy**: $S(A\|B) = \text{Tr } A(\log A - \log B)$. For this we can obtain a better estimate. If $\text{Tr } A = \text{Tr } B = 1$, then all eigenvalues are in $[0, 1]$. Analysis tells us that for some $\xi \in (x, y)$

$$-\eta(x) + \eta(y) + (x - y)\eta'(y) = -\frac{1}{2}(x - y)^2 \eta''(\xi) \geq \frac{1}{2}(x - y)^2$$

when $x, y \in [0, 1]$. According to Theorem 3.18 we have

$$\text{Tr } A(\log A - \log B) \geq \frac{1}{2}\text{Tr } (A - B)^2. \tag{3.13}$$

The **Streater inequality** (3.13) has the consequence that $A = B$ if the relative entropy is 0. Indeed, a stronger inequality called **Pinsker's inequality** is known: If $\text{Tr } A = \text{Tr } B$, then

$$\text{Tr } A(\log A - \log B) \geq \frac{1}{2}\|A - B\|_1^2,$$

where $\|A - B\|_1 := \text{Tr } |A - B|$ is the trace-norm of $A - B$, see Sect. 6.3. $\square$

3.3 Derivation

This section introduces derivatives of scalar-valued and matrix-valued functions. From the latter, a scalar-valued function can be obtained, for example, by taking the trace.

Example 3.20 Assume that $A \in \mathbb{M}_n$ is invertible. Then $A + tT$ is invertible as well for $T \in \mathbb{M}_n$ and for a small real number t. The identity

$$(A + tT)^{-1} - A^{-1} = (A + tT)^{-1}(A - (A + tT))A^{-1} = -t(A + tT)^{-1}TA^{-1}$$

gives

$$\lim_{t \to 0} \frac{1}{t}\left((A + tT)^{-1} - A^{-1}\right) = -A^{-1}TA^{-1}.$$

The derivative was computed at $t = 0$, but if $A + tT$ is invertible, then

$$\frac{d}{dt}(A + tT)^{-1} = -(A + tT)^{-1}T(A + tT)^{-1}$$

by a similar computation. We can continue the derivation:

$$\frac{d^2}{dt^2}(A + tT)^{-1} = 2(A + tT)^{-1}T(A + tT)^{-1}T(A + tT)^{-1},$$

$$\frac{d^3}{dt^3}(A + tT)^{-1} = -6(A + tT)^{-1}T(A + tT)^{-1}T(A + tT)^{-1}T(A + tT)^{-1}.$$

So the Taylor expansion is

$$(A + tT)^{-1} = A^{-1} - tA^{-1}TA^{-1} + t^2A^{-1}TA^{-1}TA^{-1}$$
$$- t^3A^{-1}TA^{-1}TA^{-1}TA^{-1} + \cdots$$
$$= \sum_{n=0}^{\infty}(-t)^n A^{-1/2}(A^{-1/2}TA^{-1/2})^n A^{-1/2}.$$

Since

$$(A + tT)^{-1} = A^{-1/2}(I + tA^{-1/2}TA^{-1/2})^{-1}A^{-1/2},$$

we can also obtain Taylor expansion from the Neumann series of $(I + tA^{-1/2}TA^{-1/2})^{-1}$, see Example 1.8.   □

Example 3.21 There is an interesting formula which relates the functional calculus and derivation:

$$f\left(\begin{bmatrix} A & B \\ 0 & A \end{bmatrix}\right) = \begin{bmatrix} f(A) & \frac{d}{dt}f(A + tB) \\ 0 & f(A) \end{bmatrix}.$$

If f is a polynomial, then it is easy to check this formula.   □

Example 3.22 Assume that $A \in M_n$ is positive invertible. Then $A + tT$ is also positive invertible for $T \in M_n^{sa}$ and for a small real number t. Therefore $\log(A+tT)$ is defined and it is expressed as

$$\log(A + tT) = \int_0^\infty (x + 1)^{-1} I - (xI + A + tT)^{-1} \, dx.$$

This is a convenient formula for the derivation (with respect to $t \in \mathbb{R}$):

$$\frac{d}{dt} \log(A + tT) = \int_0^\infty (xI + A)^{-1} T (xI + A)^{-1} \, dx$$

from the derivative of the inverse. The derivation can be continued, yielding the Taylor expansion

$$\log(A + tT) = \log A + t \int_0^\infty (x + A)^{-1} T (x + A)^{-1} \, dx$$

$$- t^2 \int_0^\infty (x + A)^{-1} T (x + A)^{-1} T (x + A)^{-1} \, dx + \cdots$$

$$= \log A - \sum_{n=1}^\infty (-t)^n \int_0^\infty (x + A)^{-1/2}$$

$$\times ((x + A)^{-1/2} T (x + A)^{-1/2})^n (x + A)^{-1/2} \, dx.$$

$\square$

Theorem 3.23 Let $A, B \in M_n(\mathbb{C})$ be self-adjoint matrices and $t \in \mathbb{R}$. Assume that $f : (\alpha, \beta) \to \mathbb{R}$ is a continuously differentiable function defined on an interval and assume that the eigenvalues of $A + tB$ are in (α, β) for small $t - t_0$. Then

$$\frac{d}{dt} \mathrm{Tr} \, f(A + tB) \Big|_{t=t_0} = \mathrm{Tr} \, (Bf'(A + t_0 B)) .$$

Proof: One can verify the formula for a polynomial f by an easy direct computation: $\mathrm{Tr} \, (A + tB)^n$ is a polynomial in the real variable t. We are interested in the coefficient of t which is

$$\mathrm{Tr} \, (A^{n-1} B + A^{n-2} BA + \cdots + ABA^{n-2} + BA^{n-1}) = n \mathrm{Tr} \, A^{n-1} B.$$

We have the result for polynomials and the formula can be extended to a more general f by means of polynomial approximation. $\square$

Example 3.24 Let $f : (\alpha, \beta) \to \mathbb{R}$ be a continuous increasing function and assume that the spectrum of the self-adjoint matrices A and C lie in (α, β). We use the previous theorem to show that

$$A \leq C \quad \text{implies} \quad \text{Tr } f(A) \leq \text{Tr } f(C). \tag{3.14}$$

We may assume that f is smooth and it is enough to show that the derivative of $\text{Tr } f(A + tB)$ is positive when $B \geq 0$. (To conclude (3.14), one takes $B = C - A$.) The derivative is $\text{Tr } (Bf'(A + tB))$ and this is the trace of the product of two positive operators. Therefore, it is positive.

Another (simpler) way to show this is to use Theorem 2.16. For the eigenvalues of A, C we have $\lambda_k(A) \leq \lambda_k(C)$ $(1 \leq k \leq n)$ and hence $\text{Tr } f(A) = \sum_k f(\lambda_k(A)) \leq \sum_k f(\lambda_k(C)) = \text{Tr } f(C)$.   □

For a holomorphic function f, we can compute the derivative of $f(A + tB)$ by using (3.10), where Γ is a positively oriented simple contour satisfying the properties required above. The derivation is reduced to the differentiation of the resolvent $(zI - (A + tB))^{-1}$ and we obtain

$$X := \frac{d}{dt} f(A + tB)\Big|_{t=0} = \frac{1}{2\pi i} \int_\Gamma f(z)(zI - A)^{-1} B(zI - A)^{-1} \, dz. \tag{3.15}$$

When A is self-adjoint, then there is no loss of generality in assuming that it is diagonal, $A = \text{Diag}(t_1, t_2, \ldots, t_n)$, and we compute the entries of the matrix (3.15) using the Frobenius formula

$$\frac{1}{2\pi i} \int_\Gamma \frac{f(z)}{(z - t_i)(z - t_j)} \, dz = \frac{f(t_i) - f(t_j)}{t_i - t_j}$$

(this means $f'(t_i)$ if $t_i = t_j$). Therefore,

$$X_{ij} = \frac{1}{2\pi i} \int_\Gamma f(z) \frac{1}{z - t_i} B_{ij} \frac{1}{z - t_j} \, dz = \frac{f(t_i) - f(t_j)}{t_i - t_j} B_{ij}.$$

A C^1-function, together with its derivative, can be approximated by polynomials. Hence we have the following result.

Theorem 3.25 *Assume that $f : (\alpha, \beta) \to \mathbb{R}$ is a C^1-function and $A = \text{Diag}(t_1, t_2, \ldots, t_n)$ with $\alpha < t_i < \beta$ $(1 \leq i \leq n)$. If $B = B^*$, then the derivative $t \mapsto f(A + tB)$ is a Schur product:*

$$\frac{d}{dt} f(A + tB)\Big|_{t=0} = D \circ B, \tag{3.16}$$

where D is the divided difference matrix:

$$D_{ij} = \begin{cases} \dfrac{f(t_i) - f(t_j)}{t_i - t_j} & \text{if } t_i - t_j \neq 0, \\[2mm] f'(t_i) & \text{if } t_i - t_j = 0. \end{cases}$$

Let $f : (\alpha, \beta) \to \mathbb{R}$ be a continuous function. It is called **matrix monotone** if

$$A \leq C \quad \text{implies} \quad f(A) \leq f(C)$$

when the spectra of the self-adjoint matrices B and C lie in (α, β).

Theorem 2.15 tells us that $f(x) = -1/x$ is a matrix monotone function. Matrix monotonicity means that $f(A + tB)$ is an increasing function when $B \geq 0$. The increasing property is equivalent to the positivity of the derivative. We use the previous theorem to show that the function $f(x) = \sqrt{x}$ is matrix monotone.

Example 3.26 Assume that $A > 0$ is diagonal: $A = \text{Diag}(t_1, t_2, \ldots, t_n)$. Then the derivative of the function $\sqrt{A + tB}$ is $D \circ B$, where

$$
D_{ij} =
\begin{cases}
\dfrac{1}{\sqrt{t_i} + \sqrt{t_j}} & \text{if } t_i - t_j \neq 0, \\[2mm]
\dfrac{1}{2\sqrt{t_i}} & \text{if } t_i - t_j = 0.
\end{cases}
$$

This is a Cauchy matrix (see Example 1.41) and it is positive. If B is positive, then so is the Schur product. We have shown that the derivative is positive, hence $f(x) = \sqrt{x}$ is matrix monotone.

An idea for another proof appears in Exercise 28. $\qquad\qquad\square$

A subset $K \subset \mathbb{M}_n$ is **convex** if for any $A, B \in K$ and for any real number $0 < \lambda < 1$

$$\lambda A + (1 - \lambda)B \in K.$$

The functional $F : K \to \mathbb{R}$ is convex if for $A, B \in K$ and for any real number $0 < \lambda < 1$ the inequality

$$F(\lambda A + (1 - \lambda)B) \leq \lambda F(A) + (1 - \lambda)F(B)$$

holds. This inequality is equivalent to the convexity of the function

$$G : [0, 1] \to \mathbb{R}, \quad G(\lambda) := F(B + \lambda(A - B)).$$

It is well-known in analysis that convexity is related to the second derivative.

Theorem 3.27 *Let K be the set of self-adjoint $n \times n$ matrices with spectrum in the interval (α, β). Assume that the function $f : (\alpha, \beta) \to \mathbb{R}$ is a convex C^2-function. Then the functional $A \mapsto \text{Tr} f(A)$ is convex on K.*

Proof: The stated convexity is equivalent to the convexity of the numerical functions

$$t \mapsto \operatorname{Tr} f(tX_1 + (1-t)X_2) = \operatorname{Tr} (X_2 + t(X_1 - X_2)) \qquad (t \in [0,1]).$$

It is enough to prove that the second derivative of $t \mapsto \operatorname{Tr} f(A + tB)$ is positive at $t = 0$.

The first derivative of the functional $t \mapsto \operatorname{Tr} f(A + tB)$ is $\operatorname{Tr} f'(A + tB)B$. To compute the second derivative we differentiate $f'(A + tB)$. We can assume that A is diagonal and we differentiate at $t = 0$. Using (3.16) we get

$$\left[\frac{d}{dt} f'(A + tB) \Big|_{t=0} \right]_{ij} = \frac{f'(t_i) - f'(t_j)}{t_i - t_j} B_{ij}.$$

Therefore,

$$\frac{d^2}{dt^2} \operatorname{Tr} f(A + tB) \Big|_{t=0} = \operatorname{Tr} \left[\frac{d}{dt} f'(A + tB) \Big|_{t=0} \right] B$$

$$= \sum_{i,k} \left[\frac{d}{dt} f'(A + tB) \Big|_{t=0} \right]_{ik} B_{ki}$$

$$= \sum_{i,k} \frac{f'(t_i) - f'(t_k)}{t_i - t_k} B_{ik} B_{ki}$$

$$= \sum_{i,k} f''(s_{ik}) |B_{ik}|^2,$$

where s_{ik} is between t_i and t_k. The convexity of f means $f''(s_{ik}) \geq 0$, hence we conclude the positivity.   □

In the above theorem one can remove the C^2 assumption for f by using the so-called regularization (or smoothing) technique; for this technique, see [20] or [36]. Note that another less analytic proof is sketched in Exercise 22.

Example 3.28 The function

$$\eta(x) = \begin{cases} -x \log x & \text{if } x > 0, \\ 0 & \text{if } x = 0 \end{cases}$$

is continuous and concave on $\mathbb{R}^+$. For a positive matrix $D \geq 0$

$$S(D) := \operatorname{Tr} \eta(D)$$

is called the **von Neumann entropy**. It follows from the previous theorem that $S(D)$ is a concave function of D. If we are being very rigorous, then we cannot apply the theorem, since η is not differentiable at 0. Therefore we should apply the theorem to $f(x) := \eta(x + \varepsilon)$, where $\varepsilon > 0$ and take the limit $\varepsilon \to 0$.   □

Example 3.29 Let a self-adjoint matrix H be fixed. The state of a quantum system is described by a density matrix D which has the properties $D \geq 0$ and $\mathrm{Tr}\, D = 1$. The equilibrium state minimizes the energy

$$F(D) = \mathrm{Tr}\, DH - \frac{1}{\beta} S(D),$$

where β is a positive number. To find the minimizer, we solve the equation

$$\frac{\partial}{\partial t} F(D + tX)\Big|_{t=0} = 0$$

for self-adjoint matrices X satisfying $\mathrm{Tr}\, X = 0$. The equation is

$$\mathrm{Tr}\, X \left(H + \frac{1}{\beta} \log D + \frac{1}{\beta} I \right) = 0$$

and

$$H + \frac{1}{\beta} \log D + \frac{1}{\beta} I$$

must be cI. Hence the minimizer is

$$D = \frac{e^{-\beta H}}{\mathrm{Tr}\, e^{-\beta H}},$$

which is called the **Gibbs state**.   □

Example 3.30 Next we restrict ourselves to the self-adjoint case $A, B \in \mathbb{M}_n(\mathbb{C})^{sa}$ in the analysis of (3.15).

The space $\mathbb{M}_n(\mathbb{C})^{sa}$ can be decomposed as $\mathcal{M}_A \oplus \mathcal{M}_A^{\perp}$, where $\mathcal{M}_A := \{C \in \mathbb{M}_n(\mathbb{C})^{sa} : CA = AC\}$ is the commutant of A and $\mathcal{M}_A^{\perp}$ is its orthogonal complement. If we consider the operator $\mathbf{L}_A : X \mapsto i[A, X] := i(AX - XA)$, then $\mathcal{M}_A$ is precisely the kernel of $\mathbf{L}_A$, while $\mathcal{M}_A^{\perp}$ is its range.

If $B \in \mathcal{M}_A$, then

$$\frac{1}{2\pi i} \int_\Gamma f(z)(zI - A)^{-1} B(zI - A)^{-1} \, dz = \frac{B}{2\pi i} \int_\Gamma f(z)(zI - A)^{-2} \, dz = Bf'(A)$$

and we have

$$\frac{d}{dt} f(A + tB)\Big|_{t=0} = Bf'(A).$$

If $B = \mathrm{i}[A, X] \in \mathcal{M}_A^\perp$, then we use the identity

$$(zI - A)^{-1}[A, X](zI - A)^{-1} = [(zI - A)^{-1}, X]$$

and we conclude

$$\frac{d}{dt} f(A + t\mathrm{i}[A, X])\Big|_{t=0} = \mathrm{i}[f(A), X].$$

To compute the derivative in an arbitrary direction B we should decompose B as $B_1 \oplus B_2$ with $B_1 \in \mathcal{M}_A$ and $B_2 \in \mathcal{M}_A^\perp$. Then

$$\frac{d}{dt} f(A + tB)\Big|_{t=0} = B_1 f'(A) + \mathrm{i}[f(A), X],$$

where X is the solution to the equation $B_2 = \mathrm{i}[A, X]$.   □

The next lemma was used in the proof of Theorem 3.12.

Lemma 3.31 *Let $A_0, B_0 \in \mathbb{M}_n^{sa}$ and assume $A_0 > 0$. Define $A = A_0^{-1}$ and $B = A_0^{-1/2} B_0 A_0^{-1/2}$, and let $t \in \mathbb{R}$. For all $p, r \in \mathbb{N}$*

$$\frac{d^r}{dt^r} \mathrm{Tr}\, (A_0 + tB_0)^{-p}\Big|_{t=0} = \frac{p}{p+r}(-1)^r \frac{d^r}{dt^r} \mathrm{Tr}\, (A + tB)^{p+r}\Big|_{t=0}. \qquad (3.17)$$

Proof: By induction it is easy to show that

$$\frac{d^r}{dt^r}(A + tB)^{p+r} = r! \sum_{\substack{0 \le i_1, \dots, i_{r+1} \le p \\ \sum_j i_j = p}} (A + tB)^{i_1} B(A + tB)^{i_2} \cdots B(A + tB)^{i_{r+1}}.$$

By taking the trace at $t = 0$ we obtain

$$\kappa_1 := \frac{d^r}{dt^r} \mathrm{Tr}\, (A + tB)^{p+r}\Big|_{t=0} = r! \sum_{\substack{0 \le i_1, \dots, i_{r+1} \le p \\ \sum_j i_j = p}} \mathrm{Tr}\, A^{i_1} B A^{i_2} \cdots B A^{i_{r+1}}.$$

Moreover, by similar arguments,

$$\frac{d^r}{dt^r}(A_0 + tB_0)^{-p}$$
$$= (-1)^r r! \sum_{\substack{1 \le i_1, \dots, i_{r+1} \le p \\ \sum_j i_j = p+r}} (A_0 + t B_0)^{-i_1} B_0(A_0 + tB_0)^{-i_2} \cdots B_0(A_0 + tB_0)^{-i_{r+1}}.$$

By taking the trace at $t = 0$ and using cyclicity, we get

$$\kappa_2 := \frac{d^r}{dt^r} \text{Tr} \, (A_0 + t B_0)^{-p} \Big|_{t=0} = (-1)^r r! \sum_{\substack{0 \le i_1,\dots,i_{r+1} \le p-1 \\ \sum_j i_j = p-1}} \text{Tr} \, A \, A^{i_1} B A^{i_2} \cdots B A^{i_{r+1}} .$$

We have to show that

$$\kappa_2 = \frac{p}{p+r}(-1)^r \kappa_1 .$$

To see this we rewrite κ_1 in the following way. Define $p+r$ matrices M_j by

$$M_j = \begin{cases} B & \text{for } 1 \le j \le r \\ A & \text{for } r+1 \le j \le r+p . \end{cases}$$

Let S_n denote the permutation group on $\{1, \dots, n\}$. Then

$$\kappa_1 = \frac{1}{p!} \sum_{\pi \in S_{p+r}} \text{Tr} \prod_{j=1}^{p+r} M_{\pi(j)} .$$

Because of the cyclicity of the trace we can always arrange the product so that M_{p+r} has the first position in the trace. Since there are $p+r$ possible locations for M_{p+r} to appear in the product above, and all products are equally weighted, we get

$$\kappa_1 = \frac{p+r}{p!} \sum_{\pi \in S_{p+r-1}} \text{Tr} \, A \prod_{j=1}^{p+r-1} M_{\pi(j)} .$$

On the other hand,

$$\kappa_2 = (-1)^r \frac{1}{(p-1)!} \sum_{\pi \in S_{p+r-1}} \text{Tr} \, A \prod_{j=1}^{p+r-1} M_{\pi(j)} ,$$

so we arrive at the desired equality.   □

3.4 Fréchet Derivatives

Let f be a real-valued function on $(a, b) \subset \mathbb{R}$, and denote by $\mathbb{M}_n^{sa}(a, b)$ the set of all matrices $A \in \mathbb{M}_n^{sa}$ with $\sigma(A) \subset (a, b)$. In this section we discuss the differentiability properties of the matrix functional calculus $A \mapsto f(A)$ when $A \in \mathbb{M}_n^{sa}(a, b)$.

The case $n = 1$ corresponds to differentiation in classical analysis. There the **divided differences** are important and will also appear here. Let $x_1, x_2, \dots$ be distinct points in (a, b). Then we define

$$f^{[0]}[x_1] := f(x_1), \qquad f^{[1]}[x_1, x_2] := \frac{f(x_1) - f(x_2)}{x_1 - x_2}$$

and recursively for $n = 2, 3, \ldots$,

$$f^{[n]}[x_1, x_2, \ldots, x_{n+1}] := \frac{f^{[n-1]}[x_1, x_2, \ldots, x_n] - f^{[n-1]}[x_2, x_3, \ldots, x_{n+1}]}{x_1 - x_{n+1}}.$$

The functions $f^{[1]}$, $f^{[2]}$ and $f^{[n]}$ are called the **first**, the **second** and the nth **divided differences**, respectively, of f.

From the recursive definition the symmetry is not clear. If f is a C^n-function, then

$$f^{[n]}[x_0, x_1, \ldots, x_n] = \int_S f^{(n)}(t_0 x_0 + t_1 x_1 + \cdots + t_n x_n) \, dt_1 dt_2 \cdots dt_n, \quad (3.18)$$

where the integral is on the set $S := \{(t_1, \ldots, t_n) \in \mathbb{R}^n : t_i \geq 0, \ \sum_i t_i \leq 1\}$ and $t_0 = 1 - \sum_{i=1}^n t_i$. From this formula the symmetry is clear and if $x_0 = x_1 = \cdots = x_n = x$, then

$$f^{[n]}[x_0, x_1, \ldots, x_n] = \frac{f^{(n)}(x)}{n!}.$$

Next we introduce the notion of Fréchet differentiability. Assume that a mapping $F : \mathbb{M}_m \to \mathbb{M}_n$ is defined in a neighbourhood of $A \in \mathbb{M}_m$. The derivative $\partial f(A) : \mathbb{M}_m \to \mathbb{M}_n$ is a linear mapping such that

$$\frac{\|F(A + X) - F(A) - \partial F(A)(X)\|_2}{\|X\|_2} \longrightarrow 0 \quad \text{as } X \in \mathbb{M}_m \text{ and } X \to 0,$$

where $\| \cdot \|_2$ is the Hilbert–Schmidt norm in (1.4). This is the general definition. In the next theorem $F(A)$ will be defined via the matrix functional calculus as $f(A)$ when $f : (a, b) \to \mathbb{R}$ and $A \in \mathbb{M}_n^{sa}(a, b)$. Then the Fréchet derivative is a linear mapping $\partial f(A) : \mathbb{M}_n^{sa} \to \mathbb{M}_n^{sa}$ such that

$$\frac{\|f(A + X) - f(A) - \partial f(A)(X)\|_2}{\|X\|_2} \longrightarrow 0 \quad \text{as } X \in \mathbb{M}_n^{sa} \text{ and } X \to 0,$$

or equivalently

$$f(A + X) = f(A) + \partial f(A)(X) + o(\|X\|_2).$$

Since Fréchet differentiability implies Gâtaux (or directional) differentiability, one can differentiate $f(A + tX)$ with respect to the real parameter t and

$$\frac{f(A + tX) - f(A)}{t} \to \partial f(A)(X) \quad \text{as } t \to 0.$$

This notion of Fréchet differentiability for $f(A)$ is inductively extended to the general higher degree. To do this, we denote by $B((\mathbb{M}_n^{sa})^m, \mathbb{M}_n^{sa})$ the set of all m-multilinear maps from $(\mathbb{M}_n^{sa})^m := \mathbb{M}_n^{sa} \times \cdots \times \mathbb{M}_n^{sa}$ (m times) to $\mathbb{M}_n^{sa}$, and introduce the norm of $\Phi \in B((\mathbb{M}_n^{sa})^m, \mathbb{M}_n^{sa})$ as

$$\|\Phi\| := \sup\left\{\|\Phi(X_1, \ldots, X_m)\|_2 : X_i \in \mathbb{M}_n^{sa}, \ \|X_i\|_2 \leq 1, \ 1 \leq i \leq m\right\}. \quad (3.19)$$

Now assume that $m \in \mathbb{N}$ with $m \geq 2$ and the $(m-1)$th Fréchet derivative $\partial^{m-1} f(B) \in B((\mathbb{M}_n^{sa})^{m-1}, \mathbb{M}_n^{sa})$ exists for all $B \in \mathbb{M}_n^{sa}(a, b)$ in a neighborhood of $A \in \mathbb{M}_n^{sa}(a, b)$. We say that $f(B)$ is m **times Fréchet differentiable** at A if $\partial^{m-1} f(B)$ is once more Fréchet differentiable at A, i.e., there exists a

$$\partial^m f(A) \in B(\mathbb{M}_n^{sa}, B((\mathbb{M}_n^{sa})^{m-1}, \mathbb{M}_n^{sa})) = B((\mathbb{M}_n^{sa})^m, \mathbb{M}_n^{sa})$$

such that

$$\frac{\|\partial^{m-1} f(A+X) - \partial^{m-1} f(A) - \partial^m f(A)(X)\|}{\|X\|_2} \longrightarrow 0 \text{ as } X \in \mathbb{M}_n^{sa} \text{ and } X \to 0,$$

with respect to the norm (3.19) of $B((\mathbb{M}_n^{sa})^{m-1}, \mathbb{M}_n^{sa})$. Then $\partial^m f(A)$ is called the m**th Fréchet derivative** of f at A. Note that the norms of $\mathbb{M}_n^{sa}$ and $B((\mathbb{M}_n^{sa})^m, \mathbb{M}_n^{sa})$ are irrelevant to the definition of Fréchet derivatives since the norms on a finite-dimensional vector space are all equivalent; we can use the Hilbert–Schmidt norm just for convenience.

Example 3.32 Let $f(x) = x^k$ with $k \in \mathbb{N}$. Then $(A + X)^k$ can be expanded and $\partial f(A)(X)$ consists of the terms containing exactly one factor of X:

$$\partial f(A)(X) = \sum_{u=0}^{k-1} A^u X A^{k-1-u}.$$

To obtain the second derivative, we put $A + Y$ in place of A in $\partial f(A)(X)$ and again we take the terms containing exactly one factor of Y:

$$\partial^2 f(A)(X, Y)$$
$$= \sum_{u=0}^{k-1} \left(\sum_{v=0}^{u-1} A^v Y A^{u-1-v}\right) X A^{k-1-u} + \sum_{u=0}^{k-1} A^u X \left(\sum_{v=0}^{k-2-u} A^v Y A^{k-2-u-v}\right).$$

The formulation

$$\partial^2 f(A)(X_1, X_2) = \sum_{u+v+w=n-2} \sum_{\pi} A^u X_{\pi(1)} A^v X_{\pi(2)} A^w$$

is more convenient, where $u, v, w \geq 0$ and π denotes the permutations of $\{1, 2\}$. $\square$

Theorem 3.33 *Let $m \in \mathbb{N}$ and assume that $f : (a, b) \to \mathbb{R}$ is a C^m-function. Then the following properties hold:*

(1) *$f(A)$ is m times Fréchet differentiable at every $A \in \mathbb{M}_n^{sa}(a, b)$. If the diagonalization of $A \in \mathbb{M}_n^{sa}(a, b)$ is $A = U\mathrm{Diag}(\lambda_1, \ldots, \lambda_n)U^*$, then the mth Fréchet derivative $\partial^m f(A)$ is given by*

$$\partial^m f(A)(X_1, \ldots, X_m) = U\left[\sum_{k_1, \ldots, k_{m-1}=1}^{n} f^{[m]}[\lambda_i, \lambda_{k_1}, \ldots, \lambda_{k_{m-1}}, \lambda_j] \right.$$

$$\left. \times \sum_{\pi \in S_m} (X'_{\pi(1)})_{ik_1} (X'_{\pi(2)})_{k_1 k_2} \cdots (X'_{\pi(m-1)})_{k_{m-2}k_{m-1}} (X'_{\pi(m)})_{k_{m-1}j} \right]_{i,j=1}^{n} U^*$$

for all $X_i \in \mathbb{M}_n^{sa}$ with $X'_i = U^* X_i U$ $(1 \leq i \leq m)$. (S_m is the set of permutations on $\{1, \ldots, m\}$.)

(2) *The map $A \mapsto \partial^m f(A)$ is a norm-continuous map from $\mathbb{M}_n^{sa}(a, b)$ to $B((\mathbb{M}_n^{sa})^m, \mathbb{M}_n^{sa})$.*

(3) *For every $A \in \mathbb{M}_n^{sa}(a, b)$ and every $X_1, \ldots, X_m \in \mathbb{M}_n^{sa}$,*

$$\partial^m f(A)(X_1, \ldots, X_m) = \frac{\partial^m}{\partial t_1 \cdots \partial t_m} f(A + t_1 X_1 + \cdots + t_m X_m)\Big|_{t_1=\cdots=t_m=0}.$$

Proof: (Sketch) When $f(x) = x^k$, it is easily verified by a direct computation that $\partial^m f(A)$ exists and

$$\partial^m f(A)(X_1, \ldots, X_m)$$
$$= \sum_{\substack{u_0, u_1, \ldots, u_m \geq 0 \\ u_0 + u_1 + \cdots + u_m = k-m}} \sum_{\pi \in S_m} A^{u_0} X_{\pi(1)} A^{u_1} X_{\pi(2)} A^{u_2} \cdots A^{u_{m-1}} X_{\pi(m)} A^{u_m},$$

see Example 3.32. (If $m > k$, then $\partial^m f(A) = 0$.) The above expression is further written as

$$\sum_{\substack{u_0, u_1, \ldots, u_m \geq 0 \\ u_0 + u_1 + \cdots + u_m = k-m}} \sum_{\pi \in S_m} U\left[\sum_{k_1, \ldots, k_{m-1}=1} \lambda_i^{u_0} \lambda_{k_1}^{u_1} \cdots \lambda_{k_{m-1}}^{u_{m-1}} \lambda_j^{u_m} \right.$$

$$\left. \times (X'_{\pi(1)})_{ik_1} (X'_{\pi(2)})_{k_1 k_2} \cdots (X'_{\pi(m-1)})_{k_{m-2}k_{m-1}} (X'_{\pi(m)})_{k_{m-1}j} \right]_{i,j=1}^{n} U^*$$

$$= U\left[\sum_{k_1, \ldots, k_{m-1}=1}^{n} \left(\sum_{\substack{u_0, u_1, \ldots, u_m \geq 0 \\ u_0 + u_1 + \cdots + u_m = k-m}} \lambda_i^{u_0} \lambda_{k_1}^{u_1} \cdots \lambda_{k_{m-1}}^{u_{m-1}} \lambda_j^{u_m} \right) \right.$$

$$
\times \sum_{\pi \in S_m} (X'_{\pi(1)})_{ik_1} (X'_{\pi(2)})_{k_1 k_2} \cdots (X'_{\pi(m-1)})_{k_{m-2} k_{m-1}} (X'_{\pi(m)})_{k_{m-1} j} \Bigg]_{i,j=1}^{n} U^*
$$

$$
= U \Bigg[\sum_{k_1, \ldots, k_{m-1}=1}^{n} f^{[m]}[\lambda_i, \lambda_{k_1}, \ldots, \lambda_{k_{m-1}}, \lambda_j]
$$

$$
\times \sum_{\pi \in S_m} (X'_{\pi(1)})_{ik_1} (X'_{\pi(2)})_{k_1 k_2} \cdots (X'_{\pi(m-1)})_{k_{m-2} k_{m-1}} (X'_{\pi(m)})_{k_{m-1} j} \Bigg]_{i,j=1}^{n} U^*
$$

by Exercise 31. Hence it follows that $\partial^m f(A)$ exists and the expression in (1) is valid for all polynomials f. We can prove this and the continuity assertion in (2) for all C^m functions f on (a, b) by a method based on induction on m and approximation by polynomials. The details are not given here.

Formula (3) follows from the fact that Fréchet differentiability implies Gâteaux (or directional) differentiability. One can differentiate $f(A + t_1 X_1 + \cdots + t_m X_m)$ as

$$
\frac{\partial^m}{\partial t_1 \cdots \partial t_m} f(A + t_1 X_1 + \cdots + t_m X_m) \Big|_{t_1 = \cdots = t_m = 0}
$$

$$
= \frac{\partial^m}{\partial t_1 \cdots \partial t_{m-1}} \partial f(A + t_1 X_1 + \cdots + t_{m-1} X_{m-1})(X_m) \Big|_{t_1 = \cdots = t_{m-1} = 0}
$$

$$
= \cdots = \partial^m f(A)(X_1, \ldots, X_m).
$$

$\square$

Example 3.34 In particular, when f is C^1 on (a, b) and $A = \mathrm{Diag}(\lambda_1, \ldots, \lambda_n)$ is diagonal in $\mathbb{M}_n^{sa}(a, b)$, then the Fréchet derivative $\partial f(A)$ at A is written as

$$
\partial f(A)(X) = \big[f^{[1]}(\lambda_i, \lambda_j) \big]_{i,j=1}^{n} \circ X,
$$

where $\circ$ denotes the Schur product. This was Theorem 3.25.

When f is C^2 on (a, b), the second Fréchet derivative $\partial^2 f(A)$ at $A = \mathrm{Diag}(\lambda_1, \ldots, \lambda_n) \in \mathbb{M}_n^{sa}(a, b)$ is written as

$$
\partial^2 f(A)(X, Y) = \Bigg[\sum_{k=1}^{n} f^{[2]}(\lambda_i, \lambda_k, \lambda_j)(X_{ik} Y_{kj} + Y_{ik} X_{kj}) \Bigg]_{i,j=1}^{n}.
$$

$\square$

Example 3.35 If f is a holomorphic function then the **Taylor expansion**

$$
f(A + X) = f(A) + \sum_{k=1}^{\infty} \frac{1}{k!} \partial^k f(A) \underbrace{(X, \ldots, X)}_{m}
$$

has a simple computation, see (3.10):

$$f(A + X) = \frac{1}{2\pi i} \int_\Gamma f(z)(zI - A - X)^{-1} dz.$$

Since

$$zI - A - X = (zI - A)^{1/2}(I - (zI - A)^{-1/2}X(zI - A)^{-1/2})(zI - A)^{1/2},$$

we have the expansion

$$
\begin{aligned}
(zI - A - X)^{-1} \\
&= (zI - A)^{-1/2}(I - (zI - A)^{-1/2}X(zI - A)^{-1/2})^{-1}(zI - A)^{-1/2} \\
&= (zI - A)^{-1/2} \sum_{n=0}^{\infty} \left((zI - A)^{-1/2}X(zI - A)^{-1/2}\right)^n (zI - A)^{-1/2} \\
&= (zI - A)^{-1} + (zI - A)^{-1}X(zI - A)^{-1} \\
&\quad + (zI - A)^{-1}X(zI - A)^{-1}X(zI - A)^{-1} + \cdots.
\end{aligned}
$$

Hence

$$
\begin{aligned}
f(A + X) &= \frac{1}{2\pi i} \int_\Gamma f(z)(zI - A)^{-1} dz \\
&\quad + \frac{1}{2\pi i} \int_\Gamma f(z)(zI - A)^{-1}X(zI - A)^{-1} dz + \cdots \\
&= f(A) + \partial f(A)(X) + \frac{1}{2!}\partial^2 f(A)(X, X) + \cdots,
\end{aligned}
$$

which is the Taylor expansion.

When f satisfies the C^m assumption as in Theorem 3.33, we have the **Taylor formula**

$$f(A + X) = f(A) + \sum_{k=1}^{\infty} \frac{1}{k!}\partial^k f(A)(X^{(1)}, \ldots, X^{(k)}) + o(\|X\|_2^m)$$

as $X \in \mathbb{M}_n^{sa}$ and $X \to 0$. The details are not given here.   □

3.5 Notes and Remarks

Formula (3.7) is due to Masuo **Suzuki**, Generalized Trotter's formula and systematic approximants of exponential operators and inner derivations with applications to many-body problems, Commun. Math. Phys. **51**(1976), 183–190.

The **Bessis–Moussa–Villani conjecture** (or BMV conjecture) was published in the paper D. Bessis, P. Moussa and M. Villani, Monotonic converging variational approximations to the functional integrals in quantum statistical mechanics, J. Math. Phys. **16** (1975), 2318–2325. Theorem 3.12 is from the paper [64] of E. H. Lieb and R. Seiringer. A proof appeared in the paper H. R. Stahl, Proof of the BMV conjecture, arXiv:1107.4875v3.

The contour integral representation (3.10) was found by Henri Poincaré in 1899. Formula (3.18) is called the Hermite–Genocchi formula.

Formula (3.11) first appeared in the work of J. J. Sylvester in 1833 and (3.12) is due to H. Richter (1949). It is remarkable that J. von Neumann proved in 1929 that $\|A - I\|$, $\|B - I\|$, $\|AB - I\| < 1$ and $AB = BA$ implies $\log AB = \log A + \log B$.

Theorem 3.33 is essentially due to Ju. L. Daleckii and S. G. Krein, Integration and differentiation of functions of Hermitian operators and applications to the theory of perturbations, Amer. Math. Soc. Transl., Ser. 2 **47** (1965), 1–30. There the higher Gâteaux derivatives of the function $t \mapsto f(A + tX)$ were obtained for self-adjoint operators in an infinite-dimensional Hilbert space. As to the version of Fréchet derivatives in Theorem 3.33, the proof for the case $m = 1$ is in the book [20] of Rajendra **Bhatia** and the proof for the higher degree case is in Fumio **Hiai**, Matrix Analysis: Matrix Monotone Functions, Matrix Means, and Majorization, Interdisciplinary Information Sciences **16** (2010), 139–248.

3.6 Exercises

1. Prove that

$$\frac{\partial}{\partial t} e^{tA} = e^{tA} A.$$

2. Compute the exponential of the matrix

$$\begin{bmatrix} 0 & 0 & 0 & 0 & 0 \\ 1 & 0 & 0 & 0 & 0 \\ 0 & 2 & 0 & 0 & 0 \\ 0 & 0 & 3 & 0 & 0 \\ 0 & 0 & 0 & 4 & 0 \end{bmatrix}.$$

   What is the extension to the $n \times n$ case?
3. Use formula (3.3) to prove Theorem 3.4.
4. Let P and Q be ortho-projections. Give an elementary proof of the inequality

$$\mathrm{Tr}\, e^{P+Q} \le \mathrm{Tr}\, e^P e^Q.$$

5. Prove the Golden–Thompson inequality using the trace inequality

$$\mathrm{Tr}\,(CD)^n \leq \mathrm{Tr}\,C^n D^n \qquad (n \in \mathbb{N})$$

for $C, D \geq 0$.

6. Give a counterexample for the inequality

$$|\mathrm{Tr}\,e^A e^B e^C| \leq \mathrm{Tr}\,e^{A+B+C}$$

with Hermitian matrices. (Hint: Use the Pauli matrices.)

7. Solve the equation

$$e^A = \begin{bmatrix} \cos t & -\sin t \\ \sin t & \cos t \end{bmatrix}$$

where $t \in \mathbb{R}$ is given.

8. Show that

$$\exp\left(\begin{bmatrix} A & B \\ 0 & A \end{bmatrix}\right) = \begin{bmatrix} e^A & \int_0^1 e^{tA} B e^{(1-t)A}\,dt \\ 0 & e^A \end{bmatrix}.$$

9. Let A and B be self-adjoint matrices. Show that

$$|\mathrm{Tr}\,e^{A+iB}| \leq \mathrm{Tr}\,e^A.$$

10. Prove the estimate

$$\|e^{A+B} - (e^{A/n}e^{B/n})^n\|_2 \leq \frac{1}{2n}\|AB - BA\|_2 \exp(\|A\|_2 + \|B\|_2).$$

11. Show that $\|A - I\|, \|B - I\|, \|AB - I\| < 1$ and $AB = BA$ implies $\log AB = \log A + \log B$ for matrices A and B.

12. Give an example of a pair of commuting matrices A and B such that $\log AB \neq \log A + \log B$.

13. Let

$$C = c_0 I + c(c_1\sigma_1 + c_2\sigma_2 + c_3\sigma_3) \quad \text{with} \quad c_1^2 + c_2^2 + c_3^2 = 1,$$

where σ_1, σ_2, σ_3 are the Pauli matrices and $c_0, c_1, c_2, c_3 \in \mathbb{R}$. Show that

$$e^C = e^{c_0}\left((\cosh c)I + (\sinh c)(c_1\sigma_1 + c_2\sigma_2 + c_3\sigma_3)\right).$$

14. Let $A \in \mathbb{M}_3$ have eigenvalues λ, λ, μ with $\lambda \neq \mu$. Show that

$$e^{tA} = e^{\lambda t}(I + t(A - \lambda I)) + \frac{e^{\mu t} - e^{\lambda t}}{(\mu - \lambda)^2}(A - \lambda I)^2 - \frac{t e^{\lambda t}}{\mu - \lambda}(A - \lambda I)^2.$$

15. Assume that $A \in \mathbb{M}_3$ has distinct eigenvalues λ, μ, ν. Show that e^{tA} is

$$e^{\lambda t}\frac{(A-\mu I)(A-\nu I)}{(\lambda-\mu)(\lambda-\nu)} + e^{\mu t}\frac{(A-\lambda I)(A-\nu I)}{(\mu-\lambda)(\mu-\nu)} + e^{\nu t}\frac{(A-\lambda I)(A-\mu I)}{(\nu-\lambda)(\nu-\mu)}.$$

16. Assume that $A \in \mathbb{M}_n$ is diagonalizable and let $f(t) = t^m$ with $m \in \mathbb{N}$. Show that (3.8) and (3.10) are the same matrices.

17. Prove Corollary 3.11 directly in the case $B = AX - XA$.

18. Let $0 < D \in \mathbb{M}_n$ be a fixed invertible positive matrix. Show that the inverse of the linear mapping

$$\mathbb{J}_D : \mathbb{M}_n \to \mathbb{M}_n, \qquad \mathbb{J}_D(B) := \tfrac{1}{2}(DB + BD)$$

is the mapping

$$\mathbb{J}_D^{-1}(A) = \int_0^\infty e^{-tD/2} A e^{-tD/2}\, dt\,.$$

19. Let $0 < D \in \mathbb{M}_n$ be a fixed invertible positive matrix. Show that the inverse of the linear mapping

$$\mathbb{J}_D : \mathbb{M}_n \to \mathbb{M}_n, \qquad \mathbb{J}_D(B) := \int_0^1 D^t B D^{1-t}\, dt$$

is the mapping

$$\mathbb{J}_D^{-1}(A) = \int_0^\infty (D+tI)^{-1} A (D+tI)^{-1}\, dt. \tag{3.20}$$

20. Prove (3.16) directly for the case $f(t) = t^n, n \in \mathbb{N}$.

21. Let $f : [\alpha, \beta] \to \mathbb{R}$ be a convex function. Show that

$$\mathrm{Tr}\, f(B) \geq \sum_i f(\mathrm{Tr}\, B\, p_i) \tag{3.21}$$

for a pairwise orthogonal family (p_i) of minimal projections with $\sum_i p_i = I$ and for a self-adjoint matrix B with spectrum in $[\alpha, \beta]$. (Hint: Use the spectral decomposition of B.)

22. Prove Theorem 3.27 using formula (3.21). (Hint: Take the spectral decomposition of $B = \lambda B_1 + (1 - \lambda)B_2$ and show that

$$\lambda \mathrm{Tr}\, f(B_1) + (1 - \lambda)\mathrm{Tr}\, f(B_2) \geq \mathrm{Tr}\, f(B).)$$

23. Let A and B be positive matrices. Show that

$$A^{-1} \log(AB^{-1}) = A^{-1/2} \log(A^{1/2} B^{-1} A^{1/2}) A^{-1/2}.$$

(Hint: Use (3.17).)

24. Show that

$$\frac{d^2}{dt^2} \log(A + tK)\Big|_{t=0} = -2 \int_0^\infty (A + sI)^{-1} K (A + sI)^{-1} K (A + sI)^{-1} \, ds.$$

25. Show that

$$\partial^2 \log A(X_1, X_2) = - \int_0^\infty (A + sI)^{-1} X_1 (A + sI)^{-1} X_2 (A + sI)^{-1} \, ds$$

$$- \int_0^\infty (A + sI)^{-1} X_2 (A + sI)^{-1} X_1 (A + sI)^{-1} \, ds$$

for a positive invertible matrix A.

26. Prove the BMV conjecture for 2×2 matrices.

27. Show that

$$\partial^2 A^{-1}(X_1, X_2) = A^{-1} X_1 A^{-1} X_2 A^{-1} + A^{-1} X_2 A^{-1} X_1 A^{-1}$$

for an invertible variable A.

28. Differentiate the equation

$$\sqrt{A + tB} \sqrt{A + tB} = A + tB$$

and show that for positive A and B

$$\frac{d}{dt} \sqrt{A + tB}\Big|_{t=0} \geq 0.$$

29. For a real number $0 < \alpha \neq 1$ the **Rényi entropy** is defined as

$$S_\alpha(D) := \frac{1}{1 - \alpha} \log \operatorname{Tr} D^\alpha$$

for a positive matrix D such that $\operatorname{Tr} D = 1$. Show that $S_\alpha(D)$ is a decreasing function of α. What is the limit $\lim_{\alpha \to 1} S_\alpha(D)$? Show that $S_\alpha(D)$ is a concave functional of D for $0 < \alpha < 1$.

30. Fix a positive invertible matrix $D \in \mathbb{M}_n$ and define a linear mapping $\mathbb{M}_n \to \mathbb{M}_n$ by $\mathbb{K}_D(A) := DAD$. Consider the differential equation

$$\frac{\partial}{\partial t} D(t) = \mathbb{K}_{D(t)} T, \qquad D(0) = \rho_0,$$

where ρ_0 is positive invertible and T is self-adjoint in $\mathbb{M}_n$. Show that $D(t) = (\rho_0^{-1} - tT)^{-1}$ is the solution of the equation.

31. If $f(x) = x^k$ with $k \in \mathbb{N}$, verify that

$$f^{[n]}[x_1, x_2, \ldots, x_{n+1}] = \sum_{\substack{u_1, u_2, \ldots, u_{n+1} \geq 0 \\ u_1 + u_2 + \ldots + u_{n+1} = k-n}} x_1^{u_1} x_2^{u_2} \cdots x_n^{u_n} x_{n+1}^{u_{n+1}}.$$

32. Show that for a matrix $A > 0$ the integral

$$\log(I + A) = \int_1^\infty A(tI + A)^{-1} t^{-1} \, dt$$

holds. (Hint: Use (3.12).)

Chapter 4
Matrix Monotone Functions and Convexity

Let $(a, b) \subset \mathbb{R}$ be an interval. A function $f : (a, b) \to \mathbb{R}$ is said to be monotone for $n \times n$ matrices if $f(A) \le f(B)$ whenever A and B are self-adjoint $n \times n$ matrices, $A \le B$ and their eigenvalues are in (a, b). If a function is monotone for every matrix size, then it is called **matrix monotone** or **operator monotone**. (One can see by an approximation argument that if a function is matrix monotone for every matrix size, then $A \le B$ also implies $f(A) \le f(B)$ for operators on an infinite-dimensional Hilbert space.) On the other hand, a function $f : (a, b) \to \mathbb{R}$ is said to be **matrix convex** if

$$f(tA + (1 - t)B) \le tf(A) + (1 - t)f(B)$$

for all self-adjoint matrices A, B with eigenvalues in (a, b) and for all $0 \le t \le 1$. When $-f$ is matrix convex, then f is called **matrix concave**.

The theory of operator/matrix monotone functions was initiated by Karel Löwner, which was soon followed by Fritz Kraus' theory of operator/matrix convex functions. After further developments due to other authors (for instance, Bendat and Sherman, Korányi), Hansen and Pedersen established a modern treatment of matrix monotone and convex functions. A remarkable feature of Löwner's theory is that we have several characterizations of matrix monotone and matrix convex functions from several different points of view. The importance of complex analysis in studying matrix monotone functions is well understood from their characterization in terms of analytic continuation as Pick functions. Integral representations for matrix monotone and matrix convex functions are essential ingredients of the theory both theoretically and in applications. The notion of divided differences has played a vital role in the theory from its very beginning.

In real analysis, monotonicity and convexity are not directly related, but in matrix analysis the situation is very different. For example, a matrix monotone function on $(0, \infty)$ is matrix concave. Matrix monotone and matrix convex functions have several applications, but for a concrete function it is not so easy to verify its matrix monotonicity or matrix convexity. Such functions are typically described in terms of integral formulas.

F. Hiai and D. Petz, *Introduction to Matrix Analysis and Applications*, Universitext, DOI: 10.1007/978-3-319-04150-6_4, © Hindustan Book Agency 2014

4.1 Some Examples of Functions

Example 4.1 Let $t > 0$ be a parameter. The function $f(x) = -(t + x)^{-1}$ is matrix monotone on $[0, \infty)$.

Let A and B be positive matrices of the same order. Then $A_t := tI + A$ and $B_t := tI + B$ are invertible, and

$$A_t \leq B_t \iff B_t^{-1/2} A_t B_t^{-1/2} \leq I \iff \|B_t^{-1/2} A_t B_t^{-1/2}\| \leq 1$$
$$\iff \|A_t^{1/2} B_t^{-1/2}\| \leq 1.$$

Since the adjoint preserves the operator norm, the latter condition is equivalent to $\|B_t^{-1/2} A_t^{1/2}\| \leq 1$, which implies that $B_t^{-1} \leq A_t^{-1}$. □

Example 4.2 The function $f(x) = \log x$ is matrix monotone on $(0, \infty)$.

This follows from the formula

$$\log x = \int_0^\infty \frac{1}{1+t} - \frac{1}{x+t} \, dt,$$

which is easy to verify. The integrand

$$f_t(x) := \frac{1}{1+t} - \frac{1}{x+t}$$

is matrix monotone according to the previous example. It follows that

$$\sum_{i=1}^n c_i f_{t(i)}(x)$$

is matrix monotone for any $t(i)$ and positive $c_i \in \mathbb{R}$. The integral is the limit of such functions. Therefore it is a matrix monotone function as well.

There are several other ways to demonstrate the matrix monotonicity of the logarithm. □

Example 4.3 The function

$$f_+(x) = \sum_{n=-\infty}^0 \left(\frac{1}{(n - 1/2)\pi - x} - \frac{n\pi}{n^2\pi + 1} \right)$$

is matrix monotone on the interval $(-\pi/2, +\infty)$ and

$$f_-(x) = \sum_{n=1}^\infty \left(\frac{1}{(n - 1/2)\pi - x} - \frac{n\pi}{n^2\pi + 1} \right)$$

is matrix monotone on the interval $(-\infty, \pi/2)$. Therefore,

$$\tan x = f_+(x) + f_-(x) = \sum_{n=-\infty}^{\infty} \left(\frac{1}{(n-1/2)\pi - x} - \frac{n\pi}{n^2\pi + 1} \right)$$

is matrix monotone on the interval $(-\pi/2, \pi/2)$. □

Example 4.4 To show that the square root function is matrix monotone, consider the function

$$F(t) := \sqrt{A + tX}$$

defined for $t \in [0, 1]$ and for fixed positive matrices A and X. If F is increasing, then $F(0) = \sqrt{A} \leq \sqrt{A+X} = F(1)$.

In order to show that F is increasing, it is enough to verify that the eigenvalues of $F'(t)$ are positive. Differentiating the equality $F(t)F(t) = A + tX$, we have

$$F'(t)F(t) + F(t)F'(t) = X.$$

Being a limit of self-adjoint matrices, F' is self-adjoint. Let $F'(t) = \sum_i \lambda_i E_i$ be its spectral decomposition. (Of course, both the eigenvalues and the projections depend on the value of t.) Then

$$\sum_i \lambda_i (E_i F(t) + F(t) E_i) = X$$

and after multiplication by E_j on the left and on the right, we have for the trace

$$2\lambda_j \operatorname{Tr} E_j F(t) E_j = \operatorname{Tr} E_j X E_j.$$

Since both traces are positive, λ_j must be positive as well.

More generally, for every $0 < t < 1$, matrix monotonicity holds: $0 \leq A \leq B$ implies $A^t \leq B^t$. This is often called the **Löwner–Heinz inequality**. A proof will be given in Example 4.45, and another approach is in Theorem 5.3 based on the geometric mean.

Next we consider the case $t > 1$. Take the matrices

$$A = \begin{bmatrix} \frac{3}{2} & 0 \\ 0 & \frac{3}{4} \end{bmatrix} \quad \text{and} \quad B = \frac{1}{2} \begin{bmatrix} 1 & 1 \\ 1 & 1 \end{bmatrix}.$$

Then $A \geq B \geq 0$ can be checked. Since B is an orthogonal projection, for each $p > 1$ we have $B^p = B$ and

$$A^p - B^p = \begin{bmatrix} \left(\frac{3}{2}\right)^p - \frac{1}{2} & -\frac{1}{2} \\ -\frac{1}{2} & \left(\frac{3}{4}\right)^p - \frac{1}{2} \end{bmatrix}.$$

We can compute

$$\det(A^p - B^p) = \frac{1}{2}\left(\frac{3}{8}\right)^p (2 \cdot 3^p - 2^p - 4^p).$$

If $A^p \geq B^p$ then we must have $\det(A^p - B^p) \geq 0$ so that $2 \cdot 3^p - 2^p - 4^p \geq 0$, which is not true when $p > 1$. Hence $A^p \geq B^p$ does not hold for any $p > 1$.   □

The previous example contained an important idea. To determine whether a function f is matrix monotone, one has to investigate the derivative of $f(A + tX)$.

Theorem 4.5 *A smooth function $f : (a, b) \to \mathbb{R}$ is matrix monotone for $n \times n$ matrices if and only if the divided difference matrix $D \in \mathbb{M}_n$ defined as*

$$D_{ij} = \begin{cases} \frac{f(t_i)-f(t_j)}{t_i-t_j} & if \quad t_i - t_j \neq 0, \\ f'(t_i) & if \quad t_i - t_j = 0 \end{cases}$$

is positive semidefinite for $t_1, t_2, \ldots, t_n \in (a, b)$.

Proof: Let A be a self-adjoint and B be a positive semidefinite $n \times n$ matrix. When f is matrix monotone, the function $t \mapsto f(A + tB)$ is an increasing function of the real variable t. Therefore, the derivative, which is a matrix, must be positive semidefinite. To compute the derivative, we use formula (3.16) of Theorem 3.25. Schur's theorem implies that the derivative is positive if the divided difference matrix is positive.

To show the converse, let the matrix B have all entries equal to 1. Then the (positive) derivative $D \circ B$ is equal to D.   □

The smoothness assumption in the previous theorem is not essential. At the beginning of the theory Löwner proved that if a function $f : (a, b) \to \mathbb{R}$ has the property that $A \leq B$ for $A, B \in \mathbb{M}_2$ implies $f(A) \leq f(B)$, then f must be a C^1-function.

The previous theorem can be reformulated in terms of a **positive definite kernel**. The divided difference

$$\psi(x, y) = \begin{cases} \frac{f(x)-f(y)}{x-y} & if \quad x \neq y, \\ f'(x) & if \quad x = y \end{cases}$$

is an $(a, b) \times (a, b) \to \mathbb{R}$ kernel function. f is matrix monotone if and only if ψ is a positive definite kernel.

Example 4.6 The function $f(x) := \exp x$ is not matrix monotone, since the divided difference matrix

$$\begin{bmatrix} \exp x & \dfrac{\exp x - \exp y}{x - y} \\ \dfrac{\exp y - \exp x}{y - x} & \exp y \end{bmatrix}$$

does not have positive determinant (for $x = 0$ and for large y). $\qquad\square$

Example 4.7 We study the monotone function

$$f(x) = \begin{cases} \sqrt{x} & \text{if } 0 \le x \le 1, \\ (1 + x)/2 & \text{if } 1 \le x. \end{cases}$$

This is matrix monotone in the intervals $[0, 1]$ and $[1, \infty)$. Theorem 4.5 can be used to show that this is monotone on $[0, \infty)$ for 2×2 matrices. We need to show that for $0 < x < 1$ and $1 < y$

$$\begin{bmatrix} f'(x) & \dfrac{f(x)-f(y)}{x-y} \\ \dfrac{f(x)-f(y)}{x-y} & f'(y) \end{bmatrix} = \begin{bmatrix} f'(x) & f'(z) \\ f'(z) & f'(y) \end{bmatrix} \qquad \text{(for some } z \in [x, y])$$

is a positive matrix. This is true, however f is not monotone for larger matrices. $\quad\square$

Example 4.8 The function $f(x) = x^2$ is matrix convex on the whole real line. This follows from the obvious inequality

$$\left(\frac{A + B}{2}\right)^2 \le \frac{A^2 + B^2}{2}. \qquad\square$$

Example 4.9 The function $f(x) = (x + t)^{-1}$ is matrix convex on $[0, \infty)$ when $t > 0$. It is enough to show that

$$\left(\frac{A + B}{2}\right)^{-1} \le \frac{A^{-1} + B^{-1}}{2},$$

which is equivalent to

$$\left(\frac{B^{-1/2}AB^{-1/2} + I}{2}\right)^{-1} \le \frac{(B^{-1/2}AB^{-1/2})^{-1} + I}{2}.$$

This holds, since

$$\left(\frac{X + I}{2}\right)^{-1} \le \frac{X^{-1} + I}{2}$$

is true for any invertible matrix $X \ge 0$.

Note that this convexity inequality is equivalent to the relation between the arithmetic and harmonic means.   □

4.2 Convexity

Let V be a vector space (over the real scalars). Then $u, v \in V$ are called the endpoints of the line-segment

$$[u, v] := \{\lambda u + (1 - \lambda)v \, : \, \lambda \in \mathbb{R}, \, 0 \le \lambda \le 1\}.$$

A subset $\mathcal{A} \subset V$ is **convex** if for any $u, v \in \mathcal{A}$ the line-segment $[u, v]$ is contained in $\mathcal{A}$. A set $\mathcal{A} \subset V$ is convex if and only if for every finite subset $v_1, v_2, \ldots, v_n$ and for every family of real positive numbers $\lambda_1, \lambda_2, \ldots, \lambda_n$ with sum 1

$$\sum_{i=1}^{n} \lambda_i v_i \in \mathcal{A}.$$

For example, if $\| \cdot \| : V \to \mathbb{R}^+$ is a norm, then

$$\{v \in V \, : \, \|v\| \le 1\}$$

is a convex set. The intersection of convex sets is a convex set.

In the vector space $\mathbb{M}_n$ the self-adjoint matrices and the positive matrices form a convex set. Let (a, b) a real interval. Then

$$\{A \in \mathbb{M}_n^{sa} \, : \, \sigma(A) \subset (a, b)\}$$

is a convex set.

Example 4.10 Let

$$\mathcal{S}_n := \{D \in \mathbb{M}_n^{sa} \, : \, D \ge 0 \quad \text{and} \quad \operatorname{Tr} D = 1\}.$$

This is a convex set, since it is the intersection of convex sets. (In quantum theory this set is called the state space.)

If $n = 2$, then a popular parametrization of the matrices in $\mathcal{S}_2$ is

$$\frac{1}{2} \begin{bmatrix} 1 + \lambda_3 & \lambda_1 - i\lambda_2 \\ \lambda_1 + i\lambda_2 & 1 - \lambda_3 \end{bmatrix} = \frac{1}{2}(I + \lambda_1 \sigma_1 + \lambda_2 \sigma_2 + \lambda_3 \sigma_3),$$

where $\sigma_1, \sigma_2, \sigma_3$ are the Pauli matrices and the necessary and sufficient condition to be in $\mathcal{S}_2$ is

$$\lambda_1^2 + \lambda_2^2 + \lambda_3^2 \leq 1.$$

This shows that the convex set $\mathcal{S}_2$ can be viewed as the unit ball in $\mathbb{R}^3$. If $n > 2$, then the geometric picture of $\mathcal{S}_n$ is not so clear. □

If $\mathcal{A}$ is a subset of the vector space V, then its **convex hull**, denoted by $\operatorname{co} \mathcal{A}$, is the smallest convex set containing $\mathcal{A}$, i.e.,

$$\operatorname{co} \mathcal{A} := \left\{ \sum_{i=1}^{n} \lambda_i v_i \ : \ v_i \in \mathcal{A}, \ \lambda_i \geq 0, \ 1 \leq i \leq n, \ \sum_{i=1}^{n} \lambda_i = 1, \ n \in \mathbb{N} \right\}.$$

Let $\mathcal{A} \subset V$ be a convex set. The vector $v \in \mathcal{A}$ is an **extreme point** of $\mathcal{A}$ if the conditions

$$v_1, v_2 \in \mathcal{A}, \quad 0 < \lambda < 1, \quad \lambda v_1 + (1 - \lambda) v_2 = v$$

imply that $v_1 = v_2 = v$.

In the convex set $\mathcal{S}_2$ the extreme points correspond to the parameters satisfying $\lambda_1^2 + \lambda_2^2 + \lambda_3^2 = 1$. (If $\mathcal{S}_2$ is viewed as a ball in $\mathbb{R}^3$, then the extreme points are in the boundary of the ball.) For extreme points of $\mathcal{S}_n$, see Exercise 14.

Let $J \subset \mathbb{R}$ be an interval. A function $f : J \to \mathbb{R}$ is said to be **convex** if

$$f(ta + (1 - t)b) \leq tf(a) + (1 - t)f(b) \tag{4.1}$$

for all $a, b \in J$ and $0 \leq t \leq 1$. This inequality is equivalent to the positivity of the **second divided difference**

$$f^{[2]}[a, b, c] = \frac{f(a)}{(a - b)(a - c)} + \frac{f(b)}{(b - a)(b - c)} + \frac{f(c)}{(c - a)(c - b)}$$

$$= \frac{1}{c - b} \left(\frac{f(c) - f(a)}{c - a} - \frac{f(b) - f(a)}{b - a} \right)$$

for all distinct $a, b, c \in J$. If $f \in C^2(J)$, then for $x \in J$ we have

$$\lim_{a, b, c \to x} f^{[2]}[a, b, c] = \frac{f''(x)}{2}.$$

Hence the convexity is equivalent to the positivity of the second derivative. For a convex function f the **Jensen inequality**

$$f \left(\sum_i t_i a_i \right) \leq \sum_i t_i f(a_i)$$

holds whenever $a_i \in J$, $t_i \geq 0$ and $\sum_i t_i = 1$. This inequality has an integral form

$$f\left(\int g(x)\,d\mu(x)\right) \le \int f \circ g(x)\,d\mu(x).$$

For a finite discrete probability measure μ this is exactly the Jensen inequality, but it also holds for any probability measure μ on J and for a bounded Borel function g with values in J.

Definition (4.1) makes sense if J is a convex subset of a vector space and f is a real functional defined on it.

A functional f is **concave** if $-f$ is convex.

Let V be a finite-dimensional vector space and $\mathcal{A} \subset V$ be a convex subset. The functional $F : \mathcal{A} \to \mathbb{R} \cup \{+\infty\}$ is called **convex** if

$$F(\lambda x + (1 - \lambda)y) \le \lambda F(x) + (1 - \lambda)F(y)$$

for every $x, y \in \mathcal{A}$ and real number $0 < \lambda < 1$. Let $[u, v] \subset \mathcal{A}$ be a line-segment and define the function

$$F_{[u,v]}(\lambda) = F(\lambda u + (1 - \lambda)v)$$

on the interval $[0, 1]$. F is convex if and only if all functions $F_{[u,v]} : [0, 1] \to \mathbb{R}$ are convex when $u, v \in \mathcal{A}$.

Example 4.11 We show that the functional

$$A \mapsto \log \operatorname{Tr} e^A$$

is convex on the self-adjoint matrices, see Example 4.13.

The statement is equivalent to the convexity of the function

$$f(t) = \log \operatorname{Tr}\left(e^{A+tB}\right) \qquad (t \in \mathbb{R}) \tag{4.2}$$

for every $A, B \in \mathbb{M}_n^{sa}$. To show this we prove that $f''(0) \ge 0$. It follows from Theorem 3.23 that

$$f'(t) = \frac{\operatorname{Tr} e^{A+tB} B}{\operatorname{Tr} e^{A+tB}} \,.$$

In the computation of the second derivative we use Dyson's expansion

$$e^{A+tB} = e^A + t \int_0^1 e^{uA} B e^{(1-u)(A+tB)}\, du \,.$$

In order to write $f''(0)$ in a convenient form we introduce the inner product

$$\langle X, Y \rangle_{\mathrm{Bo}} := \int_0^1 \operatorname{Tr} e^{tA} X^* e^{(1-t)A} Y\, dt \,.$$

(This is often called the Bogoliubov inner product.) Now

$$f''(0) = \frac{\langle I, I \rangle_{\mathrm{Bo}} \langle B, B \rangle_{\mathrm{Bo}} - \langle I, B \rangle_{\mathrm{Bo}}^2}{(\mathrm{Tr}\, e^A)^2},$$

which is positive by the Schwarz inequality. □

Let V be a finite-dimensional vector space with dual V^*. Assume that the duality is given by a bilinear pairing $\langle \cdot, \cdot \rangle$. For a convex function $F : V \to \mathbb{R} \cup \{+\infty\}$ the **conjugate convex function** $F^* : V^* \to \mathbb{R} \cup \{+\infty\}$ is given by the formula

$$F^*(v^*) = \sup\{\langle v, v^* \rangle - F(v) : v \in V\}.$$

F^* is sometimes called the **Legendre transform** of F. Since F^* is the supremum of continuous linear functionals, it is convex and lower semi-continuous. The following duality theorem is basic in convex analysis.

Theorem 4.12 *If $F : V \to \mathbb{R}\cup\{+\infty\}$ is a lower semi-continuous convex functional, then $F^{**} = F$.*

Example 4.13 The negative von Neumann entropy $-S(D) = -\mathrm{Tr}\, \eta(D) = \mathrm{Tr}\, D \log D$ is continuous and convex on the density matrices. Let

$$F(X) = \begin{cases} \mathrm{Tr}\, X \log X & \text{if } X \geq 0 \text{ and } \mathrm{Tr}\, X = 1, \\ +\infty & \text{otherwise.} \end{cases}$$

This is a lower semi-continuous convex functional on the linear space of all self-adjoint matrices. The duality is given by $\langle X, H \rangle = \mathrm{Tr}\, XH$. The conjugate functional is

$$
\begin{aligned}
F^*(H) &= \sup\{\mathrm{Tr}\, XH - F(X) : X \in \mathbb{M}_n^{sa}\} \\
&= -\inf\{-\mathrm{Tr}\, XH - S(D) : D \in \mathbb{M}_n^{sa}, D \geq 0, \mathrm{Tr}\, D = 1\}.
\end{aligned}
$$

According to Example 3.29 the minimizer is $D = e^H / \mathrm{Tr}\, e^H$, and therefore

$$F^*(H) = \log \mathrm{Tr}\, e^H.$$

This is a continuous convex function of $H \in \mathbb{M}_n^{sa}$. So Example 4.11 is recovered. The duality theorem gives that

$$\mathrm{Tr}\, X \log X = \sup\{\mathrm{Tr}\, XH - \log \mathrm{Tr}\, e^H : H = H^*\}$$

when $X \geq 0$ and $\mathrm{Tr}\, X = 1$. □

Example 4.14 Fix a **density matrix** $\rho = e^H$ (with a self-adjoint H) and consider the functional F defined on the self-adjoint matrices by

$$F(X) := \begin{cases} \operatorname{Tr} X (\log X - H) & \text{if } X \geq 0 \text{ and } \operatorname{Tr} X = 1, \\ +\infty & \text{otherwise.} \end{cases}$$

F is essentially the **relative entropy** with respect to ρ:

$$S(X \| \rho) := \operatorname{Tr} X (\log X - \log \rho).$$

The duality is $\langle X, B \rangle = \operatorname{Tr} XB$ if X and B are self-adjoint matrices. We want to show that the functional $B \mapsto \log \operatorname{Tr} e^{H+B}$ is the Legendre transform or the conjugate function of F:

$$\log \operatorname{Tr} e^{B+H} = \max\{\operatorname{Tr} XB - S(X \| e^H) : X \text{ is positive, } \operatorname{Tr} X = 1\}.$$

We introduce the notation

$$f(X) = \operatorname{Tr} XB - S(X \| e^H)$$

for a density matrix X. When $P_1, \ldots, P_n$ are projections of rank one with $\sum_{i=1}^n P_i = I$, we write

$$f\left(\sum_{i=1}^n \lambda_i P_i\right) = \sum_{i=1}^n (\lambda_i \operatorname{Tr} P_i B + \lambda_i \operatorname{Tr} P_i H - \lambda_i \log \lambda_i),$$

where $\lambda_i \geq 0$, $\sum_{i=1}^n \lambda_i = 1$. Since

$$\frac{\partial}{\partial \lambda_i} f\left(\sum_{i=1}^n \lambda_i P_i\right)\bigg|_{\lambda_i = 0} = +\infty,$$

we see that $f(X)$ attains its maximum at a matrix $X_0 > 0$, $\operatorname{Tr} X_0 = 1$. Then for any self-adjoint Z, $\operatorname{Tr} Z = 0$, we have

$$0 = \frac{d}{dt} f(X_0 + tZ)\bigg|_{t=0} = \operatorname{Tr} Z(B + H - \log X_0),$$

so that $B + H - \log X_0 = cI$ with $c \in \mathbb{R}$. Therefore $X_0 = e^{B+H} / \operatorname{Tr} e^{B+H}$ and $f(X_0) = \log \operatorname{Tr} e^{B+H}$ by a simple computation.

On the other hand, if X is positive invertible with $\operatorname{Tr} X = 1$, then

$$S(X \| e^H) = \max\{\operatorname{Tr} XB - \log \operatorname{Tr} e^{H+B} : B \text{ is self-adjoint}\}$$

by the duality theorem.   □

Theorem 4.15 *Let* $\alpha : \mathbb{M}_n \to \mathbb{M}_m$ *be a positive unital linear mapping and* $f : \mathbb{R} \to \mathbb{R}$ *be a convex function. Then*

$$\mathrm{Tr}\, f(\alpha(A)) \leq \mathrm{Tr}\,\alpha(f(A))$$

for every $A \in \mathbb{M}_n^{sa}$.

Proof: Take the spectral decompositions

$$A = \sum_j \nu_j Q_j \quad \text{and} \quad \alpha(A) = \sum_i \mu_i P_i.$$

So we have

$$\mu_i = \mathrm{Tr}\,(\alpha(A)P_i)/\mathrm{Tr}\,P_i = \sum_j \nu_j \mathrm{Tr}\,(\alpha(Q_j)P_i)/\mathrm{Tr}\,P_i,$$

whereas the convexity of f yields

$$f(\mu_i) \leq \sum_j f(\nu_j)\mathrm{Tr}\,(\alpha(Q_j)P_i)/\mathrm{Tr}\,P_i.$$

Therefore,

$$\mathrm{Tr}\, f(\alpha(A)) = \sum_i f(\mu_i)\mathrm{Tr}\,P_i \leq \sum_{i,j} f(\nu_j)\mathrm{Tr}\,(\alpha(Q_j)P_i) = \mathrm{Tr}\,\alpha(f(A)),$$

which was to be proven.   □

It was stated in Theorem 3.27 that for a convex function $f : (a, b) \to \mathbb{R}$, the functional $A \mapsto \mathrm{Tr}\, f(A)$ is convex. It is rather surprising that in the equivalent statement of the convexity of this functional the numerical coefficient $0 < t < 1$ can be replaced by a matrix.

Theorem 4.16 *Let* $f : (a, b) \to \mathbb{R}$ *be a convex function and* $C_i, A_i \in \mathbb{M}_n$ *be such that*

$$\sigma(A_i) \subset (a, b) \quad \text{and} \quad \sum_{i=1}^k C_i C_i^* = I.$$

Then

$$\mathrm{Tr}\, f\left(\sum_{i=1}^k C_i A_i C_i^*\right) \leq \sum_{i=1}^k \mathrm{Tr}\, C_i f(A_i) C_i^*.$$

Proof: We prove only the case

$$\operatorname{Tr} f(CAC^* + DBD^*) \le \operatorname{Tr} Cf(A)C^* + \operatorname{Tr} Df(B)D^*,$$

when $CC^* + DD^* = I$. (The more general version can be treated similarly.)

Set $F := CAC^* + DBD^*$ and consider the spectral decomposition of A and B as integrals:

$$X = \sum_i \mu_i^X P_i^X = \int \lambda dE^X(\lambda) \qquad (X = A, B),$$

where μ_i^X are eigenvalues, P_i^X are eigenprojections and the operator-valued measure E^X is defined on the Borel subsets S of $\mathbb{R}$ as

$$E^X(S) = \sum \{P_i^X : \mu_i^X \in S\}.$$

Assume that $A, B, C, D \in \mathbb{M}_n$ and for a vector $\xi \in \mathbb{C}^n$ we define a measure μ_ξ:

$$\mu_\xi(S) = \langle(CE^A(S)C^* + DE^B(S)D^*)\xi, \xi\rangle$$
$$= \langle E^A(S)C^*\xi, C^*\xi\rangle + \langle E^B(S)D^*\xi, D^*\xi\rangle.$$

The motivation for defining this measure is the formula

$$\langle F\xi, \xi\rangle = \int \lambda d\mu_\xi(\lambda).$$

If ξ is a unit eigenvector of F (and $f(F)$), then

$$\langle f(CAC^* + DBD^*)\xi, \xi\rangle = \langle f(F)\xi, \xi\rangle = f(\langle F\xi, \xi\rangle) = f\left(\int \lambda d\mu_\xi(\lambda)\right)$$

$$\le \int f(\lambda) d\mu_\xi(\lambda)$$

$$= \langle(Cf(A)C^* + Df(B)D^*)\xi, \xi\rangle.$$

(The inequality follows from the convexity of the function f.) To obtain the inequality in the statement of the theorem we take a sum of the above inequalities where ξ ranges over an orthonormal basis of eigenvectors of F.   □

Example 4.17 This example concerns a positive block matrix A and a concave function $f : \mathbb{R}^+ \to \mathbb{R}$. The inequality

$$\operatorname{Tr} f\left(\begin{bmatrix} A_{11} & A_{12} \\ A_{12}^* & A_{22} \end{bmatrix}\right) \le \operatorname{Tr} f(A_{11}) + \operatorname{Tr} f(A_{22})$$

is called the **subadditivity** of Tr f. We can take ortho-projections P_1 and P_2 such that $P_1 + P_2 = I$ and the subadditivity of Tr f

$$\text{Tr } f(A) \leq \text{Tr } f(P_1 A P_1) + \text{Tr } f(P_2 A P_2)$$

follows from the previous theorem. A stronger version of this inequality is less trivial.

Let P_1, P_2 and P_3 be ortho-projections such that $P_1 + P_2 + P_3 = I$. We adopt the notation $P_{12} := P_1 + P_2$ and $P_{23} := P_2 + P_3$. The **strong subadditivity** of Tr f is the inequality

$$\text{Tr } f(A) + \text{Tr } f(P_2 A P_2) \leq \text{Tr } f(P_{12} A P_{12}) + \text{Tr } f(P_{23} A P_{23}). \qquad (4.3)$$

Further details on this will come later, see Theorems 4.50 and 4.51. $\qquad \square$

Example 4.18 The log function is concave. If $A \in \mathbb{M}_n$ is positive definite and we define the projections $P_i := E(ii)$, then from the previous theorem we have

$$\text{Tr } \log \sum_{i=1}^{n} P_i A P_i \geq \sum_{i=1}^{n} \text{Tr } P_i (\log A) P_i.$$

This means

$$\sum_{i=1}^{n} \log A_{ii} \geq \text{Tr } \log A$$

and the exponential is

$$\prod_{i=1}^{n} A_{ii} \geq \exp(\text{Tr } \log A) = \det A.$$

This is the well-known **Hadamard inequality** for the determinant, see Theorem 1.30. $\qquad \square$

When $\mathcal{K}, \mathcal{L}$ are two convex sets of matrices and $F : \mathcal{K} \times \mathcal{L} \to \mathbb{R} \cup \{+\infty\}$ is a function of two matrix variables, F is called **jointly concave** if

$$F(\lambda A_1 + (1 - \lambda) A_2, \lambda B_1 + (1 - \lambda) B_2) \geq \lambda F(A_1, B_1) + (1 - \lambda) F(A_2, B_2)$$

for every $A_i \in \mathcal{K}$, $B_i \in \mathcal{L}$ and $0 < \lambda < 1$. The function $(A, B) \in \mathcal{K} \times \mathcal{L} \mapsto F(A, B)$ is jointly concave if and only if the function

$$A \oplus B \in \mathcal{K} \oplus \mathcal{L} \mapsto F(A, B)$$

is concave. In this way joint convexity and concavity can be conveniently studied.

Lemma 4.19 *If $(A, B) \in \mathcal{K} \times \mathcal{L} \mapsto F(A, B)$ is jointly concave, then*

$$f(A) := \sup\{F(A, B) : B \in \mathcal{L}\}$$

is concave on $\mathcal{K}$.

Proof: Assume that $f(A_1), f(A_2) < +\infty$. Let $\varepsilon > 0$ be a small number. We have B_1 and B_2 such that

$$f(A_1) \le F(A_1, B_1) + \varepsilon \quad \text{and} \quad f(A_2) \le F(A_2, B_2) + \varepsilon.$$

Then

$$\begin{aligned}
\lambda f(A_1) + (1 - \lambda) f(A_2) &\le \lambda F(A_1, B_1) + (1 - \lambda) F(A_2, B_2) + \varepsilon \\
&\le F(\lambda A_1 + (1 - \lambda) A_2, \lambda B_1 + (1 - \lambda) B_2) + \varepsilon \\
&\le f(\lambda A_1 + (1 - \lambda) A_2) + \varepsilon
\end{aligned}$$

and this gives the proof.

The case of $f(A_1) = +\infty$ or $f(A_2) = +\infty$ has a similar proof.   □

Example 4.20 The quantum relative entropy of $X \ge 0$ with respect to $Y \ge 0$ is defined as

$$S(X \| Y) := \mathrm{Tr}\,(X \log X - X \log Y) - \mathrm{Tr}\,(X - Y).$$

It is known (see Example 3.19) that $S(X \| Y) \ge 0$ and equality holds if and only if $X = Y$. (We assumed $X, Y > 0$ in Example 3.19 but this is true for general $X, Y \ge 0$.) A different formulation is

$$\mathrm{Tr}\,Y = \max\{\mathrm{Tr}\,(X \log Y - X \log X + X) : X \ge 0\}.$$

For a positive definite D, selecting $Y = \exp(L + \log D)$ we obtain

$$\begin{aligned}
\mathrm{Tr}\,\exp(L + \log D) &= \max\{\mathrm{Tr}\,(X(L + \log D) - X \log X + X) : X \ge 0\} \\
&= \max\{\mathrm{Tr}\,(XL) - S(X \| D) + \mathrm{Tr}\,D : X \ge 0\}.
\end{aligned}$$

Since the quantum relative entropy is a jointly convex function, the function

$$F(X, D) := \mathrm{Tr}\,(XL) - S(X \| D) + \mathrm{Tr}\,D$$

is jointly concave as well. It follows that the maximization in X is concave and we obtain that the functional

$$D \mapsto \mathrm{Tr}\,\exp(L + \log D) \tag{4.4}$$

is concave on positive definite matrices. (This result is due to Lieb, but the present proof is taken from [81].)   □

In the next lemma the operators

$$\mathbb{J}_D X = \int_0^1 D^t X D^{1-t} \, dt, \qquad \mathbb{J}_D^{-1} K = \int_0^\infty (t+D)^{-1} K (t+D)^{-1} \, dt$$

for $D, X, K \in \mathbb{M}_n, D > 0$, are used (see Exercise 19 of Chap. 3). Lieb's concavity theorem (see Example 7.9) says that $D > 0 \mapsto \operatorname{Tr} X^* D^t X D^{1-t}$ is concave for every $X \in \mathbb{M}_n$. Thus, $D > 0 \mapsto \langle X, \mathbb{J}_D X \rangle$ is concave. By using this we prove the following:

Theorem 4.21 *The functional*

$$(D, K) \mapsto Q(D, K) := \langle K, \mathbb{J}_D^{-1} K \rangle$$

is jointly convex on the domain $\{D \in \mathbb{M}_n : D > 0\} \times \mathbb{M}_n$.

Proof: $\mathbb{M}_n$ is a Hilbert space $\mathcal{H}$ with the Hilbert–Schmidt inner product. The mapping $K \mapsto Q(D, K)$ is a quadratic form. If $\mathcal{K} := \mathcal{H} \oplus \mathcal{H}$ and $D = \lambda D_1 + (1 - \lambda) D_2$, then

$$\mathcal{M}(K_1 \oplus K_2) := \lambda Q(D_1, K_1) + (1 - \lambda) Q(D_2, K_2)$$
$$\mathcal{N}(K_1 \oplus K_2) := Q(D, \lambda K_1 + (1 - \lambda) K_2)$$

are quadratic forms on $\mathcal{K}$. Note that $\mathcal{M}$ is non-degenerate. In terms of $\mathcal{M}$ and $\mathcal{N}$ the dominance $\mathcal{N} \leq \mathcal{M}$ is to be shown.

Let m and n be the corresponding sesquilinear forms on $\mathcal{K}$, that is,

$$\mathcal{M}(\xi) = m(\xi, \xi), \quad \mathcal{N}(\xi) = n(\xi, \xi) \qquad (\xi \in \mathcal{K}).$$

There exists a positive operator X on $\mathcal{K}$ such that

$$n(\xi, \eta) = m(\xi, X\eta) \qquad (\xi, \eta \in \mathcal{K})$$

and our aim is to show that its eigenvalues are ≥ 1. If $X(K \oplus L) = \gamma(K \oplus L)$ for $0 \neq K \oplus L \in \mathcal{H} \oplus \mathcal{H}$, we have

$$n(K \oplus L', K \oplus L) = \gamma m(K' \oplus L', K \oplus L)$$

for every $K', L' \in \mathcal{H}$. This is rewritten in terms of the Hilbert–Schmidt inner product as

$$\langle \lambda K' + (1 - \lambda) L', \mathbb{J}_D^{-1} (\lambda K + (1 - \lambda) L) \rangle = \gamma \lambda \langle K', \mathbb{J}_{D_1}^{-1} K \rangle + \gamma (1 - \lambda) \langle L', \mathbb{J}_{D_2}^{-1} L \rangle,$$

which is equivalent to the equations

$$\mathbb{J}_D^{-1}(\lambda K + (1 - \lambda)L) = \gamma \mathbb{J}_{D_1}^{-1} K$$

and

$$\mathbb{J}_D^{-1}(\lambda K + (1 - \lambda)L) = \gamma \mathbb{J}_{D_2}^{-1} L.$$

We infer

$$\gamma \mathbb{J}_D M = \lambda \mathbb{J}_{D_1} M + (1 - \lambda) \mathbb{J}_{D_2} M$$

with the new notation $M := \mathbb{J}_D^{-1}(\lambda K + (1 - \lambda)L)$. It follows that

$$\gamma \langle M, \mathbb{J}_D M \rangle = \lambda \langle M, \mathbb{J}_{D_1} M \rangle + (1 - \lambda) \langle M, \mathbb{J}_{D_2} M \rangle.$$

On the other hand, the concavity of $D \mapsto \langle M, \mathbb{J}_D M \rangle$ yields the inequality

$$\langle M, \mathbb{J}_D M \rangle \geq \lambda \langle M, \mathbb{J}_{D_1} M \rangle + (1 - \lambda) \langle M, \mathbb{J}_{D_2} M \rangle$$

and we arrive at $\gamma \leq 1$ if $M \neq 0$. Otherwise, if $M = 0$, then we must have $\gamma K = \gamma L = 0$ so that $\gamma = 0$. $\qquad\square$

Let $J \subset \mathbb{R}$ be an interval. As introduced at the beginning of the chapter, a function $f : J \to \mathbb{R}$ is said to be **matrix convex** if

$$f(tA + (1 - t)B) \leq tf(A) + (1 - t)f(B) \tag{4.5}$$

for all self-adjoint matrices A and B whose spectra are in J and for all numbers $0 \leq t \leq 1$. (The function f is matrix convex if the functional $A \mapsto f(A)$ is convex.) f is matrix concave if $-f$ is matrix convex.

The classical result concerns matrix convex functions on the interval $(-1, 1)$. Such a function has an integral decomposition

$$f(x) = \beta_0 + \beta_1 x + \frac{\beta_2}{2} \int_{-1}^{1} \frac{x^2}{1 - \lambda x} \, d\mu(\lambda), \tag{4.6}$$

where μ is a probability measure and $\beta_2 \geq 0$. (In particular, f must be an analytic function.) The details will be given in Theorem 4.40.

Since self-adjoint operators on an infinite-dimensional Hilbert space may be approximated by self-adjoint matrices, (4.5) holds for operators when it holds for all matrices. The next theorem shows that in the convex combination $tA + (1 - t)B$ the numbers t and $1 - t$ can be replaced by matrices.

Theorem 4.22 *Let $f : [a, b] \to \mathbb{R}$ be a matrix convex function and $C_i, A_i = A_i^* \in \mathbb{M}_n$ be such that*

$$\sigma(A_i) \subset [a, b] \quad and \quad \sum_{i=1}^{k} C_i C_i^* = I.$$

Then

$$f\left(\sum_{i=1}^{k} C_i A_i C_i^*\right) \le \sum_{i=1}^{k} C_i f(A_i) C_i^*. \tag{4.7}$$

Proof: We are content to prove the case $k = 2$:

$$f(CAC^* + DBD^*) \le Cf(A)C^* + Df(B)D^*,$$

when $CC^* + DD^* = I$. The essential idea is contained in this case.

The condition $CC^* + DD^* = I$ implies that we can find a unitary block matrix

$$U := \begin{bmatrix} C & D \\ X & Y \end{bmatrix}$$

when the entries X and Y are chosen properly. (Indeed, since $|D^*| = (I - CC^*)^{1/2}$, we have the polar decomposition $D^* = W|D^*|$ with a unitary W. Then it is an exercise to show that the choice of $X = (I - C^*C)^{1/2}$ and $Y = -C^*W^*$ satisfies the requirements.) Then

$$U \begin{bmatrix} A & 0 \\ 0 & B \end{bmatrix} U^* = \begin{bmatrix} CAC^* + DBD^* & CAX^* + DBY^* \\ XAC^* + YBD^* & XAX^* + YBY^* \end{bmatrix} =: \begin{bmatrix} A_{11} & A_{12} \\ A_{21} & A_{22} \end{bmatrix}.$$

It is easy to check that

$$\frac{1}{2} V \begin{bmatrix} A_{11} & A_{12} \\ A_{21} & A_{22} \end{bmatrix} V + \frac{1}{2} \begin{bmatrix} A_{11} & A_{12} \\ A_{21} & A_{22} \end{bmatrix} = \begin{bmatrix} A_{11} & 0 \\ 0 & A_{22} \end{bmatrix}$$

for

$$V = \begin{bmatrix} -I & 0 \\ 0 & I \end{bmatrix}.$$

It follows that the matrix

$$Z := \frac{1}{2} VU \begin{bmatrix} A & 0 \\ 0 & B \end{bmatrix} U^* V + \frac{1}{2} U \begin{bmatrix} A & 0 \\ 0 & B \end{bmatrix} U^*$$

is diagonal, $Z_{11} = CAC^* + DBD^*$ and $f(Z)_{11} = f(CAC^* + DBD^*)$.

Next we use the matrix convexity of the function f:

$$f(Z) \leq \frac{1}{2} f\left(VU \begin{bmatrix} A & 0 \\ 0 & B \end{bmatrix} U^* V\right) + \frac{1}{2} f\left(U \begin{bmatrix} A & 0 \\ 0 & B \end{bmatrix} U^*\right)$$

$$= \frac{1}{2} VU f\left(\begin{bmatrix} A & 0 \\ 0 & B \end{bmatrix}\right) U^* V + \frac{1}{2} U f\left(\begin{bmatrix} A & 0 \\ 0 & B \end{bmatrix}\right) U^*$$

$$= \frac{1}{2} VU \begin{bmatrix} f(A) & 0 \\ 0 & f(B) \end{bmatrix} U^* V + \frac{1}{2} U \begin{bmatrix} f(A) & 0 \\ 0 & f(B) \end{bmatrix} U^*.$$

The right-hand side is diagonal with $Cf(A)C^* + Df(B)D^*$ as $(1, 1)$ entry. The inequality implies the inequality between the $(1, 1)$ entries and this is precisely the inequality (4.7) for $k = 2$. $\qquad \Box$

In the proof of (4.7) for $n \times n$ matrices, the ordinary matrix convexity was used for $(2n) \times (2n)$ matrices. This is an important trick. The next theorem is due to Hansen and Pedersen [40].

Theorem 4.23 *Let $f : [a, b] \to \mathbb{R}$ and $a \leq 0 \leq b$.*

If f is a matrix convex function, $\|V\| \leq 1$ and $f(0) \leq 0$, then $f(V^* AV) \leq V^* f(A)V$ holds if $A = A^*$ and $\sigma(A) \subset [a, b]$.

If $f(PAP) \leq Pf(A)P$ holds for an orthogonal projection P and $A = A^*$ with $\sigma(A) \subset [a, b]$, then f is a matrix convex function and $f(0) \leq 0$.

Proof: If f is matrix convex, we can apply Theorem 4.22. Choose $B = 0$ and W such that $V^* V + W^* W = I$. Then

$$f(V^* AV + W^* BW) \leq V^* f(A)V + W^* f(B)W$$

holds and gives our statement.

Let A and B be self-adjoint matrices with spectrum in $[a, b]$ and $0 < \lambda < 1$. Define

$$C := \begin{bmatrix} A & 0 \\ 0 & B \end{bmatrix}, \quad U := \begin{bmatrix} \sqrt{\lambda}I & -\sqrt{1-\lambda}I \\ \sqrt{1-\lambda}I & \sqrt{\lambda}I \end{bmatrix}, \quad P := \begin{bmatrix} I & 0 \\ 0 & 0 \end{bmatrix}.$$

Then $C = C^*$ with $\sigma(C) \subset [a, b]$, U is a unitary and P is an orthogonal projection. Since

$$PU^* CUP = \begin{bmatrix} \lambda A + (1 - \lambda)B & 0 \\ 0 & 0 \end{bmatrix},$$

the assumption implies

$$\begin{bmatrix} f(\lambda A + (1-\lambda)B) & 0 \\ 0 & f(0)I \end{bmatrix} = f(PU^*CUP)$$

$$\leq Pf(U^*CU)P = PU^*f(C)UP$$

$$= \begin{bmatrix} \lambda f(A) + (1-\lambda)f(B) & 0 \\ 0 & 0 \end{bmatrix}.$$

This implies that $f(\lambda A + (1-\lambda)B) \leq \lambda f(A) + (1-\lambda)f(B)$ and $f(0) \leq 0$. $\qquad\square$

Example 4.24 From the previous theorem we can deduce that if $f : [0, b] \to \mathbb{R}$ is a matrix convex function and $f(0) \leq 0$, then $f(x)/x$ is matrix monotone on the interval $(0, b]$.

Assume that $0 < A \leq B$. Then $B^{-1/2}A^{1/2} =: V$ is a contraction, since

$$\|V\|^2 = \|VV^*\| = \|B^{-1/2}AB^{-1/2}\| \leq \|B^{-1/2}BB^{-1/2}\| = 1.$$

Therefore the theorem gives

$$f(A) = f(V^*BV) \leq V^*f(B)V = A^{1/2}B^{-1/2}f(B)B^{-1/2}A^{1/2},$$

which is equivalent to $A^{-1}f(A) \leq B^{-1}f(B)$.

Now assume that $g : [0, b] \to \mathbb{R}$ is matrix monotone. We want to show that $f(x) := xg(x)$ is matrix convex. By the previous theorem we need to show

$$PAPg(PAP) \leq PAg(A)P$$

for an orthogonal projection P and $A \geq 0$. From the monotonicity

$$g(A^{1/2}PA^{1/2}) \leq g(A)$$

and this implies

$$PA^{1/2}g(A^{1/2}PA^{1/2})A^{1/2}P \leq PA^{1/2}g(A)A^{1/2}P.$$

Since $g(A^{1/2}PA^{1/2})A^{1/2}P = A^{1/2}Pg(PAP)$ and $A^{1/2}g(A)A^{1/2} = Ag(A)$, the proof is complete. $\qquad\square$

Example 4.25 Heuristically we can say that Theorem 4.22 replaces all the numbers in the Jensen inequality $f(\sum_i t_i a_i) \leq \sum_i t_i f(a_i)$ by matrices. Therefore

$$f\left(\sum_i a_i A_i\right) \leq \sum_i f(a_i)A_i \tag{4.8}$$

holds for a matrix convex function f if $\sum_i A_i = I$ for positive matrices $A_i \in \mathbb{M}_n$ and for numbers $a_i \in (a, b)$.

We want to show that the property (4.8) is equivalent to the matrix convexity

$$f(tA + (1 - t)B) \le tf(A) + (1 - t)f(B).$$

Let

$$A = \sum_i \lambda_i P_i \quad \text{and} \quad B = \sum_j \mu_j Q_j$$

be the spectral decompositions. Then

$$\sum_i tP_i + \sum_j (1 - t)Q_j = I$$

and from (4.8) we obtain

$$f(tA + (1 - t)B) = f\left(\sum_i t\lambda_i P_i + \sum_j (1 - t)\mu_j Q_j\right)$$

$$\le \sum_i f(\lambda_i)tP_i + \sum_j f(\mu_j)(1 - t)Q_j$$

$$= tf(A) + (1 - t)f(B).$$

This inequality was the aim.   □

An operator $Z \in B(\mathcal{H})$ is called a **contraction** if $Z^*Z \le I$ and an **expansion** if $Z^*Z \ge I$. For an $A \in \mathbb{M}_n(\mathbb{C})^{sa}$ let $\lambda(A) = (\lambda_1(A), \ldots, \lambda_n(A))$ denote the eigenvalue vector of A in decreasing order with multiplicities.

Theorem 4.23 says that, for a function $f : [a, b] \to \mathbb{R}$ with $a \le 0 \le b$, the matrix inequality $f(Z^*AZ) \le Z^*f(A)Z$ for every $A = A^*$ with $\sigma(A) \subset [a, b]$ and every contraction Z characterizes the matrix convexity of f with $f(0) \le 0$. Now we consider some similar inequalities in the weaker senses of eigenvalue dominance under the simple convexity or concavity condition of f.

The first theorem presents the eigenvalue dominance involving a contraction when f is a monotone convex function with $f(0) \le 0$.

Theorem 4.26 *Assume that f is a monotone convex function on $[a, b]$ with $a \le 0 \le b$ and $f(0) \le 0$. Then, for every $A \in \mathbb{M}_n(\mathbb{C})^{sa}$ with $\sigma(A) \subset [a, b]$ and for every contraction $Z \in \mathbb{M}_n(\mathbb{C})$, there exists a unitary U such that*

$$f(Z^*AZ) \le U^*Z^*f(A)ZU,$$

or equivalently,

$$\lambda_k(f(Z^*AZ)) \leq \lambda_k(Z^*f(A)Z) \quad (1 \leq k \leq n).$$

Proof: We may assume that f is increasing; the other case is covered by taking $f(-x)$ and $-A$. First, note that for every $B \in \mathbb{M}_n(\mathbb{C})^{sa}$ and for every vector x with $\|x\| \leq 1$ we have

$$f(\langle x, Bx \rangle) \leq \langle x, f(B)x \rangle. \tag{4.9}$$

Indeed, taking the spectral decomposition $B = \sum_{i=1}^n \lambda_i |u_i\rangle\langle u_i|$ we have

$$f(\langle x, Bx \rangle) = f\left(\sum_{i=1}^n \lambda_i |\langle x, u_i \rangle|^2\right) \leq \sum_{i=1}^n f(\lambda_i)|\langle x, u_i \rangle|^2 + f(0)(1 - \|x\|^2)$$

$$\leq \sum_{i=1}^n f(\lambda_i)|\langle x, u_i \rangle|^2 = \langle x, f(B)x \rangle$$

thanks to the convexity of f and $f(0) \leq 0$. By the **minimax expression** in Theorem 1.27 there exists a subspace $\mathcal{M}$ of $\mathbb{C}^n$ with $\dim \mathcal{M} = k - 1$ such that

$$\lambda_k(Z^*f(A)Z) = \max_{x \in \mathcal{M}^\perp, \|x\|=1} \langle x, Z^*f(A)Zx \rangle = \max_{x \in \mathcal{M}^\perp, \|x\|=1} \langle Zx, f(A)Zx \rangle.$$

Since Z is a contraction and f is non-decreasing, we apply (4.9) to obtain

$$\lambda_k(Z^*f(A)Z) \geq \max_{x \in \mathcal{M}^\perp, \|x\|=1} f(\langle Zx, AZx \rangle) = f\left(\max_{x \in \mathcal{M}^\perp, \|x\|=1} \langle x, Z^*AZx \rangle\right)$$

$$\geq f(\lambda_k(Z^*AZ)) = \lambda_k(f(Z^*AZ)).$$

In the second inequality above we have used the minimax expression again. $\square$

The following corollary was originally proved by Brown and Kosaki [23] in the von Neumann algebra setting.

Corollary 4.27 *Let f be a function on $[a, b]$ with $a \leq 0 \leq b$, and let $A \in \mathbb{M}_n(\mathbb{C})^{sa}$, $\sigma(A) \subset [a, b]$, and $Z \in \mathbb{M}_n(\mathbb{C})$ be a contraction. If f is a convex function with $f(0) \leq 0$, then*

$$\mathrm{Tr}\, f(Z^*AZ) \leq \mathrm{Tr}\, Z^*f(A)Z.$$

If f is a concave function on $\mathbb{R}$ with $f(0) \geq 0$, then

$$\mathrm{Tr}\, f(Z^*AZ) \geq \mathrm{Tr}\, Z^*f(A)Z.$$

Proof: Obviously, the two assertions are equivalent. To prove the first, by approximation we may assume that $f(x) = \alpha x + g(x)$ with $\alpha \in \mathbb{R}$ and a monotone and convex function g on $[a, b]$ with $g(0) \leq 0$. Since $\mathrm{Tr}\, g(Z^*AZ) \leq \mathrm{Tr}\, Z^*gf(A)Z$ by Theorem 4.26, we have $\mathrm{Tr}\, f(Z^*AZ) \leq \mathrm{Tr}\, Z^*f(A)Z$. $\square$

The next theorem is the eigenvalue dominance version of Theorem 4.23 for f under a simple convexity condition.

Theorem 4.28 *Assume that f is a monotone convex function on $[a, b]$. Then, for every $A_1, \ldots, A_m \in \mathbb{M}_n(\mathbb{C})^{sa}$ with $\sigma(A_i) \subset [a, b]$ and every $C_1, \ldots, C_m \in \mathbb{M}_n(\mathbb{C})$ with $\sum_{i=1}^m C_i^* C_i = I$, there exists a unitary U such that*

$$f\left(\sum_{i=1}^m C_i^* A_i C_i\right) \leq U^*\left(\sum_{i=1}^m C_i^* f(A_i) C_i\right) U.$$

Proof: Letting $f_0(x) := f(x) - f(0)$ we have

$$f\left(\sum_i C_i^* A_i C_i\right) = f(0)I + f_0\left(\sum_i C_i^* A_i C_i\right),$$

$$\sum_i C_i^* f(A_i) C_i = f(0)I + \sum_i C_i^* f_0(A_i) C_i.$$

So it may be assumed that $f(0) = 0$. Set

$$A := \begin{bmatrix} A_1 & 0 & \cdots & 0 \\ 0 & A_2 & \cdots & 0 \\ \vdots & \vdots & \ddots & \vdots \\ 0 & 0 & \cdots & A_m \end{bmatrix} \quad \text{and} \quad Z := \begin{bmatrix} C_1 & 0 & \cdots & 0 \\ C_2 & 0 & \cdots & 0 \\ \vdots & \vdots & \ddots & \vdots \\ C_m & 0 & \cdots & 0 \end{bmatrix}.$$

For the block matrices $f(Z^* A Z)$ and $Z^* f(A) Z$, we can take the $(1, 1)$ blocks: $f(\sum_i C_i^* A_i C_i)$ and $\sum_i C_i^* f(A_i) C_i$. Moreover, all other blocks are 0. Hence Theorem 4.26 implies that

$$\lambda_k\left(f\left(\sum_i C_i^* A_i C_i\right)\right) \leq \lambda_k\left(\sum_i C_i^* f(A_i) C_i\right) \qquad (1 \leq k \leq n),$$

as desired.   □

A special case of Theorem 4.28 is that if f and $A_1, \ldots, A_m$ are as above, $\alpha_1, \ldots, \alpha_m > 0$ and $\sum_{i=1}^m \alpha_i = 1$, then there exists a unitary U such that

$$f\left(\sum_{i=1}^m \alpha_i A_i\right) \leq U^*\left(\sum_{i=1}^m \alpha_i f(A_i)\right) U.$$

This inequality implies the trace inequality in Theorem 4.16, although monotonicity of f is not assumed there.

4.3 Pick Functions

Let $\mathbb{C}^+$ denote the upper half-plane:

$$\mathbb{C}^+ := \{z \in \mathbb{C} : \operatorname{Im} z > 0\} = \{re^{i\varphi} \in \mathbb{C} : r > 0, \, 0 < \varphi < \pi\}.$$

Now we concentrate on analytic functions $f : \mathbb{C}^+ \to \mathbb{C}$. Recall that the range $f(\mathbb{C}^+)$ is a connected open subset of $\mathbb{C}$ unless f is a constant. An analytic function $f : \mathbb{C}^+ \to \mathbb{C}^+$ is called a **Pick function**.

The next examples show that this concept is related to the matrix monotonicity property.

Example 4.29 Let $z = re^{i\theta}$ with $r > 0$ and $0 < \theta < \pi$. For a real parameter $p > 0$ the range of the function

$$f_p(z) = z^p := r^p e^{ip\theta}$$

is in $\mathcal{P}$ if and only if $p \le 1$.

This function $f_p(z)$ is a continuous extension of the real function $0 \le x \mapsto x^p$. The latter is matrix monotone if and only if $p \le 1$. The similarity to the Pick function concept is essential.

Recall that the real function $0 < x \mapsto \log x$ is matrix monotone as well. The principal branch of $\log z$ defined as

$$\operatorname{Log} z := \log r + i\theta$$

is a continuous extension of the real logarithm function and it is also in $\mathcal{P}$.
□

The next theorem, **Nevanlinna's theorem**, provides the integral representation of Pick functions.

Theorem 4.30 *A function $f : \mathbb{C}^+ \to \mathbb{C}$ is in $\mathcal{P}$ if and only if there exists an $\alpha \in \mathbb{R}$, a $\beta \ge 0$ and a positive finite Borel measure ν on $\mathbb{R}$ such that*

$$f(z) = \alpha + \beta z + \int_{-\infty}^{\infty} \frac{1 + \lambda z}{\lambda - z} \, d\nu(\lambda), \qquad z \in \mathbb{C}^+. \tag{4.10}$$

The integral representation (4.10) can also be written as

$$f(z) = \alpha + \beta z + \int_{-\infty}^{\infty} \left(\frac{1}{\lambda - z} - \frac{\lambda}{\lambda^2 + 1} \right) d\mu(\lambda), \qquad z \in \mathbb{C}^+, \tag{4.11}$$

where μ is a positive Borel measure on $\mathbb{R}$ given by $d\mu(\lambda) := (\lambda^2 + 1) \, d\nu(\lambda)$ and so

$$\int_{-\infty}^{\infty} \frac{1}{\lambda^2 + 1} \, d\mu(\lambda) < +\infty.$$

Proof: The proof of the "if" part is easy. Assume that f is defined on $\mathbb{C}^+$ as in (4.10). For each $z \in \mathbb{C}^+$, since

$$\frac{f(z + \Delta z) - f(z)}{\Delta z} = \beta + \int_{\mathbb{R}} \frac{\lambda^2 + 1}{(\lambda - z)(\lambda - z - \Delta z)} \, d\nu(\lambda)$$

and

$$\sup\left\{\left|\frac{\lambda^2 + 1}{(\lambda - z)(\lambda - z - \Delta z)}\right| : \lambda \in \mathbb{R}, \ |\Delta z| < \frac{\operatorname{Im} z}{2}\right\} < +\infty,$$

it follows from the Lebesgue dominated convergence theorem that

$$\lim_{\Delta \to 0} \frac{f(z + \Delta z) - f(z)}{\Delta z} = \beta + \int_{\mathbb{R}} \frac{\lambda^2 + 1}{(\lambda - z)^2} \, d\nu(\lambda).$$

Hence f is analytic in $\mathbb{C}^+$. Since

$$\operatorname{Im}\left(\frac{1 + \lambda z}{\lambda - z}\right) = \frac{(\lambda^2 + 1)\operatorname{Im} z}{|\lambda - z|^2}, \qquad z \in \mathbb{C}^+,$$

we have

$$\operatorname{Im} f(z) = \left(\beta + \int_{\mathbb{R}} \frac{\lambda^2 + 1}{|\lambda - z|^2} \, d\nu(\lambda)\right) \operatorname{Im} z \geq 0$$

for all $z \in \mathbb{C}^+$. Therefore, we have $f \in \mathcal{P}$. The equivalence between the two representations (4.10) and (4.11) is immediately seen from

$$\frac{1 + \lambda z}{\lambda - z} = (\lambda^2 + 1)\left(\frac{1}{\lambda - z} - \frac{\lambda}{\lambda^2 + 1}\right).$$

The "only if" is the significant part, the proof of which is skipped here. □

Note that α, β and ν in Theorem 4.30 are uniquely determined by f. In fact, letting $z = \mathrm{i}$ in (4.10) we have $\alpha = \operatorname{Re} f(\mathrm{i})$. Letting $z = \mathrm{i}y$ with $y > 0$ we have

$$f(\mathrm{i}y) = \alpha + \mathrm{i}\beta y + \int_{-\infty}^{\infty} \frac{\lambda(1 - y^2) + \mathrm{i}y(\lambda^2 + 1)}{\lambda^2 + y^2} \, d\nu(\lambda)$$

so that

$$\frac{\operatorname{Im} f(\mathrm{i}y)}{y} = \beta + \int_{-\infty}^{\infty} \frac{\lambda^2 + 1}{\lambda^2 + y^2} \, d\nu(\lambda).$$

By the Lebesgue dominated convergence theorem this yields

$$\beta = \lim_{y \to \infty} \frac{\operatorname{Im} f(iy)}{y}.$$

Hence α and β are uniquely determined by f. By (4.11), for $z = x + iy$ we have

$$\operatorname{Im} f(x + iy) = \beta y + \int_{-\infty}^{\infty} \frac{y}{(x - \lambda)^2 + y^2} \, d\mu(\lambda), \qquad x \in \mathbb{R}, \ y > 0. \quad (4.12)$$

Thus the uniqueness of μ (hence ν) is a consequence of the so-called **Stieltjes inversion formula**. (For details omitted here, see [36, pp. 24–26] and [20, pp. 139–141].)

For any open interval (a, b), $-\infty \le a < b \le \infty$, we denote by $\mathcal{P}(a, b)$ the set of all Pick functions which admit a continuous extension to $\mathbb{C}^+ \cup (a, b)$ with real values on (a, b).

The next theorem is a specialization of Nevanlinna's theorem to functions in $\mathcal{P}(a, b)$.

Theorem 4.31 *A function $f : \mathbb{C}^+ \to \mathbb{C}$ is in $\mathcal{P}(a, b)$ if and only if f is represented as in (4.10) with $\alpha \in \mathbb{R}$, $\beta \ge 0$ and a positive finite Borel measure ν on $\mathbb{R} \setminus (a, b)$.*

Proof: Let $f \in \mathcal{P}$ be represented as in (4.10) with $\alpha \in \mathbb{R}$, $\beta \ge 0$ and a positive finite Borel measure ν on $\mathbb{R}$. It suffices to prove that $f \in \mathcal{P}(a, b)$ if and only if $\nu((a, b)) = 0$. First, assume that $\nu((a, b)) = 0$. The function f expressed by (4.10) is analytic in $\mathbb{C}^+ \cup \mathbb{C}^-$ so that $f(\overline{z}) = \overline{f(z)}$ for all $z \in \mathbb{C}^+$. For every $x \in (a, b)$, since

$$\sup \left\{ \left| \frac{\lambda^2 + 1}{(\lambda - x)(\lambda - x - \Delta z)} \right| : \lambda \in \mathbb{R} \setminus (a, b), \ |\Delta z| < \frac{1}{2} \min\{x - a, b - x\} \right\}$$

is finite, the above proof of the "if" part of Theorem 4.30, using the Lebesgue dominated convergence theorem, works for $z = x$ as well, and so f is differentiable (in the complex variable z) at $z = x$. Hence $f \in \mathcal{P}(a, b)$.

Conversely, assume that $f \in \mathcal{P}(a, b)$. It follows from (4.12) that

$$\int_{-\infty}^{\infty} \frac{1}{(x - \lambda)^2 + y^2} \, d\mu(\lambda) = \frac{\operatorname{Im} f(x + iy)}{y} - \beta, \qquad x \in \mathbb{R}, \ y > 0.$$

For any $x \in (a, b)$, since $f(x) \in \mathbb{R}$, we have

$$\frac{\operatorname{Im} f(x + iy)}{y} = \operatorname{Im} \frac{f(x + iy) - f(x)}{y} = \operatorname{Re} \frac{f(x + iy) - f(x)}{iy} \longrightarrow \operatorname{Re} f'(x)$$

as $y \searrow 0$ and so the monotone convergence theorem yields

$$\int_{-\infty}^{\infty} \frac{1}{(x - \lambda)^2} \, d\mu(\lambda) = \operatorname{Re} f'(x), \qquad x \in (a, b).$$

Hence, for any closed interval $[c, d]$ included in (a, b), we have

$$R := \sup_{x \in [c,d]} \int_{-\infty}^{\infty} \frac{1}{(x - \lambda)^2} \, d\mu(\lambda) = \sup_{x \in [c,d]} \operatorname{Re} f'(x) < +\infty.$$

For each $m \in \mathbb{N}$ let $c_k := c + (k/m)(d - c)$ for $k = 0, 1, \ldots, m$. Then

$$\mu([c, d)) = \sum_{k=1}^{m} \mu([c_{k-1}, c_k)) \leq \sum_{k=1}^{m} \int_{[c_{k-1}, c_k)} \frac{(c_k - c_{k-1})^2}{(c_k - \lambda)^2} \, d\mu(\lambda)$$

$$\leq \sum_{k=1}^{m} \left(\frac{d - c}{m}\right)^2 \int_{-\infty}^{\infty} \frac{1}{(c_k - \lambda)^2} \, d\mu(\lambda) \leq \frac{(d - c)^2 R}{m}.$$

Letting $m \to \infty$ gives $\mu([c, d)) = 0$. This implies that $\mu((a, b)) = 0$ and therefore $\nu((a, b)) = 0$. □

Now let $f \in \mathcal{P}(a, b)$. The above theorem says that $f(x)$ on (a, b) admits the integral representation

$$f(x) = \alpha + \beta x + \int_{\mathbb{R} \setminus (a,b)} \frac{1 + \lambda x}{\lambda - x} \, d\nu(\lambda)$$

$$= \alpha + \beta x + \int_{\mathbb{R} \setminus (a,b)} (\lambda^2 + 1)\left(\frac{1}{\lambda - x} - \frac{\lambda}{\lambda^2 + 1}\right) d\nu(\lambda), \qquad x \in (a, b),$$

where α, β and ν are as in the theorem. For any $n \in \mathbb{N}$ and $A, B \in \mathbb{M}_n^{sa}$ with $\sigma(A), \sigma(B) \subset (a, b)$, if $A \geq B$ then $(\lambda I - A)^{-1} \geq (\lambda I - B)^{-1}$ for all $\lambda \in \mathbb{R} \setminus (a, b)$ (see Example 4.1) and hence we have

$$f(A) = \alpha I + \beta A + \int_{\mathbb{R} \setminus (a,b)} (\lambda^2 + 1)\left((\lambda I - A)^{-1} - \frac{\lambda}{\lambda^2 + 1} I\right) d\nu(\lambda)$$

$$\geq \alpha I + \beta B + \int_{\mathbb{R} \setminus (a,b)} (\lambda^2 + 1)\left((\lambda I - B)^{-1} - \frac{\lambda}{\lambda^2 + 1} I\right) d\nu(\lambda) = f(B).$$

Therefore, $f \in \mathcal{P}(a, b)$ is matrix monotone on (a, b). It will be shown in the next section that f is matrix monotone on (a, b) if and only if $f \in \mathcal{P}(a, b)$.

The following are examples of integral representations for typical Pick functions from Example 4.29.

Example 4.32 The principal branch $\operatorname{Log} z$ of the logarithm in Example 4.29 is in $\mathcal{P}(0, \infty)$. Its integral representation in the form (4.11) is

$$\operatorname{Log} z = \int_{-\infty}^{0} \left(\frac{1}{\lambda - z} - \frac{\lambda}{\lambda^2 + 1}\right) d\lambda, \qquad z \in \mathbb{C}^+.$$

To show this, it suffices to verify the above expression for $z = x \in (0, \infty)$, that is,

$$\log x = \int_0^\infty \left(-\frac{1}{\lambda + x} + \frac{\lambda}{\lambda^2 + 1} \right) d\lambda, \qquad x \in (0, \infty),$$

which is immediate by a direct computation.   □

Example 4.33 If $0 < p < 1$, then z^p, as defined in Example 4.29, is in $\mathcal{P}(0, \infty)$. Its integral representation in the form (4.11) is

$$z^p = \cos \frac{p\pi}{2} + \frac{\sin p\pi}{\pi} \int_{-\infty}^0 \left(\frac{1}{\lambda - z} - \frac{\lambda}{\lambda^2 + 1} \right) |\lambda|^p \, d\lambda, \qquad z \in \mathbb{C}^+.$$

For this it suffices to verify that

$$x^p = \cos \frac{p\pi}{2} + \frac{\sin p\pi}{\pi} \int_0^\infty \left(-\frac{1}{\lambda + x} + \frac{\lambda}{\lambda^2 + 1} \right) \lambda^p \, d\lambda, \qquad x \in (0, \infty), \quad (4.13)$$

which is computed as follows.

The function

$$\frac{z^{p-1}}{1 + z} := \frac{r^{p-1} e^{i(p-1)\theta}}{1 + r e^{i\theta}}, \qquad z = r e^{i\theta}, \, 0 < \theta < 2\pi,$$

is analytic in the cut plane $\mathbb{C} \setminus (-\infty, 0]$ and we integrate it along the contour

$$z = \begin{cases} r e^{i\theta} & (\varepsilon \le r \le R, \theta = +0), \\ R e^{i\theta} & (0 < \theta < 2\pi), \\ r e^{i\theta} & (R \ge r \ge \varepsilon, \theta = 2\pi - 0), \\ \varepsilon e^{i\theta} & (2\pi > \theta > 0), \end{cases}$$

where $0 < \varepsilon < 1 < R$. Apply the residue theorem and let $\varepsilon \searrow 0$ and $R \nearrow \infty$ to show that

$$\int_0^\infty \frac{t^{p-1}}{1 + t} \, dt = \frac{\pi}{\sin p\pi}. \tag{4.14}$$

For each $x > 0$, substitute λ/x for t in (4.14) to obtain

$$x^p = \frac{\sin p\pi}{\pi} \int_0^\infty \frac{x \lambda^{p-1}}{\lambda + x} \, d\lambda, \qquad x \in (0, \infty).$$

Since

$$\frac{x}{\lambda + x} = \frac{1}{\lambda^2 + 1} + \left(\frac{\lambda}{\lambda^2 + 1} - \frac{1}{\lambda + x} \right) \lambda,$$

it follows that

$$x^p = \frac{\sin p\pi}{\pi} \int_0^\infty \frac{\lambda^{p-1}}{\lambda^2+1}\, d\lambda + \frac{\sin p\pi}{\pi} \int_0^\infty \left(\frac{\lambda}{\lambda^2+1} - \frac{1}{\lambda+x}\right) \lambda^p \, d\lambda, \qquad x \in (0, \infty).$$

Substitute λ^2 for t in (4.14) with p replaced by $p/2$ to obtain

$$\int_0^\infty \frac{\lambda^{p-1}}{\lambda^2+1}\, d\lambda = \frac{\pi}{2\sin\frac{p\pi}{2}}.$$

Hence (4.13) follows.   □

4.4 Löwner's Theorem

The main aim of this section is to prove the primary result in Löwner's theory, which says that a matrix monotone (i.e., operator monotone) function on (a, b) belongs to $\mathcal{P}(a, b)$.

Operator monotone functions on a finite open interval (a, b) are transformed into those on a symmetric interval $(-1, 1)$ via an affine function. So it is essential to analyze matrix monotone functions on $(-1, 1)$. They are C^∞-functions and $f'(0) > 0$ unless f is constant. We denote by $\mathcal{K}$ the set of all matrix monotone functions on $(-1, 1)$ such that $f(0) = 0$ and $f'(0) = 1$.

Lemma 4.20 *Let $f \in \mathcal{K}$. Then:*

(1) *For every $\alpha \in [-1, 1]$, $(x + \alpha)f(x)$ is matrix convex on $(-1, 1)$.*
(2) *For every $\alpha \in [-1, 1]$, $\left(1 + \frac{\alpha}{x}\right)f(x)$ is matrix monotone on $(-1, 1)$.*
(3) *f is twice differentiable at 0 and*

$$\frac{f''(0)}{2} = \lim_{x\to 0} \frac{f(x) - f'(0)x}{x^2}.$$

Proof: (1) The proof is based on Example 4.24, but we have to change the argument of the function. Let $\varepsilon \in (0, 1)$. Since $f(x - 1 + \varepsilon)$ is matrix monotone on $[0, 2 - \varepsilon)$, it follows that $xf(x - 1 + \varepsilon)$ is matrix convex on the same interval $[0, 2 - \varepsilon)$. So $(x + 1 - \varepsilon)f(x)$ is matrix convex on $(-1 + \varepsilon, 1)$. By letting $\varepsilon \searrow 0$, $(x + 1)f(x)$ is matrix convex on $(-1, 1)$.

We repeat the same argument with the matrix monotone function $-f(-x)$ and get the matrix convexity of $(x - 1)f(x)$. Since

$$(x + \alpha)f(x) = \frac{1+\alpha}{2}(x+1)f(x) + \frac{1-\alpha}{2}(x-1)f(x),$$

this function is matrix convex as well.

(2) $(x + \alpha) f(x)$ is already known to be matrix convex, so its division by x is also matrix monotone.

(3) To prove this, we use the continuous differentiability of matrix monotone functions. Then, by (2), $\left(1 + \frac{1}{x}\right) f(x)$ as well as $f(x)$ is C^1 on $(-1, 1)$ so that the function h on $(-1, 1)$ defined by $h(x) := f(x)/x$ for $x \neq 0$ and $h(0) := f'(0)$ is C^1. This implies that

$$h'(x) = \frac{f'(x)x - f(x)}{x^2} \longrightarrow h'(0) \quad \text{as } x \to 0.$$

Therefore,

$$f'(x)x = f(x) + h'(0)x^2 + o(|x|^2)$$

so that

$$f'(x) = h(x) + h'(0)x + o(|x|) = h(0) + 2h'(0)x + o(|x|) \quad \text{as } x \to 0,$$

which shows that f is twice differentiable at 0 with $f''(0) = 2h'(0)$. Hence

$$\frac{f''(0)}{2} = h'(0) = \lim_{x \to 0} \frac{h(x) - h(0)}{x} = \lim_{x \to 0} \frac{f(x) - f'(0)x}{x^2}$$

and the proof is complete. $\qquad\qquad\qquad\qquad\qquad\qquad\qquad\qquad\qquad$ □

Lemma 4.35 *If $f \in \mathcal{K}$, then*

$$\frac{x}{1 + x} \leq f(x) \quad \text{for } x \in (-1, 0), \qquad f(x) \leq \frac{x}{1 - x} \quad \text{for } x \in (0, 1),$$

and $|f''(0)| \leq 2$.

Proof: For every $x \in (-1, 1)$, Theorem 4.5 implies that

$$\begin{bmatrix} f^{[1]}(x, x) & f^{[1]}(x, 0) \\ f^{[1]}(x, 0) & f^{[1]}(0, 0) \end{bmatrix} = \begin{bmatrix} f'(x) & f(x)/x \\ f(x)/x & 1 \end{bmatrix} \geq 0,$$

and hence

$$\frac{f(x)^2}{x^2} \leq f'(x). \tag{4.15}$$

By Lemma 4.34 (1),

$$\frac{d}{dx}(x \pm 1)f(x) = f(x) + (x \pm 1)f'(x)$$

is increasing on $(-1, 1)$. Since $f(0) \pm f'(0) = \pm 1$, we have

$$f(x) + (x - 1)f'(x) \geq -1 \quad \text{for} \quad 0 < x < 1, \qquad (4.16)$$
$$f(x) + (x + 1)f'(x) \leq 1 \quad \text{for} \; -1 < x < 0. \qquad (4.17)$$

By (4.15) and (4.16) we have

$$f(x) + 1 \geq \frac{(1 - x)f(x)^2}{x^2}.$$

If $f(x) > \frac{x}{1-x}$ for some $x \in (0, 1)$, then

$$f(x) + 1 > \frac{(1 - x)f(x)}{x^2} \cdot \frac{x}{1 - x} = \frac{f(x)}{x}$$

so that $f(x) < \frac{x}{1-x}$, a contradiction. Hence $f(x) \leq \frac{x}{1-x}$ for all $x \in [0, 1)$. A similar argument using (4.15) and (4.17) yields that $f(x) \geq \frac{x}{1+x}$ for all $x \in (-1, 0]$.

Moreover, by Lemma 4.34 (3) and the two inequalities just proved,

$$\frac{f''(0)}{2} \leq \lim_{x \searrow 0} \frac{\frac{x}{1-x} - x}{x^2} = \lim_{x \searrow 0} \frac{1}{1 - x} = 1$$

and

$$\frac{f''(0)}{2} \geq \lim_{x \nearrow 0} \frac{\frac{x}{1+x} - x}{x^2} = \lim_{x \searrow 0} \frac{-1}{1 + x} = -1$$

so that $|f''(0)| \leq 2$.   □

Lemma 4.36 *The set $\mathcal{K}$ is convex and compact if it is considered as a subset of the topological vector space consisting of real functions on $(-1, 1)$ with the locally convex topology of pointwise convergence.*

Proof: It is obvious that $\mathcal{K}$ is convex. Since $\{f(x) : f \in \mathcal{K}\}$ is bounded for each $x \in (-1, 1)$ thanks to Lemma 4.35, it follows that $\mathcal{K}$ is relatively compact. To prove that $\mathcal{K}$ is closed, let $\{f_i\}$ be a net in $\mathcal{K}$ converging to a function f on $(-1, 1)$. Then it is clear that f is matrix monotone on $(-1, 1)$ and $f(0) = 0$. By Lemma 4.34 (2), $\left(1 + \frac{1}{x}\right) f_i(x)$ is matrix monotone on $(-1, 1)$ for every i. Since $\lim_{x \to 0}\left(1 + \frac{1}{x}\right) f_i(x) = f_i'(0) = 1$, we thus have

$$\left(1 - \frac{1}{x}\right) f_i(-x) \leq 1 \leq \left(1 + \frac{1}{x}\right) f_i(x), \qquad x \in (0, 1).$$

Therefore,

$$\left(1 - \frac{1}{x}\right) f(-x) \leq 1 \leq \left(1 + \frac{1}{x}\right) f(x), \qquad x \in (0, 1).$$

Since f is C^1 on $(-1, 1)$, the above inequalities yield $f'(0) = 1$. □

Lemma 4.37 *The extreme points of $\mathcal{K}$ have the form*

$$f(x) = \frac{x}{1 - \lambda x}, \quad \text{where } \lambda = \frac{f''(0)}{2}.$$

Proof: Let f be an extreme point of $\mathcal{K}$. For each $\alpha \in (-1, 1)$ define

$$g_\alpha(x) := \left(1 + \frac{\alpha}{x}\right) f(x) - \alpha, \quad x \in (-1, 1).$$

By Lemma 4.34 (2), g_α is matrix monotone on $(-1, 1)$. Notice that

$$g_\alpha(0) = f(0) + \alpha f'(0) - \alpha = 0$$

and

$$g_\alpha'(0) = \lim_{x \to 0} \frac{\left(1 + \frac{\alpha}{x}\right) f(x) - \alpha}{x} = f'(0) + \alpha \lim_{x \to 0} \frac{f(x) - f'(0)x}{x^2} = 1 + \frac{1}{2} \alpha f''(0)$$

by Lemma 4.34 (3). Since $1 + \frac{1}{2}\alpha f''(0) > 0$ by Lemma 4.35, the function

$$h_\alpha(x) := \frac{\left(1 + \frac{\alpha}{x}\right) f(x) - \alpha}{1 + \frac{1}{2} \alpha f''(0)}$$

is in $\mathcal{K}$. Since

$$f = \frac{1}{2}\left(1 + \frac{1}{2}\alpha f''(0)\right) h_\alpha + \frac{1}{2}\left(1 - \frac{1}{2}\alpha f''(0)\right) h_{-\alpha},$$

the extremality of f implies that $f = h_\alpha$ so that

$$\left(1 + \frac{1}{2}\alpha f''(0)\right) f(x) = \left(1 + \frac{\alpha}{x}\right) f(x) - \alpha$$

for all $\alpha \in (-1, 1)$. This immediately implies that $f(x) = x/\left(1 - \frac{1}{2} f''(0)x\right)$. □

Theorem 4.38 *Let f be a matrix monotone function on $(-1, 1)$. Then there exists a probability Borel measure μ on $[-1, 1]$ such that*

$$f(x) = f(0) + f'(0) \int_{-1}^{1} \frac{x}{1 - \lambda x} \, d\mu(\lambda), \quad x \in (-1, 1). \tag{4.18}$$

Proof: The essential case is $f \in \mathcal{K}$. Let $\phi_\lambda(x) := x/(1 - \lambda x)$ for $\lambda \in [-1, 1]$. By Lemmas 4.36 and 4.37, the Krein–Milman theorem says that $\mathcal{K}$ is the closed convex hull of $\{\phi_\lambda : \lambda \in [-1, 1]\}$. Hence there exists a net $\{f_i\}$ in the convex hull of $\{\phi_\lambda : \lambda \in [-1, 1]\}$ such that $f_i(x) \to f(x)$ for all $x \in (-1, 1)$. Each f_i is written as $f_i(x) = \int_{-1}^1 \phi_\lambda(x)\, d\mu_i(\lambda)$ with a probability measure μ_i on $[-1, 1]$ with finite support. Note that the set $\mathcal{M}_1([-1, 1])$ of probability Borel measures on $[-1, 1]$ is compact in the weak* topology when considered as a subset of the dual Banach space of $C([-1, 1])$. Taking a subnet we may assume that μ_i converges in the weak* topology to some $\mu \in \mathcal{M}_1([-1, 1])$. For each $x \in (-1, 1)$, since $\phi_\lambda(x)$ is continuous in $\lambda \in [-1, 1]$, we have

$$f(x) = \lim_i f_i(x) = \lim_i \int_{-1}^1 \phi_\lambda(x)\, d\mu_i(\lambda) = \int_{-1}^1 \phi_\lambda(x)\, d\mu(\lambda).$$

To prove the uniqueness of the representing measure μ, let μ_1, μ_2 be probability Borel measures on $[-1, 1]$ such that

$$f(x) = \int_{-1}^1 \phi_\lambda(x)\, d\mu_1(\lambda) = \int_{-1}^1 \phi_\lambda(x)\, d\mu_2(\lambda), \qquad x \in (-1, 1).$$

Since $\phi_\lambda(x) = \sum_{k=0}^\infty x^{k+1} \lambda^k$ is uniformly convergent in $\lambda \in [-1, 1]$ for any $x \in (-1, 1)$ fixed, it follows that

$$\sum_{k=0}^\infty x^{k+1} \int_{-1}^1 \lambda^k\, d\mu_1(\lambda) = \sum_{k=0}^\infty x^{k+1} \int_{-1}^1 \lambda^k\, d\mu_2(\lambda), \qquad x \in (-1, 1).$$

Hence $\int_{-1}^1 \lambda^k\, d\mu_1(\lambda) = \int_{-1}^1 \lambda^k\, d\mu_2(\lambda)$ for all $k = 0, 1, 2, \ldots$, which implies that $\mu_1 = \mu_2$. $\qquad\square$

The integral representation appearing in the above theorem, which we proved directly, is a special case of Choquet's theorem. The uniqueness of the representing measure μ shows that $\{\phi_\lambda : \lambda \in [-1, 1]\}$ is actually the set of extreme points of $\mathcal{K}$. Since the pointwise convergence topology on $\{\phi_\lambda : \lambda \in [-1, 1]\}$ agrees with the usual topology on $[-1, 1]$, we see that $\mathcal{K}$ is a so-called Bauer simplex.

Theorem 4.39 (**Löwner's theorem**) *Let $-\infty \le a < b \le \infty$ and f be a real-valued function on (a, b). Then f is matrix monotone on (a, b) if and only if $f \in \mathcal{P}(a, b)$. Hence, a matrix monotone function is analytic.*

Proof: The "if" part was shown after Theorem 4.31. To prove the "only if" part, it is enough to assume that (a, b) is a finite open interval. Moreover, when (a, b) is a finite interval, by transforming f into a matrix monotone function on $(-1, 1)$ via a linear function, it suffices to prove the "only if" part when $(a, b) = (-1, 1)$. If f is a non-constant matrix monotone function on $(-1, 1)$, then by using the integral representation (4.18) one can define an analytic continuation of f by

$$f(z) = f(0) + f'(0) \int_{-1}^{1} \frac{z}{1 - \lambda z} d\mu(\lambda), \quad z \in \mathbb{C}^+.$$

Since

$$\operatorname{Im} f(z) = f'(0) \int_{-1}^{1} \frac{\operatorname{Im} z}{|1 - \lambda z|^2} d\mu(\lambda),$$

it follows that f maps $\mathbb{C}^+$ into itself. Hence $f \in \mathcal{P}(-1, 1)$. □

Theorem 4.40 *Let f be a non-linear matrix convex function on $(-1, 1)$. Then there exists a unique probability Borel measure μ on $[-1, 1]$ such that*

$$f(x) = f(0) + f'(0)x + \frac{f''(0)}{2} \int_{-1}^{1} \frac{x^2}{1 - \lambda x} d\mu(\lambda), \quad x \in (-1, 1).$$

Proof: To prove this statement, we use the result due to Kraus that if f is a matrix convex function on (a, b), then f is C^2 and $f^{[1]}[x, \alpha]$ is matrix monotone on (a, b) for every $\alpha \in (a, b)$. Then we may assume that $f(0) = f'(0) = 0$ by considering $f(x) - f(0) - f'(0)x$. Since $g(x) := f^{[1]}[x, 0] = f(x)/x$ is a non-constant matrix monotone function on $(-1, 1)$, by Theorem 4.38 there exists a probability Borel measure μ on $[-1, 1]$ such that

$$g(x) = g'(0) \int_{-1}^{1} \frac{x}{1 - \lambda x} d\mu(\lambda), \quad x \in (-1, 1).$$

Since $g'(0) = f''(0)/2$ is easily seen, we have

$$f(x) = \frac{f''(0)}{2} \int_{-1}^{1} \frac{x^2}{1 - \lambda x} d\mu(\lambda), \quad x \in (-1, 1).$$

Moreover, the uniqueness of μ follows from that of the representing measure for g. □

Theorem 4.41 *Let f be a continuous matrix monotone function on $[0, \infty)$. Then there exists a positive measure μ on $(0, \infty)$ and $\beta \geq 0$ such that*

$$f(x) = f(0) + \beta x + \int_{0}^{\infty} \frac{\lambda x}{x + \lambda} d\mu(\lambda), \quad x \in [0, \infty), \tag{4.19}$$

where

$$\int_{0}^{\infty} \frac{\lambda}{1 + \lambda} d\mu(\lambda) < +\infty.$$

Proof: Consider a function $\psi : (-1, 1) \to (0, \infty)$ defined by

$$\psi(x) := \frac{1 + x}{1 - x} = -1 + \frac{2}{1 - x},$$

which is matrix monotone. Let f be a continuous matrix monotone function on $\mathbb{R}^+$. Since $g(x) := f(\psi(x))$ is matrix monotone on $(-1, 1)$, by Theorem 4.38 there exists a probability measure ν on $[-1, 1]$ such that

$$g(x) = g(0) + g'(0) \int_{[-1,1]} \frac{x}{1 - \lambda x} \, d\nu(\lambda)$$

for every $x \in (-1, 1)$. We may assume $g'(0) > 0$ since otherwise g and hence f are constant functions. Since $g(-1) = \lim_{x \searrow 0} g(x) = f(0) > -\infty$, we have

$$\int_{[-1,1]} \frac{1}{1 + \lambda} \, d\nu(\lambda) < +\infty$$

and in particular $\nu(\{-1\}) = 0$. Therefore,

$$g(x) - g(-1) = g'(0) \int_{(-1,1]} \frac{1 + x}{(1 - \lambda x)(1 + \lambda)} \, d\mu(\lambda) = \int_{(-1,1]} \frac{1 + x}{1 - \lambda x} \, d\tilde{\mu}(\lambda),$$

where $d\tilde{\mu}(\lambda) := g'(0)(1 + \lambda)^{-1} \, d\mu(\lambda)$. Define a finite measure m on $(0, \infty)$ by $m := \tilde{\mu} \circ \psi^{-1}$. Transform the above integral expression by $x = \psi^{-1}(t)$ to obtain

$$f(t) - f(0) = t\tilde{\mu}(\{1\}) + \int_{(0,\infty)} \frac{1 + \psi^{-1}(t)}{1 - \psi^{-1}(\zeta)\psi^{-1}(t)} \, dm(\zeta)$$

$$= \beta t + \int_{(0,\infty)} \frac{t(1 + \zeta)}{t + \zeta} \, dm(\zeta),$$

where $\beta := \tilde{\mu}(\{1\})$. With the measure $d\mu(\zeta) := ((1 + \zeta)/\zeta) \, dm(\zeta)$ we have the desired integral expression of f.   □

Since the integrand

$$\frac{x}{x + \lambda} = 1 - \frac{\lambda}{x + \lambda}$$

is a matrix monotone function of x (see Example 4.1), it is obvious that a function on $[0, \infty)$ admitting the integral expression (4.19) is matrix monotone. The theorem shows that a matrix monotone function on $[0, \infty)$ is matrix concave.

Theorem 4.42 *If $f : \mathbb{R}^+ \to \mathbb{R}$ is matrix monotone, then $xf(x)$ is matrix convex.*

Proof: Let $\lambda > 0$. First we consider the function $f(x) = -(x + \lambda)^{-1}$. Then

$$xf(x) = -\frac{x}{\lambda + x} = -1 + \frac{\lambda}{\lambda + x}$$

and it is well-known that $x \mapsto (x + \lambda)^{-1}$ is matrix convex.

For a general matrix monotone f, we use the integral decomposition (4.19) and the statement follows from the previous special case. □

Theorem 4.43 *If $f : (0, \infty) \to (0, \infty)$, then the following conditions are equivalent:*

(1) *f is matrix monotone;*
(2) *$x/f(x)$ is matrix monotone;*
(3) *f is matrix concave.*

Proof: For $\varepsilon > 0$ the function $f_\varepsilon(x) := f(x + \varepsilon)$ is defined on $[0, \infty)$. If the statement is proved for this function, then the limit $\varepsilon \to 0$ gives the result. So we assume $f : [0, \infty) \to (0, \infty)$.

The implication (1) $\Rightarrow$ (3) has already been observed above.

The implication (3) $\Rightarrow$ (2) is based on Example 4.24. It says that $-f(x)/x$ is matrix monotone. Therefore $x/f(x)$ is matrix monotone as well.

(2) $\Rightarrow$ (1): Assume that $x/f(x)$ is matrix monotone on $(0, \infty)$. Let $\alpha := \lim_{x \searrow 0} x/f(x)$. Then it follows from the Löwner representation that, dividing by x, we have

$$\frac{1}{f(x)} = \frac{\alpha}{x} + \beta + \int_0^\infty \frac{\lambda}{\lambda + x} \, d\mu(\lambda).$$

This multiplied by -1 yields the matrix monotone $-1/f(x)$. Therefore $f(x)$ is matrix monotone as well. □

We have shown that the matrix monotonicity is equivalent to the positive definiteness of the divided difference kernel. Matrix concavity has a somewhat similar property.

Theorem 4.44 *Let $f : \mathbb{R}^+ \to \mathbb{R}^+$ be a smooth function. If the divided difference kernel function is conditionally negative definite, then f is matrix convex.*

Proof: By continuity it suffices to prove that the function $f(x) + \varepsilon$ is matrix convex on $\mathbb{R}^+$ for any $\varepsilon > 0$. So we may assume that $f > 0$. Example 2.42 and Theorem 4.5 give that $g(x) = x^2/f(x)$ is matrix monotone. Then $x/g(x) = f(x)/x$ is matrix monotone due to Theorem 4.43. Multiplying by x we have a matrix convex function by Theorem 4.42. □

It is not always easy to determine if a function is matrix monotone. An efficient method is based on Theorem 4.39. The theorem says that a function $\mathbb{R}^+ \to \mathbb{R}$ is matrix monotone if and only if it has a holomorphic extension to the upper half-plane $\mathbb{C}^+$ such that its range is in the closure of $\mathbb{C}^+$. It is remarkable that matrix monotone functions are very smooth and connected with functions of a complex variable.

Example 4.45 The representation

$$x^t = \frac{\sin \pi t}{\pi} \int_0^\infty \frac{\lambda^{t-1} x}{\lambda + x} \, d\lambda$$

shows that $f(x) = x^t$ is matrix monotone on $\mathbb{R}^+$ when $0 < t < 1$. In other words,

$$0 \le A \le B \quad \text{implies} \quad A^t \le B^t,$$

which is often called the **Löwner–Heinz inequality**.

We can arrive at the same conclusion by holomorphic extension as a Pick function (see Examples 4.29 and 4.33) so that $f(x) = x^t$ is matrix monotone on $\mathbb{R}^+$ for these values of the parameter but not for any other value. Another familiar example of a matrix monotone function is $\log x$ on $(0, \infty)$, see Examples 4.29 and 4.32.   □

4.5 Some Applications

If the complex extension of a function $f : [0, \infty) \to \mathbb{R}$ is rather natural, then it can be checked numerically that the upper half-plane is mapped into itself, and the function is expected to be matrix monotone. For example, $x \mapsto x^p$ has a natural complex extension. In the following we give a few more examples of matrix monotone functions on $\mathbb{R}^+$.

Theorem 4.46 *Let*

$$f_p(x) := \left(\frac{p(x-1)}{x^p - 1} \right)^{\frac{1}{1-p}} \qquad (x > 0).$$

In particular, $f_2(x) = (x+1)/2$, $f_{-1}(x) = \sqrt{x}$ and

$$f_1(x) := \lim_{p \to 1} f_p(x) = e^{-1} x^{\frac{x}{x-1}}, \qquad f_0(x) := \lim_{p \to 0} f_p(x) = \frac{x-1}{\log x}.$$

Then f_p is matrix monotone if $-2 \le p \le 2$.

Proof: Since the functions are continuous in the parameter p and the matrix monotone functions are closed under pointwise convergence, it is enough to prove the result for $p \in [-2, 2]$ such that $p \ne -2, -1, 0, 1, 2$. By Löwner's theorem (Theorem 4.39) it suffices to show that f_p has a holomorphic continuation to $\mathbb{C}^+$ mapping into itself. We define $\log z$ with $\log 1 = 0$; then in case $-2 < p < 2$, the real function $p(x-1)/(x^p-1)$ has a holomorphic continuation $p(z-1)/(z^p-1)$ to $\mathbb{C}^+$ since $z^p - 1 \ne 0$ in $\mathbb{C}^+$. Moreover, it is continuous in the closed upper half-plane $\text{Im}\, z \ge 0$. Further, since $p(z-1)/(z^p-1) \ne 0$ $(z \ne 1)$, f_p has a holomorphic continuation (denoted by the same f_p) to $\mathbb{C}^+$ and it is also continuous in $\text{Im}\, z \ge 0$.

Assume $p \in (0, 1) \cup (1, 2)$. For $R > 0$ let $K_R := \{z : |z| \le R, \text{Im}\, z \ge 0\}$, $\Gamma_R := \{z : |z| = R, \text{Im}\, z > 0\}$ and K_R° be the interior of K_R; then the boundary of K_R is $\Gamma_R \cup [-R, R]$. Note that $f_p(K_R)$ is a compact set. Recall the well-known fact that the image of a connected open set under a holomorphic function is a connected open set, if it is not a single point. This yields that $f_p(K_R^\circ)$ is open and hence the boundary of $f_p(K_R)$ is included in $f_p(\Gamma_R \cup [-R, R])$. Below let us prove that for any sufficiently small $\varepsilon > 0$, if R is sufficiently large (depending on ε) then

$$f_p(\Gamma_R \cup [-R, R]) \subset \{z : -\varepsilon \leq \arg z \leq \pi + \varepsilon\},$$

which yields that $f_p(K_R) \subset \{z : -\varepsilon \leq \arg z \leq \pi + \varepsilon\}$. Thus, letting $R \nearrow \infty$ (so $\varepsilon \searrow 0$) we conclude that $f_p(\mathbb{C}^+) \subset \{z : \operatorname{Im} z \geq 0\}$.

Clearly, $[0, \infty)$ is mapped into $[0, \infty)$ by f_p. If $z \in (-\infty, 0)$, then $\arg(z-1) = \pi$ and $p\pi \leq \arg(z^p - 1) \leq \pi$ for $0 < p < 1$ and $\pi \leq \arg(z^p - 1) \leq p\pi$ for $1 < p < 2$. Hence

$$0 \leq \arg \frac{z-1}{z^p - 1} \leq (1-p)\pi \quad \text{for } 0 < p < 1,$$

$$(1-p)\pi \leq \arg \frac{z-1}{z^p - 1} \leq 0 \quad \text{for } 1 < p < 2.$$

Thus, since

$$\arg\left(\frac{z-1}{z^p - 1}\right)^{\frac{1}{1-p}} = \frac{1}{1-p} \arg \frac{z-1}{z^p - 1},$$

it follows that $0 \leq \arg f_p(z) \leq \pi$, so $(-\infty, 0)$ is mapped into $\operatorname{Im} z \geq 0$.

Next, for any small $\varepsilon > 0$, if R is sufficiently large, then we have for every $z \in \Gamma_R$

$$|\arg(z-1) - \arg z| \leq \varepsilon, \qquad |\arg(z^p - 1) - p \log z| \leq \varepsilon$$

so that

$$\left| \arg \frac{z-1}{z^p - 1} - (1-p)\arg z \right| \leq 2\varepsilon.$$

Since

$$\left| \arg\left(\frac{z-1}{z^p-1}\right)^{\frac{1}{1-p}} - \arg z \right| = \left| \frac{1}{1-p}\left(\arg\frac{z-1}{z^p-1} - (1-p)\arg z\right) \right| \leq \frac{2\varepsilon}{|1-p|},$$

we have $f_p(\Gamma_R) \subset \{z : -2\varepsilon/|1-p| \leq \arg z \leq \pi + 2\varepsilon/|1-p|\}$. Thus, the desired assertion follows.

The case $-2 < p < 0$ can be treated similarly by noting that

$$f_p(x) = \left(\frac{|p|x^{|p|}(x-1)}{x^{|p|} - 1}\right)^{\frac{1}{1+|p|}}.$$

$\square$

Theorem 4.47 *The function*

$$f_p(x) = \left(\frac{x^p + 1}{2}\right)^{\frac{1}{p}}$$

is matrix monotone if and only if $-1 \le p \le 1$.

Proof: Observe that $f_{-1}(x) = 2x/(x+1)$ and $f_1(x) = (x+1)/2$, so f_p could be matrix monotone only if $-1 \le p \le 1$. We show that it is indeed matrix monotone. The case $p = 0$ is well-known. Further, note that if f_p is matrix monotone for $0 < p < 1$ then

$$f_{-p}(x) = \left(\left(\frac{x^{-p} + 1}{2}\right)^{\frac{1}{p}}\right)^{-1}$$

is also matrix monotone since x^{-p} is matrix monotone decreasing for $0 < p \le 1$.

So let us assume that $0 < p < 1$. Then, since $z^p + 1 \ne 0$ in the upper half plane, f_p has a holomorphic continuation to the upper half plane (by defining $\log z$ as $\log 1 = 0$). By Löwner's theorem it suffices to show that f_p maps the upper half plane into itself. If $0 < \arg z < \pi$ then $0 < \arg(z^p + 1) < \arg z^p = p \arg z$ so that

$$0 < \arg\left(\frac{z^p + 1}{2}\right)^{\frac{1}{p}} = \frac{1}{p}\arg\left(\frac{z^p + 1}{2}\right) < \arg z < \pi.$$

Thus z is mapped into the upper half plane.   □

In the special case $p = \frac{1}{n}$,

$$f_p(x) = \left(\frac{x^{\frac{1}{n}} + 1}{2}\right)^n = \frac{1}{2^n}\sum_{k=0}^{n}\binom{n}{k}x^{\frac{k}{n}},$$

and it is well-known that x^α is matrix monotone for $0 < \alpha < 1$ thus f_p is also matrix monotone.

The matrix monotone functions in the next theorem play an important role in some applications.

Theorem 4.48 *For* $-1 \le p \le 2$ *the function*

$$f_p(x) = p(1-p)\frac{(x-1)^2}{(x^p - 1)(x^{1-p} - 1)} \tag{4.20}$$

is matrix monotone.

Proof: The special cases $p = -1, 0, 1, 2$ are well-known. For $0 < p < 1$ we can use an integral representation

$$\frac{1}{f_p(x)} = \frac{\sin p\pi}{\pi} \int_0^\infty d\lambda\, \lambda^{p-1} \int_0^1 ds \int_0^1 dt \frac{1}{x((1-t)\lambda + (1-s)) + (t\lambda + s)}$$

and since the integrand is matrix monotone decreasing as a function of x, $1/f_p$ is matrix monotone decreasing as well. It follows that $f_p(x)$ is matrix monotone for $0 < p < 1$.

The proof below is based on Löwner's theorem. Since $f_p = f_{1-p}$, we may assume $p \in (0, 1) \cup (1, 2)$. It suffices to show that f_p has a holomorphic continuation to $\mathbb{C}^+$ mapping into itself. It is clear that $[0, \infty)$ is mapped into $[0, \infty)$ by f_p. First, when $0 < p < 1$, f_p has a holomorphic continuation to $\mathbb{C}^+$ in the form

$$f_p(z) = p(1-p)\frac{(z-1)^2}{(z^p - 1)(z^{1-p} - 1)}.$$

For $z \in (-\infty, 0)$, since

$$p\pi \le \arg(z^p - 1) \le \pi, \qquad (1-p)\pi \le \arg(z^{1-p} - 1) \le \pi$$

and

$$\arg f_p(z) = -\arg(z^p - 1) - \arg(z^{1-p} - 1),$$

we have $-2\pi \le \arg f_p(z) \le -\pi$ so that $\operatorname{Im} f_p(z) \ge 0$. Let K_R and Γ_R be as in the proof of Theorem 4.46. For every $\varepsilon > 0$, if $z \in \Gamma_R$ with a sufficiently large $R > 0$ (depending on ε), then we have

$$|\arg(z-1)^2 - 2\arg z| \le \varepsilon, \quad |\arg(z^p - 1) - p\arg z| \le \varepsilon,$$

$$|\arg(z^{1-p} - 1) - (1-p)\arg z| \le \varepsilon,$$

so that

$$|\arg f_p(z) - \arg z| = |\arg(z-1)^2 - \arg(z^p - 1) - \arg(z^{1-p} - 1) - \arg z| \le 3\varepsilon,$$

which yields that $f_p(\Gamma_R) \cup [-R, R]) \subset \{z : -3\varepsilon \le \arg z \le 3\varepsilon\}$. Thus, letting $R \nearrow \infty$ (so $\varepsilon \searrow 0$) we have $f_p(\mathbb{C}^+) \subset \{z : \operatorname{Im} z \ge 0\}$ as in the proof of Theorem 4.46.

Next, when $1 < p < 2$, f_p has a holomorphic continuation to $\mathbb{C}^+$ in the form

$$f_p(z) = p(p-1)\frac{z^{p-1}(z-1)^2}{(z^p - 1)(z^{p-1} - 1)}.$$

For every $\varepsilon > 0$, if $z \in \Gamma_R$ with a sufficiently large R, then

$$|\arg f_p(z) - (2-p)\log z| \le 3\varepsilon$$

as above. The assertion follows similarly. $\square$

Theorem 4.49 *If $f : \mathbb{R}^+ \to \mathbb{R}$ is a matrix monotone function and $A, B \ge 0$, then*

$$2Af(A) + 2Bf(B) \ge \sqrt{A+B}\Big(f(A) + f(B)\Big)\sqrt{A+B}.$$

This proof is left for Exercise 8.

For a function $f : (a, b) \to \mathbb{R}$ the notion of strong subadditivity is introduced via the inequality (4.3). Recall that the condition for f is

$$\mathrm{Tr}\, f(A) + \mathrm{Tr}\, f(A_{22}) \le \mathrm{Tr}\, f(B) + \mathrm{Tr}\, f(C)$$

for every matrix $A = A^*$ with $\sigma(A) \subset (a, b)$ in the form of a 3×3 block matrix

$$A = \begin{bmatrix} A_{11} & A_{12} & A_{13} \\ A_{12}^* & A_{22} & A_{23} \\ A_{13}^* & A_{23}^* & A_{33} \end{bmatrix}$$

and

$$B = \begin{bmatrix} A_{11} & A_{12} \\ A_{12}^* & A_{22} \end{bmatrix}, \quad C = \begin{bmatrix} A_{22} & A_{23} \\ A_{23}^* & A_{33} \end{bmatrix}.$$

The next theorem tells us that $f(x) = \log x$ on $(0, \infty)$ is a strong subadditive function, since $\log \det A = \mathrm{Tr}\, \log A$ for a positive definite matrix A.

Theorem 4.50 *Let*

$$S = \begin{bmatrix} S_{11} & S_{12} & S_{13} \\ S_{12}^* & S_{22} & S_{23} \\ S_{13}^* & S_{23}^* & S_{33} \end{bmatrix}$$

be a positive definite block matrix. Then

$$\det S \cdot \det S_{22} \le \det \begin{bmatrix} S_{11} & S_{12} \\ S_{12}^* & S_{22} \end{bmatrix} \cdot \det \begin{bmatrix} S_{22} & S_{23} \\ S_{23}^* & S_{33} \end{bmatrix}$$

and the condition for equality is $S_{13} = S_{12}S_{22}^{-1}S_{23}$.

Proof: Take the ortho-projections

$$P = \begin{bmatrix} I & 0 & 0 \\ 0 & 0 & 0 \\ 0 & 0 & 0 \end{bmatrix} \quad \text{and} \quad Q = \begin{bmatrix} I & 0 & 0 \\ 0 & I & 0 \\ 0 & 0 & 0 \end{bmatrix}.$$

Since $P \le Q$, we have the matrix inequality

$$[P]S \le [P]QSQ,$$

which implies the determinant inequality

$$\det[P]S \le \det[P]QSQ.$$

According to the Schur determinant formula (Theorem 2.3), this is exactly the determinant inequality of the theorem.

The equality in the determinant inequality implies $[P]S = [P]QSQ$ which is

$$S_{11} - [S_{12}, S_{13}] \begin{bmatrix} S_{22} & S_{23} \\ S_{32} & S_{33} \end{bmatrix}^{-1} \begin{bmatrix} S_{21} \\ S_{31} \end{bmatrix} = S_{11} - S_{12}S_{22}^{-1}S_{21}.$$

This can be written as

$$[S_{12}, S_{13}] \left(\begin{bmatrix} S_{22} & S_{23} \\ S_{32} & S_{33} \end{bmatrix}^{-1} - \begin{bmatrix} S_{22}^{-1} & 0 \\ 0 & 0 \end{bmatrix} \right) \begin{bmatrix} S_{21} \\ S_{31} \end{bmatrix} = 0. \qquad (4.21)$$

For the moment, let

$$\begin{bmatrix} S_{22} & S_{23} \\ S_{32} & S_{33} \end{bmatrix}^{-1} = \begin{bmatrix} C_{22} & C_{23} \\ C_{32} & C_{33} \end{bmatrix}.$$

Then

$$\begin{bmatrix} S_{22} & S_{23} \\ S_{32} & S_{33} \end{bmatrix}^{-1} - \begin{bmatrix} S_{22}^{-1} & 0 \\ 0 & 0 \end{bmatrix} = \begin{bmatrix} C_{23}C_{33}^{-1}C_{32} & C_{23} \\ C_{32} & C_{33} \end{bmatrix}$$

$$= \begin{bmatrix} C_{23}C_{33}^{-1/2} \\ C_{33}^{1/2} \end{bmatrix} \begin{bmatrix} C_{33}^{-1/2}C_{32} & C_{33}^{1/2} \end{bmatrix}.$$

Comparing this with (4.21) we arrive at

$$[S_{12}, S_{13}] \begin{bmatrix} C_{23}C_{33}^{-1/2} \\ C_{33}^{1/2} \end{bmatrix} = S_{12}C_{23}C_{33}^{-1/2} + S_{13}C_{33}^{1/2} = 0.$$

Equivalently,

$$S_{12}C_{23}C_{33}^{-1} + S_{13} = 0.$$

Since the concrete form of C_{23} and C_{33} is known, we can compute that $C_{23}C_{33}^{-1} = -S_{22}^{-1}S_{23}$ and this gives the condition stated in the theorem. $\square$

The next theorem gives a sufficient condition for the strong subadditivity (4.3) of functions on $(0, \infty)$.

Theorem 4.51 *Let $f : (0, \infty) \to \mathbb{R}$ be a function such that $-f'$ is matrix monotone. Then the inequality (4.3) holds.*

Proof: A matrix monotone function has the representation

$$a + bx + \int_0^\infty \left(\frac{\lambda}{\lambda^2 + 1} - \frac{1}{\lambda + x} \right) d\mu(\lambda),$$

where $b \geq 0$, see (V.49) in [20]. Therefore, we have the representation

$$f(t) = c - \int_1^t \left(a + bx + \int_0^\infty \left(\frac{\lambda}{\lambda^2 + 1} - \frac{1}{\lambda + x} \right) d\mu(\lambda) \right) dx.$$

By integration we have

$$f(t) = d - at - \frac{b}{2}t^2 + \int_0^\infty \left(\frac{\lambda}{\lambda^2 + 1}(1 - t) + \log \left(\frac{\lambda}{\lambda + 1} + \frac{t}{\lambda + 1} \right) \right) d\mu(\lambda).$$

The first quadratic part satisfies strong subadditivity and we have to check the integral. Since $\log x$ is a strongly subadditive function by Theorem 4.50, so is the integrand. The property is retained by integration.   □

Example 4.52 By differentiation we can see that $f(x) = -(x + t) \log(x + t)$ with $t \geq 0$ satisfies strong subadditivity. Similarly, $f(x) = -x^t$ satisfies strong subadditivity if $1 \leq t \leq 2$.

In some applications the matrix monotone functions

$$f_p(x) = p(1 - p) \frac{(x - 1)^2}{(x^p - 1)(x^{1-p} - 1)} \qquad (0 < p < 1)$$

appear.

For $p = 1/2$ this is a strongly subadditivity function. Up to a constant factor, the function is

$$(\sqrt{x} + 1)^2 = x + 2\sqrt{x} + 1$$

and all terms are known to be strongly subadditive. The function $-f'_{1/2}$ is evidently matrix monotone.

Numerical computation indicates that $-f'_p$ seems to be matrix monotone, but no proof is known.   □

For $K, L \geq 0$ and a matrix monotone function f, there is a very particular relation between $f(K)$ and $f(L)$. This is described in the next theorem.

Theorem 4.53 *Let $f : \mathbb{R}^+ \to \mathbb{R}$ be a matrix monotone function. For positive matrices K and L, let P be the projection onto the range of $(K - L)_+$. Then*

$$\mathrm{Tr}\, PL(f(K) - f(L)) \geq 0.$$

Proof: From the integral representation

$$f(x) = \int_0^\infty \frac{x(1+s)}{x+s}\, d\mu(s)$$

we have

$$\operatorname{Tr} PL(f(K) - f(L)) = \int_0^\infty (1+s)s \operatorname{Tr} PL(K+s)^{-1}(K-L)(L+s)^{-1}\, d\mu(s).$$

Hence it is sufficient to prove that

$$\operatorname{Tr} PL(K+s)^{-1}(K-L)(L+s)^{-1} \geq 0$$

for $s > 0$. Let $\Delta_0 := K - L$ and consider the integral representation

$$(K+s)^{-1}\Delta_0(L+s)^{-1} = \int_0^1 s(L+t\Delta_0+s)^{-1}\Delta_0(L+t\Delta_0+s)^{-1}\, dt.$$

So we can make another reduction: it suffices to show that

$$\operatorname{Tr} PL(L+t\Delta_0+s)^{-1}t\Delta_0(L+t\Delta_0+s)^{-1} \geq 0.$$

If $C := L + t\Delta_0$ and $\Delta := t\Delta_0$, then $L = C - \Delta$ and we have

$$\operatorname{Tr} P(C-\Delta)(C+s)^{-1}\Delta(C+s)^{-1} \geq 0. \tag{4.22}$$

We write our operators in the form of 2×2 block matrices:

$$V = (C+s)^{-1} = \begin{bmatrix} V_1 & V_2 \\ V_2^* & V_3 \end{bmatrix}, \quad P = \begin{bmatrix} I & 0 \\ 0 & 0 \end{bmatrix}, \quad \Delta = \begin{bmatrix} \Delta_+ & 0 \\ 0 & -\Delta_- \end{bmatrix}.$$

The left-hand side of the inequality (4.22) can then be rewritten as

$$\begin{aligned}
\operatorname{Tr} P(C-\Delta)(V\Delta V) &= \operatorname{Tr}[(C-\Delta)(V\Delta V)]_{11} \\
&= \operatorname{Tr}[(V^{-1}-\Delta-s)(V\Delta V)]_{11} \\
&= \operatorname{Tr}[\Delta V - (\Delta+s)(V\Delta V)]_{11} \\
&= \operatorname{Tr}(\Delta_+ V_{11} - (\Delta_+ + s)(V\Delta V)_{11}) \\
&= \operatorname{Tr}(\Delta_+(V - V\Delta V)_{11} - s(V\Delta V)_{11}). \tag{4.23}
\end{aligned}$$

Because of the positivity of L, we have $V^{-1} \geq \Delta + s$, which implies $V = VV^{-1}V \geq V(\Delta + s)V = V\Delta V + sV^2$. As the diagonal blocks of a positive operator are themselves positive, this further implies

$$V_1 - (V\Delta V)_{11} \geq s(V^2)_{11}.$$

Inserting this in (4.23) gives

$$
\begin{aligned}
\mathrm{Tr}\,[(V^{-1} - \Delta - s)(V\Delta V)]_{11} &= \mathrm{Tr}\,(\Delta_+(V - V\Delta V)_{11} - s(V\Delta V)_{11}) \\
&\geq \mathrm{Tr}\,(\Delta_+ s(V^2)_{11} - s(V\Delta V)_{11}) \\
&= s\,\mathrm{Tr}\,(\Delta_+(V^2)_{11} - (V\Delta V)_{11}) \\
&= s\,\mathrm{Tr}\,(\Delta_+(V_1 V_1 + V_2 V_2^*) - (V_1 \Delta_+ V_1 \\
&\quad - V_2 \Delta_- V_2^*)) \\
&= s\,\mathrm{Tr}\,(\Delta_+ V_2 V_2^* + V_2 \Delta_- V_2^*).
\end{aligned}
$$

This quantity is positive.   □

Theorem 4.54 *Let A and B be positive operators, then for all* $0 \leq s \leq 1$,

$$
2\mathrm{Tr}\,A^s B^{1-s} \geq \mathrm{Tr}\,(A + B - |A - B|). \tag{4.24}
$$

Proof: For a self-adjoint operator X, $X_\pm$ denotes its positive and negative parts. Decomposing $A - B = (A - B)_+ - (A - B)_-$ one gets

$$
\mathrm{Tr}\,A + \mathrm{Tr}\,B - \mathrm{Tr}\,|A - B| = 2\mathrm{Tr}\,A - 2\mathrm{Tr}\,(A - B)_+,
$$

and (4.24) is equivalent to

$$
\mathrm{Tr}\,A - \mathrm{Tr}\,B^s A^{1-s} \leq \mathrm{Tr}\,(A - B)_+.
$$

From

$$
A \leq A + (A - B)_- = B + (A - B)_+
$$

and $B \leq B + (A - B)_+$ as well as the matrix monotonicity of the function $x \mapsto x^s$, we have

$$
\begin{aligned}
\mathrm{Tr}\,A - \mathrm{Tr}\,B^s A^{1-s} &= \mathrm{Tr}\,(A^s - B^s)A^{1-s} \leq \mathrm{Tr}\,((B + (A - B)_+)^s - B^s)A^{1-s} \\
&\leq \mathrm{Tr}\,((B + (A - B)_+)^s - B^s)(B + (A - B)_+)^{1-s} \\
&= \mathrm{Tr}\,B + \mathrm{Tr}\,(A - B)_+ - \mathrm{Tr}\,B^s(B + (A - B)_+)^{1-s} \\
&\leq \mathrm{Tr}\,B + \mathrm{Tr}\,(A - B)_+ - \mathrm{Tr}\,B^s B^{1-s} \\
&= \mathrm{Tr}\,(A - B)_+
\end{aligned}
$$

and the statement is obtained.   □

The following result is Lieb's extension of the **Golden–Thompson inequality**.

Theorem 4.55 (**Golden–Thompson–Lieb**) *Let* A, B *and* C *be self-adjoint matrices. Then*

$$
\mathrm{Tr}\,e^{A+B+C} \leq \int_0^\infty \mathrm{Tr}\,e^A(t + e^{-C})^{-1} e^B(t + e^{-C})^{-1}\,dt\,.
$$

Proof: Another formulation of the statement is

$$\text{Tr } e^{A+B-\log D} \leq \text{Tr } e^A \, \mathbb{J}_D^{-1}(e^B),$$

where

$$\mathbb{J}_D^{-1} K = \int_0^\infty (t+D)^{-1} K (t+D)^{-1} \, dt$$

(which is the formulation of (3.20)). We choose $L = -\log D + A$, $\beta = e^B$ and conclude from (4.4) that the functional

$$F : \beta \mapsto -\text{Tr } e^{L+\log \beta}$$

is convex on the cone of positive definite matrices. It is also homogeneous of order 1 so that the hypothesis of Lemma 4.56 (from below) is fulfilled. So

$$-\text{Tr } e^{A+B-\log D} = -\text{Tr } \exp(L + \log \beta) = F(\beta)$$

$$\geq -\frac{d}{dx} \text{Tr } \exp(L + \log(D + x\beta)) \Big|_{x=0}$$

$$= -\text{Tr } e^A \, \mathbb{J}_D^{-1}(\beta) = -\text{Tr } e^A \, \mathbb{J}_D^{-1}(e^B).$$

This is the statement with a minus sign.   □

Lemma 4.56 *Let C be a convex cone in a vector space and $F : C \to \mathbb{R}$ be a convex function such that $F(\lambda A) = \lambda F(A)$ for every $\lambda > 0$ and $A \in C$. If $B \in C$ and the limit*

$$\lim_{x \searrow 0} \frac{F(A + xB) - F(A)}{x} =: \partial_B F(A)$$

exists, then

$$F(B) \geq \partial_B F(A).$$

If equality holds here, then $F(A + xB) = (1 - x)F(A) + xF(A + B)$ for $0 \leq x \leq 1$.

Proof: Define a function $f : [0, 1] \to \mathbb{R}$ by $f(x) := F(A + xB)$. This function is convex:

$$f(\lambda x_1 + (1 - \lambda)x_2) = F(\lambda(A + x_1 B) + (1 - \lambda)(A + x_2 B))$$
$$\leq \lambda F(A + x_1 B) + (1 - \lambda)F(A + x_2 B)$$
$$= \lambda f(x_1) + (1 - \lambda)f(x_2).$$

The assumption is the existence of the derivative $f'(0)$ (from the right). From the convexity

$$F(A + B) = f(1) \geq f(0) + f'(0) = F(A) + \partial_B F(A).$$

Actually, F is subadditive:

$$F(A + B) = 2F(A/2 + B/2) \leq F(A) + F(B),$$

and the stated inequality follows.

If $f'(0) + f(0) = f(1)$, then $f(x) - f(0)$ is linear. (This also has the alternative description: $f''(x) = 0$.) $\square$

If $C = 0$ in Theorem 4.55, then we have

$$\mathrm{Tr}\, e^{A+B} \leq \mathrm{Tr}\, e^A e^B,$$

which is the original **Golden–Thompson inequality**. If $BC = CB$, then on the right-hand side, the integral

$$\int_0^\infty (t + e^{-C})^{-2}\, dt$$

appears. This equals e^C and we have $\mathrm{Tr}\, e^{A+B+C} \leq \mathrm{Tr}\, e^A e^B e^C$. Without the assumption $BC = CB$, this inequality is not true.

The Golden–Thompson inequality is equivalent to a kind of monotonicity of the relative entropy, see [73]. An example of an application of the Golden–Thompson–Lieb inequality is the **strong subadditivity** of the von Neumann entropy.

4.6 Notes and Remarks

On the subject of convex analysis, R. Tyrell Rockafellar has a famous book: *Convex Analysis*, Princeton, Princeton University Press, 1970.

Theorem 4.46 as well as the optimality of the range $-2 \leq p \leq 2$ was proved in the paper Y. Nakamura, Classes of operator monotone functions and Stieltjes functions, in *The Gohberg Anniversary Collection, Vol. II*, H. Dym et al. (eds.), Oper. Theory Adv. Appl., vol. 41, Birkhäuser, 1989, pp. 395–404. The proof here is a modification of that in the paper Ádám Besenyei and Dénes Petz, Completely positive mappings and mean matrices, Linear Algebra Appl. **435**(2011), 984–997. Theorem 4.47 was given in the paper Fumio Hiai and Hideki Kosaki, Means for matrices and comparison of their norms, Indiana Univ. Math. J. **48**(1999), 899–936.

The matrix monotonicity of the function (4.20) for $0 < p < 1$ was recognized in [74]. The proof for $p \in [-1, 2]$ is a modification of that in the paper V. E. Sándor Szabó, A class of matrix monotone functions, Linear Algebra Appl. **420** (2007), 79–85. Related discussions are in [17], and there is an extension to

$$\frac{(x-a)(x-b)}{(f(x)-f(a))(x/f(x)-b/f(b))}$$

in the paper M. Kawasaki and M. Nagisa, Transforms on operator monotone functions, arXiv:1206.5452. ($a = b = 1$ and $f(x) = x^p$ covers (4.20).) A shorter proof of this is in the paper F. Hansen, WYD-like skew information measures, J. Stat. Phys. **151**(2013), 974–979.

The original result of Karl Löwner appeared in 1934 (when he emigrated to the US, he changed his name to Charles Loewner). Apart from Löwner's original proof, there are three other different proofs, for example by Bendat and Sherman based on the Hamburger moment problem, by Korányi based on the spectral theorem of self-adjoint operators, and by Hansen and Pedersen based on the Krein–Milman theorem. In all of them, the integral representation of operator monotone functions was obtained to prove Löwner's theorem. The proof presented here is based on [40].

The integral representation (4.6) was obtained by Julius Bendat and Seymour Sherman [16]. Theorem 4.22 is from the famous paper of Frank **Hansen** and Gert G. **Pedersen** [40], and Theorem 4.16 is from [41] by the same authors. Theorems 4.28 and 4.28 are from the paper of J. S. **Aujla** and F. C. **Silva** [15], which also contains Theorem 4.16 in a stronger form of majorization (see Theorem 6.27 in Chap. 6).

Theorem 4.51 is from the paper [14]. It is an interesting question if the converse statement is true.

Theorem 4.54 is in the paper K. M. R. **Audenaert** et al., Discriminating states: the quantum Chernoff bound, Phys. Rev. Lett. **98** (2007), 160501. A quantum information application is contained in the same paper and also in the book [73]. The present proof is due to Narutaka Ozawa, which is contained in V. Jakšić, Y. Ogata, Y. Pautrat and C.-A. Pillet, Entropic fluctuations in quantum statistical mechanics. an introduction, in the book *Quantum Theory from Small to Large Scales* (École de Physique des Houches Session XCV 2010), J. Fröhlich et al. (eds.), Oxford University Press, 2012. Theorem 4.54 from [73] is an extension of Theorem 4.53 and its proof is similar to that of Audenaert et al.

4.7 Exercises

1. Prove that the function $\kappa : \mathbb{R}^+ \to \mathbb{R}$, $\kappa(x) = -x \log x + (x+1) \log(x+1)$ is matrix monotone.
2. Prove that $f(x) = x^2$ is not matrix monotone on any positive interval.
3. Show that $f(x) = e^x$ is not matrix monotone on $[0, \infty)$.
4. Show that if $f : \mathbb{R}^+ \to \mathbb{R}$ is a matrix monotone function, then $-f$ is a completely monotone function.
5. Let f be a differentiable function on the interval (a, b) such that for some $a < c < b$ the function f is matrix monotone for 2×2 matrices on the intervals $(a, c]$ and $[c, b)$. Show that f is matrix monotone for 2×2 matrices on (a,b).
6. Show that the function

$$f(x) = \frac{ax+b}{cx+d} \qquad (a, b, c, d \in \mathbb{R}, \quad ad > bc)$$

is matrix monotone on any interval which does not contain $-d/c$.

7. Use the matrices

$$A = \begin{bmatrix} 1 & 1 \\ 1 & 1 \end{bmatrix} \quad \text{and} \quad B = \begin{bmatrix} 2 & 1 \\ 1 & 1 \end{bmatrix}$$

to show that $f(x) = \sqrt{x^2 + 1}$ is not a matrix monotone function on $\mathbb{R}^+$.

8. Let $f : \mathbb{R}^+ \to \mathbb{R}$ be a matrix monotone function. Prove the inequality

$$Af(A) + Bf(B) \le \frac{1}{2}(A + B)^{1/2}(f(A) + f(B))(A + B)^{1/2}$$

for positive matrices A and B. (Hint: Use that f is matrix concave and $xf(x)$ is matrix convex.)

9. Show that the canonical representing measure in (5.24) for the standard matrix monotone function $f(x) = (x - 1)/\log x$ is the measure

$$d\mu(\lambda) = \frac{2}{(1 + \lambda)^2} d\lambda.$$

10. The function

$$\log_\alpha(x) = \frac{x^{1-\alpha} - 1}{1 - \alpha} \qquad (x > 0, \quad \alpha > 0, \quad \alpha \ne 1)$$

is called the α-logarithmic function. Is it matrix monotone?

11. Give an example of a matrix convex function such that the derivative is not matrix monotone.

12. Show that $f(z) = \tan z := \sin z / \cos z$ is in $\mathcal{P}$, where $\cos z := (e^{iz} + e^{-iz})/2$ and $\sin z := (e^{iz} - e^{-iz})/2i$.

13. Show that $f(z) = -1/z$ is in $\mathcal{P}$.

14. Show that the extreme points of the set

$$\mathcal{S}_n := \{D \in \mathbb{M}_n^{sa} : D \ge 0 \quad \text{and} \quad \text{Tr}\, D = 1\}$$

are the orthogonal projections of trace 1. Show that for $n > 2$ not all points in the boundary are extreme.

15. Let the block matrix

$$M = \begin{bmatrix} A & B \\ B^* & C \end{bmatrix}$$

be positive and $f : \mathbb{R}^+ \to \mathbb{R}$ be a convex function. Show that

$$\text{Tr}\, f(M) \ge \text{Tr}\, f(A) + \text{Tr}\, f(C).$$

16. Show that for $A, B \in \mathbb{M}_n^{sa}$ the inequality

$$\log \operatorname{Tr} e^{A+B} \geq \log \operatorname{Tr} e^A + \frac{\operatorname{Tr} B e^A}{\operatorname{Tr} e^A}$$

holds. (Hint: Use the function (4.2).)

17. Let the block matrix

$$M = \begin{bmatrix} A & B \\ B^* & C \end{bmatrix}$$

be positive and invertible. Show that

$$\det M \leq \det A \cdot \det C.$$

18. Show that for $A, B \in \mathbb{M}_n^{sa}$ the inequality

$$|\log \operatorname{Tr} e^{A+B} - \log \operatorname{Tr} e^A| \leq \|B\|$$

holds. (Hint: Use the function (4.2).)

19. Is it true that the function

$$\eta_\alpha(x) = \frac{x^\alpha - x}{1 - \alpha} \qquad (x > 0)$$

is matrix concave if $\alpha \in (0, 2)$?

Chapter 5
Matrix Means and Inequalities

The study of numerical means has been a popular subject for centuries, and the inequalities

$$\frac{2ab}{a+b} \le \sqrt{ab} \le \frac{a+b}{2}$$

between the harmonic, geometric and arithmetic means of positive numbers are well-known. When we move from 1×1 matrices (i.e. numbers) to $n \times n$ matrices, then the study of the arithmetic mean does not require any additional theory. Historically it was the harmonic mean that was the first non-trivial matrix mean to be investigated. From the point of view of some applications the name parallel sum, introduced in the late 1960s, was popular rather than 'harmonic matrix mean'. The geometric matrix mean appeared later in 1975, and the definition is not simple.

In the period 1791 until 1828, Carl Friedrich Gauss worked on the iteration:

$$a_0 := a, \quad b_0 := b,$$
$$a_{n+1} := \frac{a_n + b_n}{2}, \quad b_{n+1} := \sqrt{a_n b_n},$$

whose (joint) limit is now called the Gauss arithmetic-geometric mean $\mathbf{AG}(a, b)$. It has a non-trivial characterization:

$$\frac{1}{\mathbf{AG}(a, b)} = \frac{2}{\pi} \int_0^\infty \frac{dt}{\sqrt{(a^2 + t^2)(b^2 + t^2)}}.$$

In this chapter, we will first generalize the geometric mean to positive matrices and several other means will be studied in terms of matrix (or operator) monotone functions. There is also a natural (limit) definition for the mean of several (i.e. more than two) matrices, but its explicit description is rather hopeless.

F. Hiai and D. Petz, *Introduction to Matrix Analysis and Applications*, Universitext, DOI: 10.1007/978-3-319-04150-6_5,

187

5.1 The Geometric Mean

The definition of the geometric mean will be motivated by a property of geodesics on Riemannian manifolds.

The positive definite matrices can be considered as the variance of multivariate normal distributions and the information geometry of Gaussians yields a natural Riemannian metric. Those distributions (with 0 expectation) are given by a positive definite matrix $A \in \mathbb{M}_n$ in the form

$$f_A(x) := \frac{1}{(2\pi)^n \det A} \exp\left(-\langle A^{-1}x, x\rangle/2\right) \qquad (x \in \mathbb{C}^n).$$

The set $\mathcal{P}_n$ of positive definite $n \times n$ matrices can be considered as an open subset of the Euclidean space $\mathbb{R}^{n^2}$ and they form a manifold. The tangent vectors at a foot point $A \in \mathcal{P}_n$ are the self-adjoint matrices $\mathbb{M}_n^{sa}$.

A standard way to construct an information geometry is to start with an **information potential function** and to introduce the Riemannian metric via the Hessian of the potential. The information potential is the **Boltzmann entropy**

$$S(f_A) := -\int f_A(x) \log f_A(x)\, dx = C + \mathrm{Tr}\, \log A \qquad (C \text{ is a constant}).$$

The **Hessian** is

$$\frac{\partial^2}{\partial s \partial t} S(f_{A+tH_1+sH_2})\Big|_{t=s=0} = \mathrm{Tr}\, A^{-1}H_1 A^{-1}H_2$$

and the inner product on the tangent space at A is

$$g_A(H_1, H_2) = \mathrm{Tr}\, A^{-1}H_1 A^{-1}H_2.$$

We note here that this geometry has many symmetries, each congruence transformation of the matrices becomes a symmetry. Namely for any invertible matrix S,

$$g_{SAS^*}(SH_1S^*, SH_2S^*) = g_A(H_1, H_2). \tag{5.1}$$

A C^1 differentiable function $\gamma : [0, 1] \to \mathcal{P}_n$ is called a **curve**, its tangent vector at t is $\gamma'(t)$ and the length of the curve is

$$\int_0^1 \sqrt{g_{\gamma(t)}(\gamma'(t), \gamma'(t))}\, dt.$$

Given $A, B \in \mathcal{P}_n$ the curve

$$\gamma(t) = A^{1/2}(A^{-1/2}BA^{-1/2})^t A^{1/2} \qquad (0 \le t \le 1) \tag{5.2}$$

connects these two points: $\gamma(0) = A$, $\gamma(1) = B$. The next lemma says that this is the shortest curve connecting the two points, and is called the **geodesic** connecting A and B.

Lemma 5.1 *The geodesic connecting $A, B \in \mathcal{P}_n$ is (5.2) and the geodesic distance is*

$$\delta(A, B) = \| \log(A^{-1/2} B A^{-1/2}) \|_2 \,,$$

where $\| \cdot \|_2$ denotes the Hilbert–Schmidt norm.

Proof : By property (5.1) we may assume that $A = I$, then $\gamma(t) = B^t$. Let $\ell(t)$ be a curve in $\mathbb{M}_n^{sa}$ such that $\ell(0) = \ell(1) = 0$. This will be used for the perturbation of the curve $\gamma(t)$ in the form $\gamma(t) + \varepsilon\ell(t)$.

We want to differentiate the length

$$\int_0^1 \sqrt{g_{\gamma(t) + \varepsilon\ell(t)}(\gamma'(t) + \varepsilon\ell'(t), \gamma'(t) + \varepsilon\ell'(t))} \, dt$$

with respect to ε at $\varepsilon = 0$. When $\gamma(t) = B^t$ ($0 \le t \le 1$), note that

$$g_{\gamma(t)}(\gamma'(t), \gamma'(t)) = \operatorname{Tr} B^{-t} B^t (\log B) B^{-t} B^t \log B = \operatorname{Tr} (\log B)^2$$

does not depend on t. The derivative of the above integral at $\varepsilon = 0$ is

$$\int_0^1 \frac{1}{2} \Big(g_{\gamma(t)}(\gamma'(t), \gamma'(t))\Big)^{-1/2} \frac{\partial}{\partial \varepsilon} g_{\gamma(t) + \varepsilon\ell(t)}(\gamma'(t) + \varepsilon\ell'(t), \gamma'(t) + \varepsilon\ell'(t))\Big|_{\varepsilon = 0} dt$$

$$= \frac{1}{2\sqrt{\operatorname{Tr}(\log B)^2}} \times \int_0^1 \frac{\partial}{\partial \varepsilon} \operatorname{Tr}$$

$$(B^t + \varepsilon\ell(t))^{-1}(B^t \log B + \varepsilon\ell'(t))(B^t + \varepsilon\ell(t))^{-1}(B^t \log B + \varepsilon\ell'(t))\Big|_{\varepsilon = 0} dt$$

$$= \frac{1}{\sqrt{\operatorname{Tr}(\log B)^2}} \int_0^1 \operatorname{Tr} \big(-B^{-t}(\log B)^2 \ell(t) + B^{-t}(\log B)\ell'(t)\big) \, dt.$$

To remove $\ell'(t)$, we integrate the second term by parts:

$$\int_0^1 \operatorname{Tr} B^{-t}(\log B)\ell'(t) \, dt = \Big[\operatorname{Tr} B^{-t}(\log B)\ell(t)\Big]_0^1 + \int_0^1 \operatorname{Tr} B^{-t}(\log B)^2 \ell(t) \, dt \,.$$

Since $\ell(0) = \ell(1) = 0$, the first term vanishes here and the derivative at $\varepsilon = 0$ is 0 for every perturbation $\ell(t)$. Thus we conclude that $\gamma(t) = B^t$ is the geodesic curve between I and B. The distance is

$$\int_0^1 \sqrt{\operatorname{Tr}(\log B)^2} \, dt = \sqrt{\operatorname{Tr}(\log B)^2} = \| \log B \|_2.$$

The lemma is proved.   □

The midpoint of the curve (5.2) will be called the **geometric mean** of $A, B \in \mathcal{P}_n$ and denoted by $A\#B$, that is,

$$A\#B := A^{1/2}(A^{-1/2}BA^{-1/2})^{1/2}A^{1/2}. \tag{5.3}$$

The motivation is the fact that in the case where $AB = BA$, the midpoint is $\sqrt{AB}$. This geodesic approach will give an idea for the geometric mean of three matrices as well.

Let $A, B \geq 0$ and assume that A is invertible. We want to study the positivity of the matrix

$$\begin{bmatrix} A & X \\ X & B \end{bmatrix} \tag{5.4}$$

for a positive X. The positivity of the block matrix implies

$$B \geq XA^{-1}X,$$

see Theorem 2.1. From the matrix monotonicity of the square root function (Example 3.26), we obtain $(A^{-1/2}BA^{-1/2})^{1/2} \geq A^{-1/2}XA^{-1/2}$, or

$$A^{1/2}(A^{-1/2}BA^{-1/2})^{1/2}A^{1/2} \geq X.$$

It is easy to see that for $X = A\#B$, the block matrix (5.4) is positive. Therefore, $A\#B$ is the largest positive matrix X such that (5.4) is positive, that is,

$$A\#B = \max\left\{ X \geq 0 : \begin{bmatrix} A & X \\ X & B \end{bmatrix} \geq 0 \right\}. \tag{5.5}$$

The definition (5.3) is for invertible A. For a non-invertible A, an equivalent possibility is

$$A\#B := \lim_{\varepsilon \searrow 0}(A + \varepsilon I)\#B.$$

(The characterization with (5.4) remains true in this general case.) If $AB = BA$, then $A\#B = A^{1/2}B^{1/2}(= (AB)^{1/2})$. The inequality between geometric and arithmetic means also holds for matrices, see Exercise 5.6.

Example 5.2 The partial ordering $\leq$ of operators has a geometric interpretation for projections. The relation $P \leq Q$ is equivalent to ran $P \subset$ ran Q, that is, P projects to a smaller subspace than Q. This implies that any two projections P and Q have a largest lower bound denoted by $P \wedge Q$. This operator is the orthogonal projection onto the (closed) subspace ran $P \cap$ ran Q.

We want to show that $P\#Q = P \wedge Q$. First we show that the block matrix

$$\begin{bmatrix} P & P \wedge Q \\ P \wedge Q & Q \end{bmatrix}$$

is positive. This is equivalent to the relation

$$\begin{bmatrix} P + \varepsilon P^\perp & P \wedge Q \\ P \wedge Q & Q \end{bmatrix} \geq 0 \qquad (5.6)$$

for every constant $\varepsilon > 0$. Since

$$(P \wedge Q)(P + \varepsilon P^\perp)^{-1}(P \wedge Q) = P \wedge Q$$

is smaller than Q, the positivity (5.6) follows from Theorem 2.1. We conclude that $P\#Q \geq P \wedge Q$.

The positivity of

$$\begin{bmatrix} P + \varepsilon P^\perp & X \\ X & Q \end{bmatrix}$$

gives the condition

$$Q \geq X(P + \varepsilon^{-1}P^\perp)X = XPX + \varepsilon^{-1}XP^\perp X.$$

Since $\varepsilon > 0$ is arbitrary, $XP^\perp X = 0$. The latter condition gives $X = XP$. Therefore, $Q \geq X^2$. Symmetrically, $P \geq X^2$ and Corollary 2.25 tells us that $P \wedge Q \geq X^2$ and so $P \wedge Q \geq X$. □

Theorem 5.3 Assume that A_1, A_2, B_1, B_2 are positive matrices and $A_1 \leq B_1$, $A_2 \leq B_2$. Then $A_1\#A_2 \leq B_1\#B_2$.

Proof : The statement is equivalent to the positivity of the block matrix

$$\begin{bmatrix} B_1 & A_1\#A_2 \\ A_1\#A_2 & B_2 \end{bmatrix}.$$

This is a sum of positive matrices:

$$\begin{bmatrix} A_1 & A_1\#A_2 \\ A_1\#A_2 & A_2 \end{bmatrix} + \begin{bmatrix} B_1 - A_1 & 0 \\ 0 & B_2 - A_2 \end{bmatrix}.$$

The proof is complete.   □

The next theorem is the **Löwner–Heinz inequality** already given in Sect. 4.4. The present proof is based on the geometric mean.

Theorem 5.4 *Assume that for matrices A and B the inequalities $0 \leq A \leq B$ hold and $0 < t < 1$ is a real number. Then $A^t \leq B^t$.*

Proof : By continuity, it is enough to prove the case $t = k/2^n$, that is, when t is a dyadic rational number. We use Theorem 5.3 to deduce from the inequalities $A \leq B$ and $I \leq I$ the inequality

$$A^{1/2} = A\#I \leq B\#I = B^{1/2}.$$

A second application of Theorem 5.3 similarly gives $A^{1/4} \leq B^{1/4}$ and $A^{3/4} \leq B^{3/4}$. The procedure can be continued to cover all dyadic rational numbers. The result for arbitrary $t \in [0, 1]$ follows by taking limits of dyadic numbers.   □

Theorem 5.5 *The geometric mean of matrices is jointly concave, that is,*

$$\frac{A_1 + A_2}{2} \# \frac{A_3 + A_4}{2} \geq \frac{A_1 \# A_3 + A_2 \# A_4}{2}.$$

Proof : The block matrices

$$\begin{bmatrix} A_1 & A_1\#A_2 \\ A_1\#A_2 & A_2 \end{bmatrix} \quad \text{and} \quad \begin{bmatrix} A_3 & A_3\#A_4 \\ A_4\#A_3 & A_4 \end{bmatrix}$$

are positive and so is the arithmetic mean,

$$\begin{bmatrix} \frac{1}{2}(A_1 + A_3) & \frac{1}{2}(A_1\#A_2 + A_3\#A_4) \\ \frac{1}{2}(A_1\#A_2 + A_3\#A_4) & \frac{1}{2}(A_2 + A_4) \end{bmatrix}.$$

Therefore the off-diagonal entry is smaller than the geometric mean of the diagonal entries.   □

Note that the joint concavity property is equivalent to the slightly simpler formula

$$(A_1 + A_2)\#(A_3 + A_4) \geq (A_1\#A_3) + (A_2\#A_4).$$

We shall use this inequality later.

The next theorem of Ando [7] is a generalization of Example 5.2. For the sake of simplicity the formulation is given in terms of block matrices.

Theorem 5.6 *Take an ortho-projection P and a positive invertible matrix R:*

$$P = \begin{bmatrix} I & 0 \\ 0 & 0 \end{bmatrix}, \quad R = \begin{bmatrix} R_{11} & R_{12} \\ R_{21} & R_{22} \end{bmatrix}.$$

The geometric mean of P and R is the following:

$$P \# R = (PR^{-1}P)^{-1/2} = \begin{bmatrix} (R_{11} - R_{12}R_{22}^{-1}R_{21})^{-1/2} & 0 \\ 0 & 0 \end{bmatrix}.$$

Proof : P and R are already given in block matrix form. By (5.5) we are looking for positive matrices

$$X = \begin{bmatrix} X_{11} & X_{12} \\ X_{21} & X_{22} \end{bmatrix}$$

such that

$$\begin{bmatrix} P & X \\ X & R \end{bmatrix} = \begin{bmatrix} I & 0 & X_{11} & X_{12} \\ 0 & 0 & X_{21} & X_{22} \\ X_{11} & X_{12} & R_{11} & R_{12} \\ X_{21} & X_{22} & R_{21} & R_{22} \end{bmatrix}$$

is positive. From the positivity it follows that $X_{12} = X_{21} = X_{22} = 0$ and the necessary and sufficient condition is

$$\begin{bmatrix} I & 0 \\ 0 & 0 \end{bmatrix} \geq \begin{bmatrix} X_{11} & 0 \\ 0 & 0 \end{bmatrix} R^{-1} \begin{bmatrix} X_{11} & 0 \\ 0 & 0 \end{bmatrix},$$

or

$$I \geq X_{11}(R^{-1})_{11}X_{11}.$$

The latter is equivalent to $I \geq ((R^{-1})_{11})^{1/2} X_{11}^2 ((R^{-1})_{11})^{1/2}$, which implies that

$$X_{11} \leq ((R^{-1})_{11})^{-1/2}.$$

The inverse of a block matrix is described in (2.4) and the proof is complete.   □

For projections P and Q, the theorem and Example 5.2 give

$$P \# Q = P \wedge Q = \lim_{\varepsilon \to +0} (P(Q + \varepsilon I)^{-1}P)^{-1/2}.$$

The arithmetic mean of several matrices is simpler: for (positive) matrices $A_1, A_2, \ldots, A_n$ it is

$$\mathbf{A}(A_1, A_2, \ldots, A_n) := \frac{A_1 + A_2 + \cdots + A_n}{n}.$$

Only the linear structure plays a role. The arithmetic mean is a good example illustrating how to move from the means of two variables to means of three variables.

Suppose we have a device which can compute the mean of two matrices. How to compute the mean of three? Assume that we aim to obtain the mean of A, B and C. In the case of the arithmetic mean, we can introduce a new device

$$W : (A, B, C) \mapsto (\mathbf{A}(A, B), \mathbf{A}(A, C), \mathbf{A}(B, C)),$$

which, applied to (A, B, C) many times, gives the mean of A, B and C:

$$W^n(A, B, C) = (A_n, B_n, C_n)$$

and

$$A_n, \ B_n, \ C_n \to \mathbf{A}(A, B, C) \quad \text{as} \quad n \to \infty.$$

Indeed, A_n, B_n, C_n are convex combinations of A, B and C, so

$$A_n = \lambda_1^{(n)} A + \lambda_2^{(n)} B + \lambda_3^{(n)} C.$$

One can compute the coefficients $\lambda_i^{(n)}$ explicitly and show that $\lambda_i^{(n)} \to 1/3$, and likewise for B_n and C_n. The idea is illustrated in the following picture and will be extended to the geometric mean (Fig. 5.1).

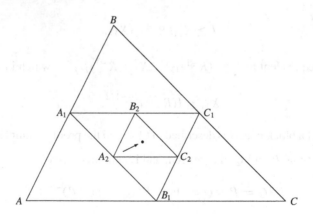

Fig. 5.1 The triangles of A_n, B_n, C_n

Theorem 5.7 *Let A, B, $C \in \mathbb{M}_n$ be positive definite matrices and define a recursion by*

$$A_0 = A, \qquad B_0 = B, \qquad C_0 = C,$$

$$A_{m+1} = A_m \# B_m, \qquad B_{m+1} = A_m \# C_m, \qquad C_{m+1} = B_m \# C_m.$$

Then the limits

$$\mathbf{G}_3(A, B, C) := \lim_m A_m = \lim_m B_m = \lim_m C_m \qquad (5.7)$$

exist.

Proof : First we assume that $A \leq B \leq C$. From the monotonicity property of the geometric mean, see Theorem 5.3, we obtain that $A_m \leq B_m \leq C_m$. It follows that the sequence (A_m) is increasing and (C_m) is decreasing. Therefore, the limits

$$L := \lim_{m \to \infty} A_m \qquad \text{and} \qquad U = \lim_{m \to \infty} C_m$$

exist, and $L \leq U$. We claim that $L = U$.

By continuity, $B_m \to L\#U =: M$, where $L \leq M \leq U$. Since

$$B_m \# C_m = C_{m+1},$$

the limit $m \to \infty$ gives $M\#U = U$. Therefore $M = U$ and so $U = L$.

The general case can be reduced to the case of an ordered triplet. If A, B, C are arbitrary, we can find numbers λ and μ such that $A \leq \lambda B \leq \mu C$ and use the formula

$$(\alpha X)\#(\beta Y) = \sqrt{\alpha\beta}(X\#Y) \qquad (5.8)$$

for positive numbers α and β.

Let

$$A_1' = A, \qquad B_1' = \lambda B, \qquad C_1' = \mu C,$$

and

$$A_{m+1}' = A_m' \# B_m', \qquad B_{m+1}' = A_m' \# C_m', \qquad C_{m+1}' = B_m' \# C_m'.$$

It is clear that for the numbers

$$a := 1, \qquad b := \lambda \qquad \text{and} \qquad c := \mu$$

the recursion provides a convergent sequence (a_m, b_m, c_m) of triplets:

$$(\lambda\mu)^{1/3} = \lim_m a_m = \lim_m b_m = \lim_m c_m.$$

Since

$$A_m = A_m'/a_m, \qquad B_m = B_m'/b_m \qquad \text{and} \qquad C_m = C_m'/c_m$$

by property (5.8) of the geometric mean, the limits stated in the theorem must exist and equal $\mathbf{G}(A', B', C')/(\lambda\mu)^{1/3}$.   □

The geometric mean of the positive definite matrices $A, B, C \in \mathbb{M}_n$ is defined as $\mathbf{G}_3(A, B, C)$ in (5.7). An explicit formula is not known and the same kind of procedure can be used to define the geometric mean of k matrices. If $P_1, P_2, \ldots, P_k$ are ortho-projections, then Example 5.2 gives the limit

$$\mathbf{G}_k(P_1, P_2, \ldots, P_k) = P_1 \wedge P_2 \wedge \cdots \wedge P_k .$$

5.2 General Theory

The first example is the parallel sum which is a constant multiple of the harmonic mean.

Example 5.8 It is a well-known result in electronics that if two resistors with resistance a and b are connected in parallel, then the total resistance q is the solution of the equation

$$\frac{1}{q} = \frac{1}{a} + \frac{1}{b}.$$

Then

$$q = (a^{-1} + b^{-1})^{-1} = \frac{ab}{a+b}$$

is the harmonic mean up to a factor 2. More generally, one can consider n-point networks, where the voltage and current vectors are connected by a positive matrix. The **parallel sum**

$$A : B = (A^{-1} + B^{-1})^{-1}$$

of two positive definite matrices represents the combined resistance of two n-port networks connected in parallel.

One can check that

$$A : B = A - A(A + B)^{-1}A.$$

Therefore $A : B$ is the **Schur complement** of $A + B$ in the block matrix

$$\begin{bmatrix} A & A \\ A & A + B \end{bmatrix},$$

see Theorem 2.4.

It is easy to see that if $0 < A \leq C$ and $0 < B \leq D$, then $A : B \leq C : D$. The parallel sum can be extended to all positive matrices:

$$A : B = \lim_{\varepsilon \searrow 0} (A + \varepsilon I) : (B + \varepsilon I).$$

Note that all matrix means can be expressed as an integral of parallel sums (see Theorem 5.11 below). $\square$

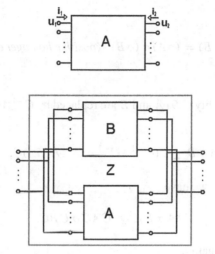

Above: *An n-point network with the input and output voltage vectors.* Below: *Two parallelly connected networks*

On the basis of the previous example, the **harmonic mean** of the positive matrices A and B is defined as

$$H(A, B) := 2(A : B).$$

Assume that for all positive matrices A, B (of the same size) the matrix $A \, \sigma \, B$ is defined. Then σ is called an **operator connection** if it satisfies the following conditions:

(i) $0 \leq A \leq C$ and $0 \leq B \leq D$ imply

$$A \, \sigma \, B \leq C \, \sigma \, D \qquad \text{(joint monotonicity)};$$

(ii) if $0 \leq A, B$ and $C = C^*$, then

$$C(A \, \sigma \, B)C \leq (CAC) \, \sigma \, (CBC) \qquad \text{(transformer inequality)}; \qquad (5.9)$$

(iii) if $0 \le A_n, B_n$ and $A_n \searrow A$, $B_n \searrow B$ then

$$(A_n \sigma B_n) \searrow (A \sigma B) \qquad \text{(upper semi-continuity)}.$$

The parallel sum is an example of an operator connection.

Lemma 5.9 *Assume that σ is an operator connection. If $C = C^*$ is invertible, then*

$$C(A \sigma B)C = (CAC) \sigma (CBC), \qquad (5.10)$$

and for every $\alpha \ge 0$

$$\alpha(A \sigma B) = (\alpha A) \sigma (\alpha B) \quad \text{(positive homogeneity)} \qquad (5.11)$$

holds.

Proof: In the inequality (5.9) A and B are replaced by $C^{-1}AC^{-1}$ and $C^{-1}BC^{-1}$, respectively:

$$A \sigma B \ge C(C^{-1}AC^{-1} \sigma C^{-1}BC^{-1})C.$$

Replacing C with C^{-1}, we have

$$C(A \sigma B)C \ge CAC \sigma CBC.$$

This and (5.9) yield equality.

When $\alpha > 0$, letting $C := \alpha^{1/2}I$ in (5.10) implies (5.11). When $\alpha = 0$, let $0 < \alpha_n \searrow 0$. Then $(\alpha_n I) \sigma (\alpha_n I) \searrow 0 \sigma 0$ by (iii) above while $(\alpha_n I) \sigma (\alpha_n I) = \alpha_n (I \sigma I) \searrow 0$. Hence $0 = 0 \sigma 0$, which is (5.11) for $\alpha = 0$. $\qquad \square$

The next fundamental theorem of Kubo and Ando says that there is a one-to-one correspondence between operator connections and matrix monotone functions on $[0, \infty)$.

Theorem 5.10 (Kubo–Ando theorem) *For each operator connection σ there exists a unique matrix monotone function $f : \mathbb{R}^+ \to \mathbb{R}^+$ such that*

$$f(t)I = I \sigma (tI) \qquad (t \in \mathbb{R}^+) \qquad (5.12)$$

and for $A > 0$ and $B \ge 0$ the formula

$$A \sigma B = A^{1/2} f(A^{-1/2}BA^{-1/2})A^{1/2} = f(BA^{-1})A \qquad (5.13)$$

holds, where the last term is defined via the analytic functional calculus.

Proof: Let σ be an operator connection. First we show that if an ortho-projection P commutes with A and B, then P commutes with $A \sigma B$ and

$$((AP)\sigma(BP))P = (A\sigma B)P. \tag{5.14}$$

Since $PAP = AP \le A$ and $PBP = BP \le B$, it follows from (ii) and (i) of the definition of σ that

$$P(A\sigma B)P \le (PAP)\sigma(PBP) = (AP)\sigma(BP) \le A\sigma B. \tag{5.15}$$

Hence $(A\sigma B - P(A\sigma B)P)^{1/2}$ exists so that

$$\left|(A\sigma B - P(A\sigma B)P)^{1/2}P\right|^2 = P(A\sigma B - P(A\sigma B)P)P = 0.$$

Therefore, $(A\sigma B - P(A\sigma B)P)^{1/2}P = 0$ and so $(A\sigma B)P = P(A\sigma B)P$. This implies that P commutes with $A\sigma B$. Similarly, P commutes with $(AP)\sigma(BP)$ as well, and (5.14) follows from (5.15). For every $t \ge 0$, since $I\sigma(tI)$ commutes with all ortho-projections, it is a scalar multiple of I. Thus, we see that there is a function $f \ge 0$ on $[0, \infty)$ satisfying (5.12). The uniqueness of such a function f is obvious, and it follows from (iii) of the definition of the operator connection that f is right-continuous for $t \ge 0$. Since $t^{-1}f(t)I = (t^{-1}I)\sigma I$ for $t > 0$ thanks to (5.11), it follows from (iii) of the definition again that $t^{-1}f(t)$ is left-continuous for $t > 0$ and so is $f(t)$. Hence f is continuous on $[0, \infty)$.

To show the operator monotonicity of f, let us prove that

$$f(A) = I\sigma A. \tag{5.16}$$

Let $A = \sum_{i=1}^{m} \alpha_i P_i$, where $\alpha_i > 0$ and P_i are projections with $\sum_{i=1}^{m} P_i = I$. Since each P_i commutes with A, using (5.14) twice we have

$$I\sigma A = \sum_{i=1}^{m}(I\sigma A)P_i = \sum_{i=1}^{m}(P_i\sigma(AP_i))P_i = \sum_{i=1}^{m}(P_i\sigma(\alpha_i P_i))P_i$$
$$= \sum_{i=1}^{m}(I\sigma(\alpha_i I))P_i = \sum_{i=1}^{m}f(\alpha_i)P_i = f(A).$$

For general $A \ge 0$ choose a sequence $0 < A_n$ of the above form such that $A_n \searrow A$. By upper semi-continuity we have

$$I\sigma A = \lim_{n\to\infty} I\sigma A_n = \lim_{n\to\infty} f(A_n) = f(A).$$

So (5.16) is shown. Hence, if $0 \le A \le B$, then

$$f(A) = I\sigma A \le I\sigma B = f(B)$$

and we conclude that f is matrix monotone.

When A is invertible, we can use (5.10):

$$A \sigma B = A^{1/2}(I \sigma A^{-1/2}BA^{-1/2})A^{1/2} = A^{1/2}f(A^{-1/2}BA^{-1/2})A^{1/2}$$

and the first part of (5.13) is obtained. The rest is a general property.   □

Note that the general formula is

$$A \sigma B = \lim_{\varepsilon \searrow 0} A_\varepsilon \sigma B_\varepsilon = \lim_{\varepsilon \searrow 0} A_\varepsilon^{1/2} f(A_\varepsilon^{-1/2} B_\varepsilon A_\varepsilon^{-1/2}) A_\varepsilon^{1/2},$$

where $A_\varepsilon := A + \varepsilon I$ and $B_\varepsilon := B + \varepsilon I$. We call f the **representing function** of σ. For scalars $s, t > 0$ we have $s \sigma t = sf(t/s)$.

The next theorem follows from the integral representation of matrix monotone functions and from the previous theorem.

Theorem 5.11 *Every operator connection σ has an integral representation*

$$A \sigma B = aA + bB + \int_{(0,\infty)} \frac{1+\lambda}{\lambda}\Big((\lambda A) : B\Big)\, d\mu(\lambda) \qquad (A, B \geq 0),$$

where μ is a positive finite Borel measure on $[0, \infty)$.

Using this integral expression, one can often derive properties of general operator connections by checking them for the parallel sum.

Lemma 5.12 *For every vector z,*

$$\inf\{\langle x, Ax \rangle + \langle y, By \rangle : x + y = z\} = \langle z, (A : B)z \rangle.$$

Proof : When A, B are invertible, we have

$$A : B = \Big(B^{-1}(A+B)A^{-1}\Big)^{-1} = \Big((A+B) - B\Big)(A+B)^{-1}B = B - B(A+B)^{-1}B.$$

For all vectors x, y we have

$$\begin{aligned}
\langle x, Ax \rangle &+ \langle z - x, B(z-x) \rangle - \langle z, (A:B)z \rangle \\
&= \langle z, Bz \rangle + \langle x, (A+B)x \rangle - 2\mathrm{Re}\,\langle x, Bz \rangle - \langle z, (A:B)z \rangle \\
&= \langle z, B(A+B)^{-1}Bz \rangle + \langle x, (A+B)x \rangle - 2\mathrm{Re}\,\langle x, Bz \rangle \\
&= \|(A+B)^{-1/2}Bz\|^2 + \|(A+B)^{1/2}x\|^2 \\
&\quad -2\mathrm{Re}\,\langle (A+B)^{1/2}x, (A+B)^{-1/2}Bz \rangle \geq 0.
\end{aligned}$$

In particular, the above is equal to 0 if $x = (A+B)^{-1}Bz$. Hence the assertion follows if $A, B > 0$. For general A, B,

$$\langle z, (A : B)z \rangle = \inf_{\varepsilon > 0} \langle z, \left((A + \varepsilon I) : (B + \varepsilon I)\right)z \rangle$$

$$= \inf_{\varepsilon > 0} \inf_{y} \left\{ \langle x, (A + \varepsilon I)x \rangle + \langle z - x, (B + \varepsilon I)(z - x) \rangle \right\}$$

$$= \inf_{y} \left\{ \langle x, Ax \rangle + \langle z - x, B(z - x) \rangle \right\}.$$

The proof is complete.   □

The next result is called the **transformer inequality**, it is a stronger version of (5.9).

Theorem 5.13 *For all $A, B \geq 0$ and general S,*

$$S^*(A \sigma B)S \leq (S^* A S) \sigma (S^* B S)$$

and equality holds if S is invertible.

Proof: For $z = x + y$ Lemma 5.12 implies

$$\langle z, S^*(A : B)Sz \rangle = \langle Sz, (A : B)Sz \rangle \leq \langle Sx, ASx \rangle + \langle Sy, BSy \rangle$$
$$= \langle x, S^* ASx \rangle + \langle y, S^* BSy \rangle.$$

Hence $S^*(A : B)S \leq (S^* AS) : (S^* BS)$ follows. The statement of the theorem is true for the parallel sum and by Theorem 5.11 we obtain it for any operator connection. The proof of the last assertion is similar to that of Lemma 5.9.   □

A very similar argument gives the **joint concavity**:

$$(A \sigma B) + (C \sigma D) \leq (A + C) \sigma (B + D).$$

The next theorem concerns a recursively defined double sequence.

Theorem 5.14 *Let σ_1 and σ_2 be operator connections dominated by the arithmetic mean. For positive matrices A and B define a recursion*

$$A_1 = A, \quad B_1 = B, \quad A_{k+1} = A_k \sigma_1 B_k, \quad B_{k+1} = A_k \sigma_2 B_k. \tag{5.17}$$

Then (A_k) and (B_k) converge to the same operator connection $A \sigma B$.

Proof: First we prove the convergence of (A_k) and (B_k). From the inequality

$$X \sigma_i Y \leq \frac{X + Y}{2}$$

we have

$$A_{k+1} + B_{k+1} = A_k \sigma_1 B_k + A_k \sigma_2 B_k \leq A_k + B_k.$$

Therefore the decreasing positive sequence has a limit:

$$A_k + B_k \to X \quad \text{as } k \to \infty.$$

Moreover,

$$a_{k+1} := \|A_{k+1}\|_2^2 + \|B_{k+1}\|_2^2 \leq \|A_k\|_2^2 + \|B_k\|_2^2 - \frac{1}{2}\|A_k - B_k\|_2^2,$$

where $\|X\|_2 = (\operatorname{Tr} X^* X)^{1/2}$, the Hilbert–Schmidt norm. The numerical sequence a_k is decreasing, it has a limit and it follows that

$$\|A_k - B_k\|_2^2 \to 0$$

and $A_k, B_k \to X/2$ as $k \to \infty$.

For each k, A_k and B_k are operator connections of the matrices A and B, and the limit is an operator connection as well. $\qquad\square$

Example 5.15 At the end of the eighteenth century J.-L. Lagrange and C.F. Gauss became interested in the arithmetic-geometric mean of positive numbers. Gauss worked on this subject in the period 1791 until 1828.

As already mentioned in the introduction to this chapter, with the initial conditions

$$a_1 = a, \qquad b_1 = b$$

define the recursion

$$a_{n+1} = \frac{a_n + b_n}{2}, \quad b_{n+1} = \sqrt{a_n b_n}.$$

Then the (joint) limit is the so-called **Gauss arithmetic-geometric mean** $AG(a, b)$ with the characterization

$$\frac{1}{AG(a,b)} = \frac{2}{\pi} \int_0^\infty \frac{dt}{\sqrt{(a^2 + t^2)(b^2 + t^2)}},$$

see [34]. It follows from Theorem 5.14 that the Gauss arithmetic-geometric mean can also be defined for matrices. Therefore the function $f(x) = AG(1, x)$ is a matrix monotone function. $\qquad\square$

It is interesting to note that (5.17) can be modified slightly:

$$A_1 = A, \quad B_1 = B, \qquad A_{k+1} = A_k \,\sigma_1\, B_k, \quad B_{k+1} = A_{k+1} \,\sigma_2\, B_k. \qquad (5.18)$$

A similar proof gives the existence of the limit. (5.17) is called the **Gaussian double-mean process**, while (5.18) is the **Archimedean double-mean process**.

The symmetric **matrix means** are binary operations on positive matrices. They are operator connections with the properties $A \sigma A = A$ and $A \sigma B = B \sigma A$. For matrix means we shall use the notation $m(A, B)$. We repeat the main properties:

(1) $m(A, A) = A$ for every A;
(2) $m(A, B) = m(B, A)$ for every A and B;
(3) if $A \leq B$, then $A \leq m(A, B) \leq B$;
(4) if $A \leq A'$ and $B \leq B'$, then $m(A, B) \leq m(A', B')$;
(5) m is upper semi-continuous;
(6) $C m(A, B) C^* \leq m(CAC^*, CBC^*)$.

It follows from the Kubo–Ando theorem (Theorem 5.10) that the operator means are in a one-to-one correspondence with matrix monotone functions $\mathbb{R}^+ \to \mathbb{R}^+$ satisfying conditions $f(1) = 1$ and $tf(t^{-1}) = f(t)$. Such a matrix monotone function on $\mathbb{R}^+$ is said to be **standard**. Given a matrix monotone function f, the corresponding mean is

$$m_f(A, B) = A^{1/2} f(A^{-1/2} B A^{-1/2}) A^{1/2} \tag{5.19}$$

when A is invertible. (When A is not invertible, take a sequence A_n of invertible operators approximating A such that $A_n \searrow A$ and let $m_f(A, B) = \lim_n m_f(A_n, B)$.) It follows from the definition (5.19) of means that if $f \leq g$, then $m_f(A, B) \leq m_g(A, B)$.

Theorem 5.16 *If* $f : \mathbb{R}^+ \to \mathbb{R}^+$ *is a standard matrix monotone function, then*

$$\frac{2x}{x+1} \leq f(x) \leq \frac{x+1}{2}.$$

Proof: By differentiating the formula $f(x) = xf(x^{-1})$, we obtain $f'(1) = 1/2$. Since $f(1) = 1$, the concavity of the function f gives $f(x) \leq (1+x)/2$.

If f is a standard matrix monotone function, then so is $f(x^{-1})^{-1}$. The inequality $f(x^{-1})^{-1} \leq (1+x)/2$ gives $f(x) \geq 2x/(x+1)$. □

If $f(x)$ is a standard matrix monotone function with matrix mean $m(\cdot, \cdot)$, then the matrix mean corresponding to $x/f(x)$ is called the **dual** of $m(\cdot, \cdot)$ and is denoted by $m^\perp(\cdot, \cdot)$. For instance, the dual of the arithmetic mean is the harmonic mean and $\#^\perp = \#$.

The next theorem is a Trotter-like product formula for matrix means.

Theorem 5.17 *For a symmetric matrix mean m and for self-adjoint A, B we have*

$$\lim_{n \to \infty} m(e^{A/n}, e^{B/n})^n = \exp \frac{A+B}{2}.$$

Proof: It is an exercise to prove that

$$\lim_{t \to 0} \frac{m(e^{tA}, e^{tB}) - I}{t} = \frac{A+B}{2}.$$

The choice $t = 1/n$ gives

$$\exp\left(-n(I - m(e^{A/n}, e^{B/n}))\right) \to \exp\frac{A + B}{2}.$$

So it is enough to show that

$$D_n := m(e^{A/n}, e^{B/n})^n - \exp\left(-n(I - m(e^{A/n}, e^{B/n}))\right) \to 0$$

as $n \to \infty$. If A is replaced by $A + aI$ and B is replaced by $B + aI$ for some real number a, then D_n does not change. Therefore we can assume $A, B \le 0$.

We use the abbreviation $F(n) := m(e^{A/n}, e^{B/n})$, so

$$D_n = F(n)^n - \exp(-n(I - F(n))) = F(n)^n - e^{-n}\sum_{k=0}^{\infty}\frac{n^k}{k!}F(n)^k$$

$$= e^{-n}\sum_{k=0}^{\infty}\frac{n^k}{k!}F(n)^n - e^{-n}\sum_{k=0}^{\infty}\frac{n^k}{k!}F(n)^k = e^{-n}\sum_{k=0}^{\infty}\frac{n^k}{k!}\left(F(n)^n - F(n)^k\right).$$

Since $F(n) \le I$, we have

$$\|D_n\| \le e^{-n}\sum_{k=0}^{\infty}\frac{n^k}{k!}\|F(n)^n - F(n)^k\| \le e^{-n}\sum_{k=0}^{\infty}\frac{n^k}{k!}\|I - F(n)^{|k-n|}\|.$$

Since

$$0 \le I - F(n)^{|k-n|} \le |k - n|(I - F(n)),$$

it follows that

$$\|D_n\| \le e^{-n}\|I - F(n)\|\sum_{k=0}^{\infty}\frac{n^k}{k!}|k - n|.$$

The Schwarz inequality gives that

$$\sum_{k=0}^{\infty}\frac{n^k}{k!}|k - n| \le \left(\sum_{k=0}^{\infty}\frac{n^k}{k!}\right)^{1/2}\left(\sum_{k=0}^{\infty}\frac{n^k}{k!}(k - n)^2\right)^{1/2} = n^{1/2}e^n.$$

So we have

$$\|D_n\| \le n^{-1/2}\|n(I - F(n))\|.$$

Since $\|n(I - F(n))\|$ is bounded, the limit is really 0. $\square$

For the geometric mean the previous theorem gives the Lie–Trotter formula, see Theorem 3.8.

Theorem 5.7 concerns the geometric mean of several matrices and it can be extended for arbitrary symmetric means. The proof is due to Miklós Pálfia and makes use of the Hilbert–Schmidt norm $\|X\|_2 = (\operatorname{Tr} X^*X)^{1/2}$.

Theorem 5.18 *Let $m(\,\cdot\,,\,\cdot\,)$ be a symmetric matrix mean and $0 \le A, B, C \in \mathbb{M}_n$. Define a recursion:*

(1) $A^{(0)} := A, \quad B^{(0)} := B, \quad C^{(0)} := C;$
(2) $A^{(k+1)} := m(A^{(k)}, B^{(k)}), \quad B^{(k+1)} := m(A^{(k)}, C^{(k)}) \quad and \quad C^{(k+1)} := m(B^{(k)}, C^{(k)}).$

Then the limits $\lim_m A^{(m)} = \lim_m B^{(m)} = \lim_m C^{(m)}$ exist. This common value will be denoted by $m(A, B, C)$.

Proof : From the well-known inequality

$$m(X, Y) \le \frac{X + Y}{2} \tag{5.20}$$

we have

$$A^{(k+1)} + B^{(k+1)} + C^{(k+1)} \le A^{(k)} + B^{(k)} + C^{(k)}.$$

Therefore the decreasing positive sequence has a limit:

$$A^{(k)} + B^{(k)} + C^{(k)} \to X \text{ as } k \to \infty. \tag{5.21}$$

It also follows from (5.20) that

$$\|m(C, D)\|_2^2 \le \frac{\|C\|_2^2 + \|D\|_2^2}{2} - \frac{1}{4}\|C - D\|_2^2.$$

Therefore,

$$a_{k+1} := \|A^{(k+1)}\|_2^2 + \|B^{(k+1)}\|_2^2 + \|C^{(k+1)}\|_2^2$$

$$\le \|A^{(k)}\|_2^2 + \|B^{(k)}\|_2^2 + \|C^{(k)}\|_2^2$$

$$- \frac{1}{4}\Big(\|A^{(k)} - B^{(k)}\|_2^2 + \|B^{(k)} - C^{(k)}\|_2^2 + \|C^{(k)} - A^{(k)}\|_2^2\Big)$$

$$=: a_k - c_k.$$

Since the numerical sequence a_k is decreasing, it has a limit and it follows that $c_k \to 0$. Therefore,

$$A^{(k)} - B^{(k)} \to 0, \qquad A^{(k)} - C^{(k)} \to 0.$$

Combining formulas with (5.21), we have

$$A^{(k)} \to \frac{1}{3}X \text{ as } k \to \infty.$$

A similar convergence holds for $B^{(k)}$ and $C^{(k)}$.   □

Theorem 5.19 *The mean* $m(A, B, C)$ *defined in Theorem 5.18 has the following properties:*

(1) $m(A, A, A) = A$ *for every* A;
(2) $m(A, B, C) = m(B, A, C) = m(C, A, B)$ *for every* A, B *and* C;
(3) *if* $A \le B \le C$, *then* $A \le m(A, B, C) \le C$;
(4) *if* $A \le A'$, $B \le B'$ *and* $C \le C'$, *then* $m(A, B, C) \le m(A', B', C')$;
(5) m *is upper semi-continuous;*
(6) $D m(A, B, C) D^* \le m(DAD^*, DBD^*, DCD^*)$ *and equality holds if* D *is invertible.*

The above properties can be shown from convergence arguments based on Theorem 5.18. The details are omitted here.

Example 5.20 If P_1, P_2, P_3 are ortho-projections, then

$$m(P_1, P_2, P_3) = P_1 \wedge P_2 \wedge P_3$$

holds for several means, see Example 5.23.

Now we consider the geometric mean $G_3(A, A, B)$. If $A > 0$, then

$$G_3(A, A, B) = A^{1/2} G_3(I, I, A^{-1/2} B A^{-1/2}) A^{1/2}.$$

Since $I, I, A^{-1/2} B A^{-1/2}$ are commuting matrices, it is easy to compute the geometric mean. We have

$$G_3(A, A, B) = A^{1/2} (A^{-1/2} B A^{-1/2})^{1/3} A^{1/2}.$$

This is an example of a **weighted geometric mean**:

$$G_t(A, B) = A^{1/2} (A^{-1/2} B A^{-1/2})^t A^{1/2} \qquad (0 < t < 1).$$

There is a general theory of weighted geometric means.   □

5.3 Mean Examples

Recall that a matrix monotone function $f : \mathbb{R}^+ \to \mathbb{R}^+$ is called **standard** if $f(1) = 1$ and $xf(x^{-1}) = f(x)$. Standard functions are used to define matrix means in (5.19).

Here are some familiar standard matrix monotone functions:

$$\frac{2x}{x+1} \leq \sqrt{x} \leq \frac{x-1}{\log x} \leq \frac{x+1}{2}.$$

The corresponding increasing means are the harmonic, geometric, logarithmic and arithmetic means. By Theorem 5.16 we see that the harmonic mean is the smallest and the arithmetic mean is the largest among the symmetric matrix means.

First we study the **harmonic mean** $H(A, B)$. A variational expression is expressed in terms of 2×2 block matrices.

Theorem 5.21

$$H(A, B) = \max \left\{ X \geq 0 : \begin{bmatrix} 2A & 0 \\ 0 & 2B \end{bmatrix} \geq \begin{bmatrix} X & X \\ X & X \end{bmatrix} \right\}.$$

Proof: The inequality of the two block matrices is equivalently written as

$$\langle x, 2Ax \rangle + \langle y, 2By \rangle \geq \langle x + y, X(x+y) \rangle.$$

Therefore the proof is reduced to Lemma 5.12, where $x + y$ is z and $H(A, B) = 2(A : B)$. $\qquad\square$

Recall the **geometric mean**

$$G(A, B) = A\#B = A^{1/2}(A^{-1/2}BA^{-1/2})^{1/2}A^{1/2}$$

which corresponds to $f(x) = \sqrt{x}$. The mean $A\#B$ is the unique positive solution to the equation $XA^{-1}X = B$ and therefore $(A\#B)^{-1} = A^{-1}\#B^{-1}$.

Example 5.22 The function

$$f(x) = \frac{x-1}{\log x}$$

is matrix monotone due to the formula

$$\int_0^1 x^t \, dt = \frac{x-1}{\log x}.$$

The standard property is obvious. The matrix mean induced by the function $f(x)$ is called the **logarithmic mean**. The logarithmic mean of positive operators A and B is denoted by $L(A, B)$.

From the inequality

$$\frac{x-1}{\log x} = \int_0^1 x^t \, dt = \int_0^{1/2} (x^t + x^{1-t}) \, dt \geq \int_0^{1/2} 2\sqrt{x} \, dt = \sqrt{x}$$

of the real functions we have the matrix inequality

$$A\#B \leq L(A, B).$$

It can similarly be proved that $L(A, B) \leq (A + B)/2$.

From the integral formula

$$\frac{1}{L(a, b)} = \frac{\log a - \log b}{a - b} = \int_0^\infty \frac{1}{(a + t)(b + t)} \, dt$$

one can obtain

$$L(A, B)^{-1} = \int_0^\infty \frac{(tA + B)^{-1}}{t + 1} \, dt.$$

$\square$

In the next example we study the means of ortho-projections.

Example 5.23 Let P and Q be ortho-projections. It was shown in Example 5.2 that $P\#Q = P \wedge Q$. The inequality

$$\begin{bmatrix} 2P & 0 \\ 0 & 2Q \end{bmatrix} \geq \begin{bmatrix} P \wedge Q & P \wedge Q \\ P \wedge Q & P \wedge Q \end{bmatrix}$$

is true since

$$\begin{bmatrix} P & 0 \\ 0 & Q \end{bmatrix} \geq \begin{bmatrix} P \wedge Q & 0 \\ 0 & P \wedge Q \end{bmatrix}, \quad \begin{bmatrix} P & -P \wedge Q \\ -P \wedge Q & Q \end{bmatrix} \geq 0.$$

This gives that $H(P, Q) \geq P \wedge Q$ and from the other inequality $H(P, Q) \leq P\#Q$, we obtain $H(P, Q) = P \wedge Q = P\#Q$.

The general matrix mean $m_f(P, Q)$ has the integral expression

$$m_f(P, Q) = aP + bQ + \int_{(0,\infty)} \frac{1 + \lambda}{\lambda} \Big((\lambda P) : Q \Big) \, d\mu(\lambda).$$

Since

$$(\lambda P) : Q = \frac{\lambda}{1 + \lambda} (P \wedge Q),$$

we have

$$m_f(P, Q) = aP + bQ + c(P \wedge Q).$$

Note that $a = f(0)$, $b = \lim_{x \to \infty} f(x)/x$ and $c = \mu((0, \infty))$. If $a = b = 0$, then $c = 1$ (since $m(I, I) = I$) and $m_f(P, Q) = P \wedge Q$. □

Example 5.24 The **power difference means** are determined by the functions

$$f_t(x) = \frac{t-1}{t} \cdot \frac{x^t - 1}{x^{t-1} - 1} \qquad (-1 \le t \le 2), \tag{5.22}$$

where the values $t = -1, 1/2, 1, 2$ correspond to the well-known harmonic, geometric, logarithmic and arithmetic means. The functions (5.22) are standard matrix monotone [39] and it can be shown that for fixed $x > 0$ the value $f_t(x)$ is an increasing function of t. The simple case $t = n/(n-1)$ yields

$$f_t(x) = \frac{1}{n} \sum_{k=0}^{n-1} x^{k/(n-1)}$$

and the matrix monotonicity is obvious. □

Example 5.25 The **Heinz mean**

$$H_t(x, y) = \frac{x^t y^{1-t} + x^{1-t} y^t}{2} \qquad (0 \le t \le 1/2)$$

interpolates the arithmetic and geometric means. The corresponding standard function

$$f_t(x) = \frac{x^t + x^{1-t}}{2}$$

is obviously matrix monotone and a decreasing function of the parameter t. Therefore we have a Heinz matrix mean. The formula is

$$H_t(A, B) = A^{1/2} \frac{(A^{-1/2} B A^{-1/2})^t + (A^{-1/2} B A^{-1/2})^{1-t}}{2} A^{1/2}.$$

This lies between the geometric and arithmetic means:

$$A \# B \le H_t(A, B) \le \frac{A + B}{2} \qquad (0 \le t \le 1/2).$$

□

Example 5.26 For $x \neq y$ the **Stolarsky mean** is

$$m_p(x, y) = \left(p\frac{x-y}{x^p - y^p}\right)^{\frac{1}{1-p}} = \left(\frac{1}{y-x}\int_x^y t^{p-1}\,dt\right)^{\frac{1}{p-1}},$$

where the case $p = 1$ is understood as

$$\lim_{p=1} m_p(x, y) = \frac{1}{e}\left(\frac{x^x}{y^y}\right)^{\frac{1}{x-y}}.$$

If $-2 \le p \le 2$, then $f_p(x) = m_p(x, 1)$ is a matrix monotone function (see Theorem 4.46), so it can define a matrix mean. The case $p = 1$ is called the **identric mean** and the case $p = 0$ is the well-known logarithmic mean.   □

It is known that the following canonical representation holds for any standard matrix monotone function $\mathbb{R}^+ \to \mathbb{R}^+$.

Theorem 5.27 *Let $f : \mathbb{R}^+ \to \mathbb{R}^+$ be a standard matrix monotone function. Then f admits a canonical representation*

$$f(x) = \frac{1+x}{2}\exp\int_0^1 \frac{(\lambda - 1)(1 - x)^2}{(\lambda + x)(1 + \lambda x)(1 + \lambda)}h(\lambda)\,d\lambda \qquad (5.23)$$

where $h : [0, 1] \to [0, 1]$ is a measurable function.

Example 5.28 In the function (5.23) we take

$$h(\lambda) = \begin{cases} 1 & \text{if } a \le \lambda \le b, \\ 0 & \text{otherwise} \end{cases}$$

where $0 \le a \le b \le 1$.

Then an easy calculation gives

$$\frac{(\lambda - 1)(1 - x)^2}{(\lambda + x)(1 + \lambda x)(1 + \lambda)} = \frac{2}{1+\lambda} - \frac{1}{\lambda + x} - \frac{x}{1 + \lambda x}.$$

Thus

$$\int_a^b \frac{(\lambda - 1)(1 - x)^2}{(\lambda + x)(1 + \lambda x)(1 + \lambda)}\,d\lambda = \left[\log(1 + \lambda)^2 - \log(\lambda + x) - \log(1 + \lambda x)\right]_{\lambda=a}^b$$

$$= \log\frac{(1 + b)^2}{(1 + a)^2} - \log\frac{b + x}{a + x} - \log\frac{1 + bx}{1 + ax}.$$

So

$$f(x) = \frac{(b+1)^2}{2(a+1)^2} \frac{(1+x)(a+x)(1+ax)}{(b+x)(1+bx)}.$$

Choosing $h \equiv 0$ yields the largest function $f(x) = (1+x)/2$ and $h \equiv 1$ yields the smallest function $f(x) = 2x/(1+x)$. If

$$\int_0^1 \frac{h(\lambda)}{\lambda} d\lambda = +\infty,$$

then $f(0) = 0$.   □

The next theorem describes the canonical representation for the reciprocal $1/f$ of a standard matrix monotone function f ($1/f$ is a matrix monotone decreasing function).

Theorem 5.29 *If $f : \mathbb{R}^+ \to \mathbb{R}^+$ is a standard matrix monotone function, then*

$$\frac{1}{f(x)} = \int_0^1 \frac{1+\lambda}{2} \left(\frac{1}{x+\lambda} + \frac{1}{1+x\lambda} \right) d\mu(\lambda), \tag{5.24}$$

where μ is a probability measure on $[0, 1]$.

A standard matrix monotone function $f : \mathbb{R}^+ \to \mathbb{R}^+$ is called **regular** if $f(0) > 0$. The next theorem provides a bijection between the regular standard matrix monotone functions and the non-regular ones.

Theorem 5.30 *Let $f : \mathbb{R}^+ \to \mathbb{R}^+$ be a standard matrix monotone function with $f(0) > 0$. Then*

$$\tilde{f}(x) := \frac{1}{2} \left((x+1) - (x-1)^2 \frac{f(0)}{f(x)} \right)$$

is standard matrix monotone as well. Moreover, $f \mapsto \tilde{f}$ gives a bijection between $\{f \in \mathcal{F} : f(0) > 0\}$ and $\{f \in \mathcal{F} : f(0) = 0\}$, where $\mathcal{F}$ is the set of all standard matrix monotone functions $\mathbb{R}^+ \to \mathbb{R}^+$.

Example 5.31 Let $A, B \in \mathbb{M}_n$ be positive definite matrices and m be a matrix mean. The block matrix

$$\begin{bmatrix} A & m(A, B) \\ m(A, B) & B \end{bmatrix}$$

is positive if and only if $m(A, B) \leq A\#B$. Similarly,

$$\begin{bmatrix} A^{-1} & m(A, B)^{-1} \\ m(A, B)^{-1} & B^{-1} \end{bmatrix} \geq 0$$

if and only $m(A, B) \geq A \# B$.

If $\lambda_1, \lambda_2, \ldots, \lambda_n$ are positive numbers, then the matrix $A \in \mathbb{M}_n$ defined as

$$A_{ij} = \frac{1}{L(\lambda_i, \lambda_j)}$$

is positive for $n = 2$ according to the above argument. However, this is true for every n by the formula

$$\frac{1}{L(x, y)} = \int_0^\infty \frac{1}{(x + t)(y + t)} dt.$$

(Another argument appears in Example 2.55.)

From the harmonic mean we obtain the mean matrix

$$[H(\lambda_i, \lambda_j)] = \left[\frac{2\lambda_i \lambda_j}{\lambda_i + \lambda_j} \right],$$

which is positive since it is the Hadamard product of two positive matrices (one of which is the Cauchy matrix).

A general description of positive mean matrices and many examples can be found in the book [45] and the paper [46]. It is worthwhile to note that two different notions of the matrix mean $m(A, B)$ and the mean matrix $[m(\lambda_i, \lambda_j)]$ are associated with a standard matrix monotone function.   □

5.4 Mean Transformation

If $0 \leq A, B \in \mathbb{M}_n$, then a matrix mean $m_f(A, B) \in \mathbb{M}_n$ has a slightly complicated formula expressed by the function $f : \mathbb{R}^+ \to \mathbb{R}^+$ of the mean. If $AB = BA$, then the situation is simpler: $m_f(A, B) = f(AB^{-1})B$. The mean introduced here will be a linear mapping $\mathbb{M}_n \to \mathbb{M}_n$. If $n > 1$, then this is essentially different from $m_f(A, B)$.

From A and B we have the linear mappings $\mathbb{M}_n \to \mathbb{M}_n$ defined as

$$\mathbb{L}_A X := AX, \qquad \mathbb{R}_B X := XB \qquad (X \in \mathbb{M}_n).$$

So $\mathbb{L}_A$ is left-multiplication by A and $\mathbb{R}_B$ is right-multiplication by B. Obviously, they are commuting operators, $\mathbb{L}_A \mathbb{R}_B = \mathbb{R}_B \mathbb{L}_A$, and they can be considered as matrices in $\mathbb{M}_n \otimes \mathbb{M}_n = \mathbb{M}_{n^2}$.

The definition of the **mean transformation** is

$$M_f(A, B) := m_f(\mathbb{L}_A, \mathbb{R}_B).$$

Sometimes the notation $\mathbb{J}^f_{A,B}$ is used for this.

For $f(x) = \sqrt{x}$ we have the geometric mean, which is a simple example.

Example 5.32 Since $\mathbb{L}_A$ and $\mathbb{R}_B$ commute, the geometric mean is the following:

$$\mathbb{L}_A \# \mathbb{R}_B = (\mathbb{L}_A)^{1/2}(\mathbb{R}_B)^{1/2} = \mathbb{L}_{A^{1/2}}\mathbb{R}_{B^{1/2}}, \qquad X \mapsto A^{1/2}XB^{1/2}.$$

It is not true that $M(A, B)X \geq 0$ if $X \geq 0$, but as a linear mapping $M(A, B)$ is positive:

$$\langle X, M(A, B)X \rangle = \operatorname{Tr} X^* A^{1/2}XB^{1/2} = \operatorname{Tr} B^{1/4}X^* A^{1/2}XB^{1/4} \geq 0$$

for every $X \in \mathbb{M}_n$.

Let $A, B > 0$. The equality $M(A, B)A = M(B, A)A$ immediately implies that $AB = BA$. From $M(A, B) = M(B, A)$ we deduce that $A = \lambda B$ for some number $\lambda > 0$. Therefore $M(A, B) = M(B, A)$ is a very special condition for the mean transformation. $\qquad \square$

The logarithmic mean transformation is

$$M_{\log}(A, B)X = \int_0^1 A^t X B^{1-t}\, dt.$$

In the next example we have a formula for general $M(A, B)$.

Example 5.33 Assume that A and B act on a Hilbert space which has two orthonormal bases $|x_1\rangle, \ldots, |x_n\rangle$ and $|y_1\rangle, \ldots, |y_n\rangle$ such that

$$A = \sum_i \lambda_i |x_i\rangle\langle x_i|, \qquad B = \sum_j \mu_j |y_j\rangle\langle y_j|.$$

Then for $f(x) = x^k$ we have

$$f(\mathbb{L}_A \mathbb{R}_B^{-1})\mathbb{R}_B |x_i\rangle\langle y_j| = A^k |x_i\rangle\langle y_j| B^{-k+1} = \lambda_i^k \mu_j^{-k+1} |x_i\rangle\langle y_j|$$
$$= f(\lambda_i/\mu_j)\mu_j |x_i\rangle\langle y_j| = m_f(\lambda_i, \mu_j)|x_i\rangle\langle y_j|$$

and for a general f

$$M_f(A, B)|x_i\rangle\langle y_j| = m_f(\lambda_i, \mu_j)|x_i\rangle\langle y_j|.$$

This also shows that $M_f(A, B) \geq 0$ with respect to the Hilbert–Schmidt inner product.

Another formulation is also possible. Let $A = U\text{Diag}(\lambda_1, \ldots, \lambda_n)U^*$ and $B = V\text{Diag}(\mu_1, \ldots, \mu_n)V^*$ with unitaries U, V. Let $|e_1\rangle, \ldots, |e_n\rangle$ be the standard basis vectors. Then

$$M_f(A, B)X = U\left([m_f(\lambda_i, \mu_j)]_{ij} \circ (U^*XV)\right) V^*.$$

It is enough to check the case $X = |x_i\rangle\langle y_j|$. Then

$$U\left([m_f(\lambda_i, \mu_j)]_{ij} \circ (U^*|x_i\rangle\langle y_j|V)\right) V^* = U\left([m_f(\lambda_i, \mu_j)]_{ij} \circ |e_i\rangle\langle e_j|\right) V^*$$
$$= m_f(\lambda_i, \mu_j)U|e_i\rangle\langle e_j|V^* = m_f(\lambda_i, \mu_j)|x_i\rangle\langle y_j|.$$

For the matrix means we have $m(A, A) = A$, but $M(A, A)$ is rather different, it cannot be A since it is a transformation. If $A = \sum_i \lambda_i|x_i\rangle\langle x_i|$, then

$$M(A, A)|x_i\rangle\langle x_j| = m(\lambda_i, \lambda_j)|x_i\rangle\langle x_j|.$$

(This is related to the so-called mean matrix, see Example 5.31.)   □

Example 5.34 Here we find a very special inequality between the geometric mean transformation $M_G(A, B)$ and the arithmetic mean transformation $M_A(A, B)$. They are

$$M_G(A, B)X = A^{1/2}XB^{1/2}, \qquad M_A(A, B)X = \tfrac{1}{2}(AX + XB).$$

There is an integral formula

$$M_G(A, B)X = \int_{-\infty}^{\infty} A^{it} M_A(A, B)XB^{-it}\, d\mu(t), \tag{5.25}$$

where the probability measure is

$$d\mu(t) = \frac{1}{\cosh(\pi t)}dt.$$

From (5.25) it follows that

$$\|M_G(A, B)X\| \le \|M_A(A, B)X\| \tag{5.26}$$

which is an operator norm inequality. A general comparison theorem of this kind between mean transformations is given in [44].   □

The next theorem gives the **transformer inequality**.

Theorem 5.35 Let $f : [0, +\infty) \to [0, +\infty)$ be a matrix monotone function and $M(\cdot, \cdot)$ be the corresponding mean transformation. If $\beta : \mathbb{M}_n \to \mathbb{M}_m$ is a 2-positive *trace-preserving mapping and the matrices $A, B \in \mathbb{M}_n$ are positive, then*

$$\beta M(A, B)\beta^* \leq M(\beta(A), \beta(B)). \tag{5.27}$$

Proof : By approximation we may assume that $A, B, \beta(A), \beta(B) > 0$. Indeed, assume that the conclusion holds under this positive definiteness condition. For each $\varepsilon > 0$ let

$$\beta_\varepsilon(X) := \frac{\beta(X) + \varepsilon(\mathrm{Tr}\, X)I_m}{1 + m\varepsilon}, \qquad X \in \mathbb{M}_n,$$

which is 2-positive and trace-preserving. If $A, B > 0$, then $\beta_\varepsilon(A), \beta_\varepsilon(B) > 0$ as well and hence (5.27) holds for β_ε. Letting $\varepsilon \searrow 0$ implies that (5.27) for β is true for all $A, B > 0$. Then by taking the limit of $A + \varepsilon I_n, B + \varepsilon I_n$ as $\varepsilon \searrow$, we have (5.27) for all $A, B \geq 0$. Now assume $A, B, \beta(A), \beta(B) > 0$.

Based on Löwner's theorem, we may consider $f(x) = x/(\lambda + x)$ ($\lambda > 0$). Then

$$M(A, B) = \frac{\mathbb{L}_A}{\lambda I + \mathbb{L}_A \mathbb{R}_B^{-1}}, \qquad M(A, B)^{-1} = (\lambda I + \mathbb{L}_A \mathbb{R}_B^{-1})\mathbb{L}_A^{-1}$$

and similarly $M(\beta(A), \beta(B))^{-1} = (\lambda I + \mathbb{L}_{\beta(A)} \mathbb{R}_{\beta(B)}^{-1})\mathbb{L}_{\beta(A)}^{-1}$. The statement (5.27) has the equivalent form

$$\beta^* M(\beta(A), \beta(B))^{-1}\beta \leq M(A, B)^{-1},$$

which means

$$\langle \beta(X), (\lambda I + \mathbb{L}_{\beta(A)} \mathbb{R}_{\beta(B)}^{-1})\mathbb{L}_{\beta(A)}^{-1}\beta(X) \rangle \leq \langle X, (\lambda I + \mathbb{L}_A \mathbb{R}_B^{-1})\mathbb{L}_A^{-1}X \rangle$$

or

$$\lambda \mathrm{Tr}\, \beta(X^*)\beta(A)^{-1}\beta(X) + \mathrm{Tr}\, \beta(X)\beta(B)^{-1}\beta(X^*) \leq \lambda \mathrm{Tr}\, X^*A^{-1}X + \mathrm{Tr}\, XB^{-1}X^*.$$

This inequality is true due to the matrix inequality

$$\beta(X^*)\beta(Y)^{-1}\beta(X) \leq \beta(X^*Y^{-1}X) \qquad (Y > 0),$$

see Lemma 2.46. □

If β^{-1} has the same properties as β in the previous theorem, then we have equality in formula (5.27).

Theorem 5.36 *Let $f : \mathbb{R}^+ \to \mathbb{R}^+$ be a matrix monotone function with $f(1) = 1$ and $M(\cdot, \cdot)$ be the corresponding mean transformation. Assume that $0 \leq A \leq A'$ and $0 \leq B \leq B'$ in $\mathbb{M}_n$. Then $M(A, B) \leq M(A', B')$.*

Proof: By continuity we may assume that A, $B > 0$. Based on Löwner's theorem, we may consider $f(x) = x/(\lambda + x)$ $(\lambda > 0)$. Then the statement is

$$\mathbb{L}_A(\lambda I + \mathbb{L}_A \mathbb{R}_B^{-1})^{-1} \leq \mathbb{L}_{A'}(\lambda I + \mathbb{L}_{A'} \mathbb{R}_{B'}^{-1})^{-1},$$

which is equivalent to the relation

$$\lambda \mathbb{L}_{A'}^{-1} + \mathbb{R}_{B'}^{-1} = (\lambda I + \mathbb{L}_{A'} \mathbb{R}_{B'}^{-1}) \mathbb{L}_{A'}^{-1} \leq (\lambda I + \mathbb{L}_A \mathbb{R}_B^{-1}) \mathbb{L}_A^{-1} = \lambda \mathbb{L}_A^{-1} + \mathbb{R}_B^{-1}.$$

This is true, since $\mathbb{L}_{A'}^{-1} \leq \mathbb{L}_A^{-1}$ and $\mathbb{R}_{B'}^{-1} \leq \mathbb{R}_B^{-1}$ by the assumption. $\square$

Theorem 5.37 *Let f be a matrix monotone function with $f(1) = 1$ and M_f be the corresponding transformation mean. Then M_f has the following properties:*

(1) $M_f(\lambda A, \lambda B) = \lambda M_f(A, B)$ *for a number $\lambda > 0$.*
(2) $\big(M_f(A, B)X\big)^* = M_f(B, A)X^*.$
(3) $M_f(A, A)I = A.$
(4) $\operatorname{Tr} M_f(A, A)^{-1}Y = \operatorname{Tr} A^{-1}Y.$
(5) $(A, B) \mapsto \langle X, M_f(A, B)Y \rangle$ *is continuous.*
(6) *Let*

$$C := \begin{bmatrix} A & 0 \\ 0 & B \end{bmatrix} \geq 0.$$

Then

$$M_f(C, C)\begin{bmatrix} X & Y \\ Z & W \end{bmatrix} = \begin{bmatrix} M_f(A, A)X & M_f(A, B)Y \\ M_f(B, A)Z & M_f(B, B)Z \end{bmatrix}.$$

The proof of the theorem is an elementary computation. Property (6) is essential. It tells us that it is sufficient to know the mean transformation for two identical matrices.

The next theorem is an axiomatic characterization of the mean transformation.

Theorem 5.38 *Assume that for any $n \in \mathbb{N}$ and for all $0 \leq A, B \in \mathbb{M}_n$, the linear operator $L(A, B) : \mathbb{M}_n \to \mathbb{M}_n$ is defined. $L(A, B) = M_f(\mathbb{L}_A, \mathbb{R}_B)$ for a matrix monotone function f if and only if L has the following properties:*

(i) $(X, Y) \mapsto \langle X, L(A, B)Y \rangle$ *is an inner product on $\mathbb{M}_n$.*
(ii) $(A, B) \mapsto \langle X, L(A, B)Y \rangle$ *is continuous.*
(iii) *For a trace-preserving completely positive mapping $\beta : \mathbb{M}_n \to \mathbb{M}_m$,*

$$\beta L(A, B) \beta^* \leq L\big(\beta A, \beta B\big)$$

holds.

(iv) *Let*

$$C := \begin{bmatrix} A & 0 \\ 0 & B \end{bmatrix} > 0.$$

Then

$$L(C, C) \begin{bmatrix} X & Y \\ Z & W \end{bmatrix} = \begin{bmatrix} L(A, A)X & L(A, B)Y \\ L(B, A)Z & L(B, B)Z \end{bmatrix}.$$

The proof needs a few lemmas. We use the notation $\mathbb{P}_n := \{A \in \mathbb{M}_n : A > 0\}$.

Lemma 5.39 *If $U, V \in \mathbb{M}_n$ are arbitrary unitary matrices, then for all $A, B \in \mathbb{P}_n$ and $X \in \mathbb{M}_n$ we have*

$$\langle X, L(A, B)X \rangle = \langle UXV^*, L(UAU^*, VBV^*)UXV^* \rangle.$$

Proof: For a unitary matrix $U \in \mathbb{M}_n$ define $\beta(A) = U^*AU$. Then $\beta \colon \mathbb{M}_n \to \mathbb{M}_n$ is trace-preserving completely positive and $\beta^*(A) = \beta^{-1}(A) = UAU^*$. Thus by double application of (iii) we obtain

$$\begin{aligned}
\langle X, L(A, A)X \rangle &= \langle X, L(\beta\beta^{-1}A, \beta\beta^{-1}A)X \rangle \\
&\geq \langle X, \beta L(\beta^{-1}A, \beta^{-1}A)\beta^*X \rangle \\
&= \langle \beta^*X, L(\beta^{-1}A, \beta^{-1}A\beta^*X \rangle \\
&\geq \langle \beta^*X, \beta^{-1}L(A, A)(\beta^{-1})^*\beta^*X \rangle \\
&= \langle X, L(A, A)X \rangle,
\end{aligned}$$

hence

$$\langle X, L(A, A)X \rangle = \langle UAU^*, L(UAU^*, UAU^*)UXU^* \rangle.$$

Now for the matrices

$$C = \begin{bmatrix} A & 0 \\ 0 & B \end{bmatrix} \in \mathbb{P}_{2n}, \quad Y = \begin{bmatrix} 0 & X \\ 0 & 0 \end{bmatrix} \in \mathbb{M}_{2n} \quad \text{and} \quad W = \begin{bmatrix} U & 0 \\ 0 & V \end{bmatrix} \in \mathbb{M}_{2n}$$

it follows by (iv) that

$$\begin{aligned}
\langle X, L(A, B)X \rangle &= \langle Y, L(C, C)Y \rangle \\
&= \langle WYW^*, L(WCW^*, WCW^*)WYW^* \rangle \\
&= \langle UXV^*L(UAU^*, VBV^*)UXV^* \rangle
\end{aligned}$$

and we have the statement. $\square$

Lemma 5.40 *Suppose that $L(A, B)$ is defined by the axioms (i)–(iv). Then there exists a unique continuous function $d \colon \mathbb{R}^+ \times \mathbb{R}^+ \to \mathbb{R}^+$ such that*

$$d(r\lambda, r\mu) = rd(\lambda, \mu) \qquad (r, \lambda, \mu > 0)$$

and for every $A = \mathrm{Diag}(\lambda_1, \ldots, \lambda_n)$ *and* $B = \mathrm{Diag}(\mu_1, \ldots, \mu_n)$ *in* $\mathbb{P}_n$,

$$\langle X, L(A, B)X \rangle = \sum_{j,k=1}^{n} d(\lambda_j, \mu_k) |X_{jk}|^2.$$

Proof : The uniqueness of such a function d is clear. We concentrate on the existence.

Denote by $E(jk)^{(n)}$ and I_n the $n \times n$ matrix units and the $n \times n$ unit matrix, respectively. We assume that $A = \mathrm{Diag}(\lambda_1, \ldots, \lambda_n)$ and $B = \mathrm{Diag}(\mu_1, \ldots, \mu_n)$ are in $\mathbb{P}_n$.

We first show that

$$\langle E(jk)^{(n)}, L(A, A)E(lm)^{(n)} \rangle = 0 \quad \text{if} \quad (j, k) \neq (l, m). \tag{5.28}$$

Indeed, if $j \neq k, l, m$ we let $U_j = \mathrm{Diag}(1, \ldots, 1, \mathrm{i}, 1, \ldots, 1)$ where the imaginary unit is the jth entry and $j \neq k, l, m$. Then by Lemma 5.39 we have

$$\begin{aligned}
&\langle E(jk)^{(n)}, L(A, A)E(lm)^{(n)} \rangle \\
&= \langle U_j E(jk)^{(n)} U_j^*, L(U_j AU_j^*, U_j AU_j^*)U_j E(lm)^{(n)} U_j^* \rangle \\
&= \langle \mathrm{i}E(jk)^{(n)}, L(A, A)E(lm)^{(n)} \rangle = -\mathrm{i}\langle E(jk)^{(n)}, L(A, A)E(lm)^{(n)} \rangle,
\end{aligned}$$

hence $\langle E(jk)^{(n)}, L(A, A)E(lm)^{(n)} \rangle = 0$. If one of the indices j, k, l, m is different from the others then (5.28) follows analogously. Finally, applying condition (iv) we obtain that

$$\langle E(jk)^{(n)}, L(A, B)E(lm)^{(n)} \rangle = \langle E(j, k + n)^{(2n)}, m(C, C)E(l, m + n)^{(2n)} \rangle = 0$$

if $(j, k) \neq (l, m)$, because $C = \mathrm{Diag}(\lambda_1, \ldots, \lambda_n, \mu_1, \ldots, \mu_n) \in H_{2n}^+$ and one of the indices $j, k + n, l, m + n$ is different from the others.

Now we claim that $\langle E(jk)^{(n)}, L(A, B)E(jk)^{(n)} \rangle$ is determined by λ_j and μ_k. More specifically,

$$\|E(jk)^{(n)}\|_{A,B}^2 = \|E(12)^{(2)}\|_{\mathrm{Diag}(\lambda_j, \mu_k)}^2, \tag{5.29}$$

where for brevity we have introduced the notations

$$\|X\|_{A,B}^2 = \langle X, L(A, B)X \rangle \quad \text{and} \quad \|X\|_A^2 = \|X\|_{A,A}^2.$$

Indeed, if $U_{j,k+n} \in \mathbb{M}_{2n}$ denotes the unitary matrix which interchanges the first and the jth coordinates and further the second and the $(k + n)$th coordinates, then by condition (iv) and Lemma 5.39 it follows that

$$\|E(jk)^{(n)}\|^2_{A,B} = \|E(j,k+n)^{(2n)}\|^2_C$$
$$= \|U_{j,k+n}E(j,k+n)^{(2n)}U^*_{j,k+n}\|^2_{U_{j,k+n}CU^*_{j,k+n}}$$
$$= \|E(12)^{(2n)}\|^2_{\mathrm{Diag}(\lambda_j,\mu_k,\lambda_3,...,\mu_n)}.$$

Thus it suffices to prove

$$\|E(12)^{(2n)}\|^2_{\mathrm{Diag}(\eta_1,\eta_2,...,\eta_{2n})} = \|E(12)^{(2)}\|^2_{\mathrm{Diag}(\eta_1,\eta_2)}. \qquad (5.30)$$

Condition (iv) with $X = E(12)^{(n)}$ and $Y = Z = W = 0$ yields

$$\|E(12)^{(2n)}\|^2_{\mathrm{Diag}(\eta_1,\eta_2,...,\eta_{2n})} = \|E(12)^{(n)}\|^2_{\mathrm{Diag}(\eta_1,\eta_2,...,\eta_n)}. \qquad (5.31)$$

Further, consider the following mappings ($n \geq 4$): $\beta_n : \mathbb{M}_n \to \mathbb{M}_{n-1}$,

$$\beta_n(E(jk)^{(n)}) := \begin{cases} E(jk)^{(n-1)}, & \text{if } 1 \leq j,k \leq n-1, \\ E(n-1,n-1)^{(n-1)}, & \text{if } j = k = n, \\ 0, & \text{otherwise}, \end{cases}$$

and $\tilde{\beta}_n : \mathbb{M}_{n-1} \to \mathbb{M}_n$, $\tilde{\beta}_n(E(jk)^{(n-1)}) := E(jk)^{(n-1)}$ if $1 \leq j,k \leq n-2$,

$$\tilde{\beta}_n(E(n-1,n-1)^{(n-1)}) := \frac{\eta_{n-1}E(n-1,n-1)^{(n)} + \eta_n E(nn)^{(n)}}{\eta_{n-1} + \eta_n}$$

and in the other cases $\tilde{\beta}_n(E(jk)^{(n-1)}) = 0$.

Clearly, β_n and $\tilde{\beta}_n$ are trace-preserving completely positive mappings. Hence by (iii)

$$\|E(12)^{(n)}\|^2_{\mathrm{Diag}(\eta_1,...,\eta_n)} = \|E(12)^{(n)}\|^2_{\tilde{\beta}_n\beta_n\mathrm{Diag}(\eta_1,...,\eta_n)}$$
$$\geq \|\tilde{\beta}^*_n E(12)^{(n)}\|^2_{\beta_n\mathrm{Diag}(\eta_1,...,\eta_n)}$$
$$\geq \|\beta^*_n\tilde{\beta}^*_n E(12)^{(n)}\|^2_{\mathrm{Diag}(\eta_1,...,\eta_n)}$$
$$= \|E(12)^{(n)}\|^2_{\mathrm{Diag}(\eta_1,...,\eta_n)}.$$

Thus equality holds, which implies that

$$\|E(12)^{(n)}\|^2_{\mathrm{Diag}(\eta_1,...,\eta_{n-1},\eta_n)} = \|E(12)^{(n-1)}\|^2_{\mathrm{Diag}(\eta_1,...,\eta_{n-2},\eta_{n-1}+\eta_n)}. \qquad (5.32)$$

Now repeated application of (5.31) and (5.32) yields (5.30) and therefore (5.29) also follows.

For $\lambda, \mu > 0$ let
$$d(\lambda,\mu) := \|E(12)^{(2)}\|^2_{\mathrm{Diag}(\lambda,\mu)}.$$

Condition (ii) implies the continuity of d. We furthermore claim that d is homogeneous of order one, that is,

$$d(r\lambda, r\mu) = rd(\lambda, \mu) \qquad (\lambda, \mu, r > 0).$$

First let $r = k \in \mathbb{N}$. Then the mappings $\alpha_k \colon \mathbb{M}_2 \to \mathbb{M}_{2k}$, $\tilde{\alpha}_k \colon \mathbb{M}_{2k} \to \mathbb{M}_k$ defined by

$$\alpha_k(X) = \frac{1}{k} I_k \otimes X$$

and

$$\tilde{\alpha}_k \begin{bmatrix} X_{11} & X_{12} & \ldots & X_{1k} \\ X_{21} & X_{22} & \ldots & X_{2k} \\ \vdots & \vdots & \ddots & \\ X_{k1} & X_{k2} & \ldots & X_{kk} \end{bmatrix} = X_{11} + X_{22} + \ldots + X_{kk}$$

are trace-preserving completely positive, for which $\tilde{\alpha}_k^* = k\alpha_k$. So applying condition (iii) twice it follows that

$$
\begin{aligned}
\|E(12)^{(2)}\|_{\mathrm{Diag}(\lambda,\mu)}^2 &= \|E(12)^{(2)}\|_{\tilde{\alpha}_k\alpha_k\mathrm{Diag}(\lambda,\mu)}^2 \\
&\geq \|\tilde{\alpha}_k^* E(12)^{(2)}\|_{\alpha_k\mathrm{Diag}(\lambda,\mu)}^2 \\
&\geq \|\alpha_k^* \tilde{\alpha}_k^* E(12)^{(2)}\|_{\mathrm{Diag}(\lambda,\mu)}^2 \\
&= \|E(12)^{(2)}\|_{\mathrm{Diag}(\lambda,\mu)}^2.
\end{aligned}
$$

Hence equality holds, which means that

$$\|E(12)^{(2)}\|_{\mathrm{Diag}(\lambda,\mu)}^2 = \|I_k \otimes E(12)^{(2)}\|_{\frac{1}{k} I_k \otimes \mathrm{Diag}(\lambda,\mu)}^2.$$

Thus by applying (5.28) and (5.29) we obtain

$$
\begin{aligned}
d(\lambda, \mu) &= \|I_k \otimes E(12)^{(2)}\|_{\frac{1}{k} I_k \otimes \mathrm{Diag}(\lambda,\mu)}^2 \\
&= \sum_{j=1}^{k} \|E(jj)^{(k)} \otimes E(12)^{(2)}\|_{\frac{1}{k} I_k \otimes \mathrm{Diag}(\lambda,\mu)}^2 \\
&= k\|E(11)^{(k)} \otimes E(12)^{(2)}\|_{\frac{1}{k} I_k \otimes \mathrm{Diag}(\lambda,\mu)}^2 \\
&= kd\left(\frac{\lambda}{k}, \frac{\mu}{k}\right).
\end{aligned}
$$

If $r = \ell/k$ where ℓ, k are positive natural numbers, then

$$d(r\lambda, r\mu) = d\left(\frac{\ell}{k}\lambda, \frac{\ell}{k}\mu\right) = \frac{1}{k}d(\ell\lambda, \ell\mu) = \frac{\ell}{k}d(\lambda, \mu).$$

By condition (ii), the homogeneity follows for every $r > 0$.
We finish the proof by using (5.28) and (5.29), obtaining

$$\|X\|_{A,B}^2 = \sum_{j,k=1}^n d(\lambda_j, \mu_k)|X_{jk}|^2.$$

□

If we require the positivity of $M(A, B)X$ for $X \geq 0$, then from the formula

$$(M(A, B)X)^* = M(B, A)X^*$$

we need $A = B$. If $A = \sum_i \lambda_i |x_i\rangle\langle x_i|$ and $X = \sum_{i,j} |x_i\rangle\langle x_j|$ for an orthonormal basis $\{|x_i\rangle : i\}$, then

$$\left(M(A, A)X\right)_{ij} = m(\lambda_i, \lambda_j).$$

The positivity of this matrix is necessary.
Given the positive numbers $\{\lambda_i : 1 \leq i \leq n\}$, the matrix

$$K_{ij} = m(\lambda_i, \lambda_j)$$

is called an $n \times n$ **mean matrix**. From the previous argument the positivity of $M(A, A) : \mathbb{M}_n \to \mathbb{M}_n$ implies the positivity of the $n \times n$ mean matrices of the mean M. It is easy to see that if the mean matrices of any size are positive, then $M(A, A) : \mathbb{M}_n \to \mathbb{M}_n$ is a completely positive mapping.

If the mean matrix

$$\begin{bmatrix} \lambda_1 & m(\lambda_1, \lambda_2) \\ m(\lambda_1, \lambda_2) & \lambda_1 \end{bmatrix}$$

is positive, then $m(\lambda_1, \lambda_2) \leq \sqrt{\lambda_1\lambda_2}$. It follows that to have a positive mean matrix, the mean m should be smaller than the geometric mean. Indeed, the next general characterization result is known.

Theorem 5.41 *Let f be a standard matrix monotone function on $\mathbb{R}^+$ and m the corresponding mean, i.e., $m(x, y) := xf(x/y)$ for $x, y > 0$. Let $M(\cdot, \cdot)$ be the corresponding mean transformation. Then the following conditions are equivalent:*

(1) $M(A, A)X \geq 0$ *for all* $0 \leq A, X \in \mathbb{M}_n$ *and every* $n \in \mathbb{N}$;
(2) *the mean transformation $M(A, A) : \mathbb{M}_n \to \mathbb{M}_n$ is completely positive for all* $0 \leq A \in \mathbb{M}_n$ *and every* $n \in \mathbb{N}$;
(3) $\|M(A, A)X\| \leq \|A^{1/2}XA^{1/2}\|$ *for all $A, X \in \mathbb{M}_n$ with $A \geq 0$ and every $n \in \mathbb{N}$, where $\|\cdot\|$ is the operator norm;*

(4) *the mean matrix* $[m(\lambda_i, \lambda_j)]_{ij}$ *is positive semidefinite for all* $\lambda_1, \ldots, \lambda_n > 0$
   and every $n \in \mathbb{N}$;
(5) $f(e^t)e^{-t/2}$ *is a positive definite function on* $\mathbb{R}$ *in the sense of Bochner, i.e., it is
   the Fourier transform of a probability measure on* $\mathbb{R}$.

The above condition (5) is stronger than $f(x) \le \sqrt{x}$ and it is a necessary and
sufficient condition for the positivity of $M(A, A)X$ for all $A, X \ge 0$.

Example 5.42 The **power mean** or **binomial mean**

$$m_t(x, y) = \left(\frac{x^t + y^t}{2}\right)^{1/t}$$

is an increasing function of t when x and y are fixed. The limit $t \to 0$ gives the
geometric mean. Therefore the positivity of the matrix mean may appear only for
$t \le 0$. Then for $t > 0$,

$$m_{-t}(x, y) = 2^{1/t} \frac{xy}{(x^t + y^t)^{1/t}}$$

and the corresponding mean matrix is positive due to the infinitely divisible Cauchy
matrix, see Example 1.41. □

5.5 Notes and Remarks

The geometric mean of operators first appeared in the paper Wieslaw Pusz and
Stanislav L. Woronowicz, Functional calculus for sesquilinear forms and the purifi-
cation map, Rep. Math. Phys. **8**(1975), 159–170, and was studied in detail in the
papers [2,58] of Tsuyoshi **Ando** and Fumio **Kubo**. The geometric mean for more
matrices is from the paper [10]. Another approach based on differential geometry
is explained in the book [21]. A popularization of the subject is the paper Rajendra
Bhatia and John **Holbrook**, Noncommutative geometric means, Math. Intelligencer
28(2006), 32–39.

Theorem 5.18 is from the paper Miklós **Pálfia**, A multivariable extension of two-
variable matrix means, SIAM J. Matrix Anal. Appl. **32**(2011), 385–393. There is a
different definition of the geometric mean X of the positive matrices $A_1, A_2, \ldots, A_k$
as defined by the equation $\sum_{k=1}^{n} \log A_i^{-1} X = 0$. See the paper Y. Lim and M. Pálfia,
Matrix power means and the Karcher mean, J. Funct. Anal. **262**(2012), 1498–1514
and the references therein.

The mean transformations are in the paper [44] and the book [45] of Fumio
Hiai and Hideki **Kosaki**. Theorem 5.38 is from the paper [18]. There are several
examples of positive and infinite divisible mean matrices in the paper Rajendra
Bhatia and Hideki Kosaki, Mean matrices and infinite divisibility, Linear Algebra

Appl. **424**(2007), 36–54. (Infinite divisibility means the positivity of matrices $A_{ij} = m(\lambda_i, \lambda_j)^t$ for every $t > 0$.)

Lajos **Molnár** proved that if a bijection $\alpha : \mathbb{M}_n^+ \to \mathbb{M}_n^+$ preserves the geometric mean, then, for $n \geq 2$, $\alpha(A) = SAS^*$ for a linear or conjugate linear mapping S (Maps preserving the geometric mean of positive operators, Proc. Amer. Math. Soc. **137**(2009), 1763–1770.)

Theorem 5.27 is from the paper K. Audenaert, L. Cai and F. Hansen, Inequalities for quantum skew information, Lett. Math. Phys. **85**(2008), 135–146. Theorem 5.29 is from the paper F. Hansen, Metric adjusted skew information, Proc. Natl. Acad. Sci. USA **105**(2008), 9909–9916, and Theorem 5.35 is from P. Gibilisco, F. Hansen, T. Isola, On a correspondence between regular and non-regular operator monotone functions, Linear Algebra Appl. **430**(2009), 2225–2232.

The norm inequality (5.26) was obtained by R. Bhatia and C. Davis, A Cauchy–Schwarz inequality for operators with applications, Linear Algebra Appl. **223/224** (1995), 119–129. The integral expression (5.25) is due to H. Kosaki, Arithmetic-geometric mean and related inequalities for operators, J. Funct. Anal. **156**(1998), 429–451. For a systematic analysis on norm inequalities and integral expressions of this kind as well as the details on Theorem 5.41, see the papers [44,45,46].

5.6 Exercises

1. Show that for positive invertible matrices A and B the inequalities

$$2(A^{-1} + B^{-1})^{-1} \leq A\#B \leq \frac{1}{2}(A + B)$$

   hold. What is the condition for equality? (Hint: Reduce the general case to $A = I$.)

2. Show that

$$A\#B = \frac{1}{\pi} \int_0^1 \frac{(tA^{-1} + (1 - t)B^{-1})^{-1}}{\sqrt{t(1 - t)}} \, dt.$$

3. Let $A, B > 0$. Show that $A\#B = A$ implies $A = B$.

4. Let $0 < A, B \in \mathbb{M}_m$. Show that the rank of the matrix

$$\begin{bmatrix} A & A\#B \\ A\#B & B \end{bmatrix}$$

   is smaller than $2m$.

5. Show that for any matrix mean m,

$$m(A, B)\#m^{\perp}(A, B) = A\#B.$$

6. Let $A \geq 0$ and P be a projection of rank 1. Show that $A \# P = \sqrt{\operatorname{Tr} AP} \, P$.

7. Give an argument that the natural map

$$(A, B) \longmapsto \exp\left(\frac{\log A + \log B}{2}\right)$$

would not be a good definition of the geometric mean.

8. Show that for positive matrices $A : B = A - A(A + B)^{-1}A$.

9. Show that for positive matrices $A : B \leq A$.

10. Show that $0 < A \leq B$ implies $A \leq 2(A : B) \leq B$.

11. Show that $L(A, B) \leq (A + B)/2$.

12. Let $A, B > 0$. Show that if for a matrix mean $m_f(A, B) = A$, then $A = B$.

13. Let $f, g : \mathbb{R}^+ \to \mathbb{R}^+$ be matrix monotone functions. Show that their arithmetic and geometric means are matrix monotone as well.

14. Show that the matrix

$$A_{ij} = \frac{1}{H_t(\lambda_i, \lambda_j)}$$

defined by the Heinz mean is positive.

15. Show that

$$\frac{\partial}{\partial t} m(e^{tA}, e^{tB})\Big|_{t=0} = \frac{A + B}{2}$$

for a symmetric mean. (Hint: Prove the result for arithmetic and harmonic means and reduce the general case to these examples.)

16. Let A and B be positive matrices and assume that there is a unitary U such that $A^{1/2}UB^{1/2} \geq 0$. Show that $A \# B = A^{1/2}UB^{1/2}$.

17. Show that

$$S^*(A : B)S \leq (S^*AS) : (S^*BS)$$

for any invertible matrix S and $A, B \geq 0$.

18. Prove the property

$$(A : B) + (C : D) \leq (A + C) : (B + D)$$

of the parallel sum.

19. Prove the logarithmic mean formula

$$L(A, B)^{-1} = \int_0^\infty \frac{(tA + B)^{-1}}{t + 1} \, dt$$

for positive definite matrices A, B.

20. Let A and B be positive definite matrices. Set $A_0 := A$, $B_0 := B$ and define recursively

$$A_n = \frac{A_{n-1} + B_{n-1}}{2} \quad \text{and} \quad B_n = 2(A_{n-1}^{-1} + B_{n-1}^{-1})^{-1} \quad (n = 1, 2, \ldots).$$

Show that

$$\lim_{n \to \infty} A_n = \lim_{n \to \infty} B_n = A \# B.$$

21. Show that the function $f_t(x)$ defined in (5.22) has the property

$$\sqrt{x} \leq f_t(x) \leq \frac{1+x}{2}$$

when $1/2 \leq t \leq 2$.

22. Let P and Q be ortho-projections. What is their Heinz mean?

23. Show that

$$\det (A \# B) = \sqrt{\det A \det B}.$$

24. Assume that A and B are invertible positive matrices. Show that

$$(A \# B)^{-1} = A^{-1} \# B^{-1}.$$

25. Let

$$A := \begin{bmatrix} 3/2 & 0 \\ 0 & 3/4 \end{bmatrix} \quad \text{and} \quad B := \begin{bmatrix} 1/2 & 1/2 \\ 1/2 & 1/2 \end{bmatrix}.$$

Show that $A \geq B \geq 0$ and for $p > 1$ the inequality $A^p \geq B^p$ does not hold.

26. Show that

$$\det \left(\mathbf{G}(A, B, C) \right) = \left(\det A \det B \det C \right)^{1/3}.$$

27. Show that

$$\mathbf{G}(\alpha A, \beta B, \gamma C) = (\alpha \beta \gamma)^{1/3} \mathbf{G}(A, B, C)$$

for positive numbers α, β, γ.

28. Show that $A_1 \geq A_2$, $B_1 \geq B_2$, $C_1 \geq C_2$ imply

$$\mathbf{G}(A_1, B_1, C_1) \geq \mathbf{G}(A_2, B_2, C_2).$$

29. Show that

$$\mathbf{G}(A, B, C) = \mathbf{G}(A^{-1}, B^{-1}, C^{-1})^{-1}.$$

30. Show that

$$3(A^{-1} + B^{-1} + C^{-1})^{-1} \leq \mathbf{G}(A, B, C) \leq \frac{1}{3}(A + B + C).$$

31. Show that

$$f_\gamma(x) = 2^{2\gamma-1}x^\gamma(1 + x)^{1-2\gamma}$$

is a matrix monotone function for $0 < \gamma < 1$.

32. Let P and Q be ortho-projections. Prove that $L(P, Q) = P \wedge Q$.

33. Show that the function

$$f_p(x) = \left(\frac{x^p + 1}{2}\right)^{1/p}$$

is matrix monotone if and only if $-1 \leq p \leq 1$.

34. For positive numbers a and b

$$\lim_{p \to 0} \left(\frac{a^p + b^p}{2}\right)^{1/p} = \sqrt{ab}.$$

Is it true that for $0 < A, B \in \mathbb{M}_n(\mathbb{C})$

$$\lim_{p \to 0} \left(\frac{A^p + B^p}{2}\right)^{1/p}$$

is the geometric mean of A and B?

Chapter 6
Majorization and Singular Values

A citation from von Neumann: "The object of this note is the study of certain properties of complex matrices of nth order: $A = (a_{ij})_{i,j=1}^{n}$, n being a finite positive integer: $n = 1, 2, \ldots$. Together with them we shall use complex vectors of nth order (in n dimensions): $x = (x_i)_{i=1}^{n}$." This classical subject in matrix theory is exposed in Sects. 6.2 and 6.3 after discussions on vectors in Sect. 6.1. This chapter also contains several matrix norm inequalities as well as majorization results for matrices, which were mostly developed more recently.

Basic properties of singular values of matrices are given in Sect. 6.2. This section also contains several fundamental majorizations, notably the Lidskii–Wielandt and Gel'fand–Naimark theorems, for the eigenvalues of Hermitian matrices and the singular values of general matrices. Section 6.3 covers the important subject of symmetric or unitarily invariant norms for matrices. Symmetric norms are written as symmetric gauge functions of the singular values of matrices (von Neumann's theorem). So they are closely connected with majorization theory as manifestly seen from the fact that the weak majorization $s(A) \prec_w s(B)$ for the singular value vectors $s(A), s(B)$ of matrices A, B is equivalent to the inequality $|||A||| \leq |||B|||$ for all symmetric norms, as summarized in Theorem 6.23. Therefore, the majorization method is of particular use to obtain various symmetric norm inequalities for matrices.

Section 6.4 further collects several more recent results on majorization (and hence, on symmetric norm inequalities) for positive matrices involving concave or convex functions, or operator monotone functions, or certain matrix means. For instance, the symmetric norm inequalities of Golden–Thompson type and of its complementary type are presented.

6.1 Majorization of Vectors

Let $a = (a_1, \ldots, a_n)$ and $b = (b_1, \ldots, b_n)$ be vectors in $\mathbb{R}^n$. The **decreasing rearrangement** of a is $a^{\downarrow} = (a_1^{\downarrow}, \ldots, a_n^{\downarrow})$ and $b^{\downarrow} = (b_1^{\downarrow}, \ldots, b_n^{\downarrow})$ is similarly defined. The **majorization** $a \prec b$ means that

F. Hiai and D. Petz, *Introduction to Matrix Analysis and Applications*,
Universitext, DOI: 10.1007/978-3-319-04150-6_6,
© Hindustan Book Agency 2014

$$\sum_{i=1}^{k} a_i^{\downarrow} \leq \sum_{i=1}^{k} b_i^{\downarrow} \qquad (1 \leq k \leq n) \tag{6.1}$$

and equality is required for $k = n$. The **weak majorization** $a \prec_w b$ is defined by the inequality (6.1), where the equality for $k = n$ is not required. These concepts were introduced by Hardy, Littlewood and Pólya.

The majorization $a \prec b$ is equivalent to the statement that a is a convex combination of permutations of the components of the vector b. This can be written as

$$a = \sum_{U} \lambda_U U b,$$

where the summation is over the $n \times n$ permutation matrices U and $\lambda_U \geq 0$, $\sum_U \lambda_U = 1$. The $n \times n$ matrix $D = \sum_U \lambda_U U$ has the property that all entries are positive and the sums of rows and columns are 1. Such a matrix D is called **doubly stochastic**. So $a = Db$. The proof is a part of the next theorem.

Theorem 6.1 *The following conditions for $a, b \in \mathbb{R}^n$ are equivalent:*

(1) $a \prec b$;
(2) $\sum_{i=1}^{n} |a_i - r| \leq \sum_{i=1}^{n} |b_i - r|$ *for all* $r \in \mathbb{R}$;
(3) $\sum_{i=1}^{n} f(a_i) \leq \sum_{i=1}^{n} f(b_i)$ *for any convex function* f *on an interval containing all* a_i, b_i;
(4) a *is a convex combination of coordinate permutations of* b;
(5) $a = Db$ *for some doubly stochastic* $n \times n$ *matrix* D.

Proof: (1) $\Rightarrow$ (4). We show that there exist a finite number of matrices $D_1, \dots, D_N$ of the form $\lambda I + (1 - \lambda)\Pi$ where $0 \leq \lambda \leq 1$ and Π is a permutation matrix interchanging two coordinates only such that $a = D_N \cdots D_1 b$. Then (4) follows because $D_N \cdots D_1$ becomes a convex combination of permutation matrices. We may assume that $a_1 \geq \cdots \geq a_n$ and $b_1 \geq \cdots \geq b_n$. Suppose $a \neq b$ and choose the largest j such that $a_j < b_j$. Then there exists a k with $k > j$ such that $a_k > b_k$. Choose the smallest such k. Let $\lambda_1 := 1 - \min\{b_j - a_j, a_k - b_k\}/(b_j - b_k)$ and Π_1 be the permutation matrix interchanging the jth and kth coordinates. Then $0 < \lambda_1 < 1$ since $b_j > a_j \geq a_k > b_k$. Define $D_1 := \lambda_1 I + (1 - \lambda_1)\Pi_1$ and $b^{(1)} := D_1 b$. Now it is easy to check that $a \prec b^{(1)} \prec b$ and $b_1^{(1)} \geq \cdots \geq b_n^{(1)}$. Moreover the jth or kth coordinates of a and $b^{(1)}$ are equal. When $a \neq b^{(1)}$, we can apply the above argument to a and $b^{(1)}$. Repeating this finitely many times we reach the conclusion.

(4) $\Rightarrow$ (5) trivially follows from the fact that any convex combination of permutation matrices is doubly stochastic.

(5) $\Rightarrow$ (2). For every $r \in \mathbb{R}$ we have

$$\sum_{i=1}^{n} |a_i - r| = \sum_{i=1}^{n} \left| \sum_{j=1}^{n} D_{ij}(b_j - r) \right| \leq \sum_{i,j=1}^{n} D_{ij}|b_j - r| = \sum_{j=1}^{n} |b_j - r|.$$

$(2) \Rightarrow (1)$. Taking large r and small r in the inequality of (2) we have $\sum_{i=1}^{n} a_i = \sum_{i=1}^{n} b_i$. Noting that $|x| + x = 2x_+$ for $x \in \mathbb{R}$, where $x_+ = \max\{x, 0\}$, we have

$$\sum_{i=1}^{n}(a_i - r)_+ \leq \sum_{i=1}^{n}(b_i - r)_+ \qquad (r \in \mathbb{R}). \qquad (6.2)$$

We now prove that (6.2) implies that $a \prec_w b$. When $b_k^{\downarrow} \geq r \geq b_{k+1}^{\downarrow}$, $\sum_{i=1}^{k} a_i^{\downarrow} \leq \sum_{i=1}^{k} b_i^{\downarrow}$ follows since

$$\sum_{i=1}^{n}(a_i - r)_+ \geq \sum_{i=1}^{k}(a_i^{\downarrow} - r)_+ \geq \sum_{i=1}^{k} a_i^{\downarrow} - kr, \qquad \sum_{i=1}^{n}(b_i - r)_+ = \sum_{i=1}^{k} b_i^{\downarrow} - kr.$$

$(4) \Rightarrow (3)$. Suppose that $a_i = \sum_{k=1}^{N} \lambda_k b_{\pi_k(i)}$, $1 \leq i \leq n$, where $\lambda_k > 0$, $\sum_{k=1}^{N} \lambda_k = 1$, and π_k are permutations on $\{1, \ldots, n\}$. Then the convexity of f implies that

$$\sum_{i=1}^{n} f(a_i) \leq \sum_{i=1}^{n} \sum_{k=1}^{N} \lambda_k f(b_{\pi_k(i)}) = \sum_{i=1}^{n} f(b_i).$$

$(3) \Rightarrow (5)$ is trivial since $f(x) = |x - r|$ is convex. $\qquad \square$

Note that the implication $(5) \Rightarrow (4)$ can be seen directly from the well-known theorem of **Birkhoff** stating that any **doubly stochastic** matrix is a convex combination of **permutation matrices** [27].

Example 6.2 Let $D^{AB} \in \mathbb{M}_n \otimes \mathbb{M}_m$ be a density matrix which is the convex combination of tensor products of density matrices: $D^{AB} = \sum_i \lambda_i D_i^A \otimes D_i^B$. We assume that the matrices D_i^A are acting on the Hilbert space $\mathcal{H}_A$ and D_i^B acts on $\mathcal{H}_B$.

The eigenvalues of D^{AB} form a probability vector $r = (r_1, r_2, \ldots, r_{nm})$. The reduced density matrix $D^A = \sum_i \lambda_i (\text{Tr } D_i^B) D_i^A$ has n eigenvalues and we add $nm - n$ zeros to get a probability vector $q = (q_1, q_2, \ldots, q_{nm})$. We want to show that there is a doubly stochastic matrix S which transforms q into r. This means $r \prec q$.

Let

$$D^{AB} = \sum_k r_k |e_k\rangle\langle e_k| = \sum_j p_j |x_j\rangle\langle x_j| \otimes |y_j\rangle\langle y_j|$$

be decompositions of a density matrix in terms of unit vectors $|e_k\rangle \in \mathcal{H}_A \otimes \mathcal{H}_B$, $|x_j\rangle \in \mathcal{H}_A$ and $|y_j\rangle \in \mathcal{H}_B$. The first decomposition is the Schmidt decomposition and the second one is guaranteed by the assumed separability condition. For the reduced density D^A we have the Schmidt decomposition and another one:

$$D^A = \sum_l q_l |f_l\rangle\langle f_l| = \sum_j p_j |x_j\rangle\langle x_j|,$$

where f_j is an orthonormal family in $\mathcal{H}_A$. According to Lemma 1.24 we have two unitary matrices V and W such that

$$\sum_k V_{kj}\sqrt{p_j}|x_j\rangle \otimes |y_j\rangle = \sqrt{r_k}|e_k\rangle$$

$$\sum_l W_{jl}\sqrt{q_l}|f_l\rangle = \sqrt{p_j}|x_j\rangle.$$

Combining these equations we obtain

$$\sum_k V_{kj} \sum_l W_{jl}\sqrt{q_l}|f_l\rangle \otimes |y_j\rangle = \sqrt{r_k}|e_k\rangle$$

and taking the squared norm:

$$r_k = \sum_l \left(\sum_{j_1, j_2} \overline{V}_{kj_1} V_{kj_2} \overline{W}_{j_1l} W_{j_2l}\langle y_{j_1}, y_{j_2}\rangle \right) q_l.$$

We verify that the matrix

$$S_{kl} = \left(\sum_{j_1, j_2} \overline{V}_{kj_1} V_{kj_2} \overline{W}_{j_1l} W_{j_2l}\langle y_{j_1}, y_{j_2}\rangle \right)$$

is doubly stochastic.   □

The **weak majorization** $a \prec_w b$ is defined by the inequality (6.1). A matrix S is called a **doubly substochastic** $n \times n$ matrix if $\sum_{j=1}^n S_{ij} \le 1$ for $1 \le i \le n$ and $\sum_{i=1}^n S_{ij} \le 1$ for $1 \le j \le n$.

While the previous theorem concerned majorization, the next theorem is about weak majorization.

Theorem 6.3 *The following conditions for $a, b \in \mathbb{R}^n$ are equivalent:*

(1) *$a \prec_w b$;*
(2) *there exists a $c \in \mathbb{R}^n$ such that $a \le c \prec b$, where $a \le c$ means that $a_i \le c_i$, $1 \le i \le n$;*
(3) *$\sum_{i=1}^n (a_i - r)_+ \le \sum_{i=1}^n (b_i - r)_+$ for all $r \in \mathbb{R}$;*
(4) *$\sum_{i=1}^n f(a_i) \le \sum_{i=1}^n f(b_i)$ for any increasing convex function f on an interval containing all a_i, b_i.*

Moreover, if $a, b \ge 0$, then the above conditions are equivalent to the following:

(5) *$a = Sb$ for some doubly substochastic $n \times n$ matrix S.*

Proof: (1) $\Rightarrow$ (2). By induction on n. We may assume that $a_1 \geq \cdots \geq a_n$ and $b_1 \geq \cdots \geq b_n$. Let $\alpha := \min_{1 \leq k \leq n} (\sum_{i=1}^k b_i - \sum_{i=1}^k a_i)$ and define $\tilde{a} := (a_1 + \alpha, a_2, \ldots, a_n)$. Then $a \leq \tilde{a} \prec_w b$ and $\sum_{i=1}^k \tilde{a}_i = \sum_{i=1}^k b_i$ for some $1 \leq k \leq n$. When $k = n$, $a \leq \tilde{a} \prec b$. When $k < n$, we have $(\tilde{a}_1, \ldots, \tilde{a}_k) \prec (b_1, \ldots, b_k)$ and $(\tilde{a}_{k+1}, \ldots, \tilde{a}_n) \prec_w (b_{k+1}, \ldots, b_n)$. Hence the induction assumption implies that $(\tilde{a}_{k+1}, \ldots, \tilde{a}_n) \leq (c_{k+1}, \ldots, c_n) \prec (b_{k+1}, \ldots, b_n)$ for some $(c_{k+1}, \ldots, c_n) \in \mathbb{R}^{n-k}$. Then $a \leq (\tilde{a}_1, \ldots, \tilde{a}_k, c_{k+1}, \ldots, c_n) \prec b$ is immediate from $\tilde{a}_k \geq b_k \geq b_{k+1} \geq c_{k+1}$.

(2) $\Rightarrow$ (4). Let $a \leq c \prec b$. If f is increasing and convex on an interval $[\alpha, \beta]$ containing a_i, b_i, then $c_i \in [\alpha, \beta]$ and

$$\sum_{i=1}^n f(a_i) \leq \sum_{i=1}^n f(c_i) \leq \sum_{i=1}^n f(b_i)$$

by Theorem 6.1.

(4) $\Rightarrow$ (3) is trivial and (3) $\Rightarrow$ (1) was already shown in the proof (2) $\Rightarrow$ (1) of Theorem 6.1.

We now assume $a, b \geq 0$ and prove that (2) $\Leftrightarrow$ (5). If $a \leq c \prec b$, then we have, by Theorem 6.1, $c = Db$ for some doubly stochastic matrix D and $a_i = \alpha_i c_i$ for some $0 \leq \alpha_i \leq 1$. So $a = \mathrm{Diag}(\alpha_1, \ldots, \alpha_n) Db$ and $\mathrm{Diag}(\alpha_1, \ldots, \alpha_n) D$ is a doubly substochastic matrix. Conversely if $a = Sb$ for a doubly substochastic matrix S, then a doubly stochastic matrix D exists so that $S \leq D$ entrywise, the proof of which is left as Exercise 1, and hence $a \leq Db \prec b$. $\qquad\square$

Example 6.4 Let $a, b \in \mathbb{R}^n$ and f be a convex function on an interval containing all a_i, b_i. We use the notation $f(a) := (f(a_1), \ldots, f(a_n))$ and similarly $f(b)$. Assume that $a \prec b$. Since f is a convex function, so is $(f(x) - r)_+$ for any $r \in \mathbb{R}$. Hence $f(a) \prec_w f(b)$ follows from Theorems 6.1 and 6.3.

Next assume that $a \prec_w b$ and f is an increasing convex function, then $f(a) \prec_w f(b)$ can be proved similarly. $\qquad\square$

Let $a, b \in \mathbb{R}^n$ and $a, b \geq 0$. We define the **weak log-majorization** $a \prec_{w(\log)} b$ when

$$\prod_{i=1}^k a_i^\downarrow \leq \prod_{i=1}^k b_i^\downarrow \qquad (1 \leq k \leq n) \tag{6.3}$$

and the **log-majorization** $a \prec_{(\log)} b$ when $a \prec_{w(\log)} b$ and equality holds for $k = n$ in (6.3). It is obvious that if a and b are strictly positive, then $a \prec_{(\log)} b$ (resp., $a \prec_{w(\log)} b$) if and only if $\log a \prec \log b$ (resp., $\log a \prec_w \log b$), where $\log a := (\log a_1, \ldots, \log a_n)$.

Theorem 6.5 Let $a, b \in \mathbb{R}^n$ with $a, b \geq 0$ and assume that $a \prec_{w(\log)} b$. If f is a continuous increasing function on $[0, \infty)$ such that $f(e^x)$ is convex, then $f(a) \prec_w f(b)$. In particular, $a \prec_{w(\log)} b$ implies $a \prec_w b$.

Proof: First assume that $a, b \in \mathbb{R}^n$ are strictly positive and $a \prec_{w(\log)} b$, so that $\log a \prec_w \log b$. Since $g \circ h$ is convex when g and h are convex with g increasing, the function $(f(e^x) - r)_+$ is increasing and convex for any $r \in \mathbb{R}$. Hence by Theorem 6.3 we have

$$\sum_{i=1}^n (f(a_i) - r)_+ \leq \sum_{i=1}^n (f(b_i) - r)_+,$$

which implies $f(a) \prec_w f(b)$ by Theorem 6.3 again. When $a, b \geq 0$ and $a \prec_{w(\log)} b$, we can choose $a^{(m)}, b^{(m)} > 0$ such that $a^{(m)} \prec_{w(\log)} b^{(m)}$, $a^{(m)} \to a$, and $b^{(m)} \to b$. Since $f(a^{(m)}) \prec_w f(b^{(m)})$ and f is continuous, we obtain $f(a) \prec_w f(b)$. □

6.2 Singular Values

In this section we discuss the majorization theory for eigenvalues and singular values of matrices. Our goal is to prove the Lidskii–Wielandt and Gel'fand–Naimark theorems for singular values of matrices. These are the most fundamental majorizations for matrices.

When A is self-adjoint, the vector of the eigenvalues of A in decreasing order with counting multiplicities is denoted by $\lambda(A)$. The majorization relation of self-adjoint matrices also appears in quantum theory.

Example 6.6 In quantum theory the states are described by density matrices (positive matrices with trace 1). Let D_1 and D_2 be density matrices. The relation $\lambda(D_1) \prec \lambda(D_2)$ has the interpretation that D_1 is **more mixed** than D_2. Among the $n \times n$ density matrices the "most mixed" has all eigenvalues equal to $1/n$.

Let $f : \mathbb{R}^+ \to \mathbb{R}^+$ be an increasing convex function with $f(0) = 0$. We show that

$$\lambda(D) \prec \lambda(f(D)/\mathrm{Tr}\, f(D)) \tag{6.4}$$

for a density matrix D.

Set $\lambda(D) = (\lambda_1, \lambda_2, \ldots, \lambda_n)$. Under the hypothesis on f the inequality $f(y)x \geq f(x)y$ holds for $0 \leq x \leq y$. Hence for $i \leq j$ we have $\lambda_j f(\lambda_i) \geq \lambda_i f(\lambda_j)$ and

$$\big(f(\lambda_1) + \cdots + f(\lambda_k)\big)(\lambda_{k+1} + \cdots + \lambda_n)$$
$$\geq (\lambda_1 + \cdots + \lambda_k)\big(f(\lambda_{k+1}) + \cdots + f(\lambda_n)\big).$$

Adding to both sides the term $(f(\lambda_1) + \cdots + f(\lambda_k))(\lambda_1 + \cdots + \lambda_k)$ we arrive at

$$(f(\lambda_1) + \cdots + f(\lambda_k)) \sum_{i=1}^{n} \lambda_i \geq (\lambda_1 + \cdots + \lambda_k) \sum_{i=1}^{n} f(\lambda_i).$$

This shows that the sum of the k largest eigenvalues of $f(D)/\mathrm{Tr}\, f(D)$ must exceed that of D (which is $\lambda_1 + \cdots + \lambda_k$).

The canonical (Gibbs) state at inverse temperature $\beta = (kT)^{-1}$ possesses the density $e^{-\beta H}/\mathrm{Tr}\, e^{-\beta H}$. Choosing $f(x) = x^{\beta'/\beta}$ with $\beta' > \beta$ the formula (6.4) tells us that

$$e^{-\beta H}/\mathrm{Tr}\, e^{-\beta H} \prec e^{-\beta' H}/\mathrm{Tr}\, e^{-\beta' H}$$

that is, at higher temperatures the canonical density is more mixed.   □

Let $\mathcal{H}$ be an n-dimensional Hilbert space and $A \in B(\mathcal{H})$. Let $s(A) = (s_1(A), \ldots, s_n(A))$ denote the vector of the **singular values** of A in decreasing order, i.e., $s_1(A) \geq \cdots \geq s_n(A)$ are the eigenvalues of $|A| = (A^*A)^{1/2}$, counting multiplicities.

The basic properties of the singular values are summarized in the next theorem, which includes the definition of the **minimax expression**, see Theorem 1.27. Recall that $\|\cdot\|$ is the operator norm.

Theorem 6.7 *Let $A, B, X, Y \in B(\mathcal{H})$ and $k, m \in \{1, \ldots, n\}$. Then:*

(1) $s_1(A) = \|A\|$.
(2) $s_k(\alpha A) = |\alpha| s_k(A)$ *for* $\alpha \in \mathbb{C}$.
(3) $s_k(A) = s_k(A^*)$.
(4) *Minimax expression:*

$$s_k(A) = \min\{\|A(I - P)\| : P \text{ is a projection, rank } P = k - 1\}. \qquad (6.5)$$

If $A \geq 0$ then

$$s_k(A) = \min \left\{ \max\{\langle x, Ax \rangle : x \in \mathcal{M}^\perp, \|x\| = 1\} : \right.$$
$$\left. \mathcal{M} \text{ is a subspace of } \mathcal{H}, \dim \mathcal{M} = k - 1 \right\}. \qquad (6.6)$$

(5) *Approximation number expression:*

$$s_k(A) = \inf\{\|A - X\| : X \in B(\mathcal{H}), \text{ rank } X < k\}. \qquad (6.7)$$

(6) *If $0 \leq A \leq B$ then $s_k(A) \leq s_k(B)$.*
(7) $s_k(XAY) \leq \|X\| \|Y\| s_k(A)$.
(8) $s_{k+m-1}(A + B) \leq s_k(A) + s_m(B)$ *if $k + m - 1 \leq n$.*
(9) $s_{k+m-1}(AB) \leq s_n(A)s_m(B)$ *if $k + m - 1 \leq n$.*
(10) $|s_k(A) - s_k(B)| \leq \|A - B\|$.
(11) $s_k(f(A)) = f(s_k(A))$ *if $A \geq 0$ and $f : \mathbb{R}^+ \to \mathbb{R}^+$ is an increasing function.*

Proof: First, we recall some basic decompositions of $A \in B(\mathcal{H})$. Let $A = U|A|$ be the polar decomposition of A and write the Schmidt decomposition of $|A|$ as

$$|A| = \sum_{i=1}^{n} s_i(A)|u_i\rangle\langle u_i|,$$

where U is a unitary and $\{u_1, \ldots, u_n\}$ is an orthonormal basis of $\mathcal{H}$. From the polar decomposition of A and the diagonalization of $|A|$ one has the expression

$$A = U_1\mathrm{Diag}(s_1(A), \ldots, s_n(A))U_2 \tag{6.8}$$

with unitaries $U_1, U_2 \in B(\mathcal{H})$, called the **singular value decomposition** of A, see Theorem 1.46.

(1) follows since $s_1(A) = \| |A| \| = \|A\|$. (2) is clear from $|\alpha A| = |\alpha| |A|$. Also, (3) immediately follows since the Schmidt decomposition of $|A^*|$ is given by

$$|A^*| = U|A|U^* = \sum_{i=1}^{n} s_i(A)|Uu_i\rangle\langle Uu_i|.$$

(4) Let α_k be the right-hand side of (6.5). For $1 \leq k \leq n$ define $P_k := \sum_{i=1}^{k} |u_i\rangle\langle u_i|$, which is a projection of rank k. We have

$$\alpha_k \leq \|A(I - P_{k-1})\| = \left\| \sum_{i=k}^{n} s_i(A)|u_i\rangle\langle u_i| \right\| = s_k(A).$$

Conversely, for any $\varepsilon > 0$ choose a projection P with rank $P = k - 1$ such that $\|A(I - P)\| < \alpha_k + \varepsilon$. Then there exists a $y \in \mathcal{H}$ with $\|y\| = 1$ such that $P_k y = y$ but $Py = 0$. Since $y = \sum_{i=1}^{k}\langle u_i, y\rangle u_i$, we have

$$\alpha_k + \varepsilon > \| |A|(I - P)y\| = \| |A|y\| = \left\| \sum_{i=1}^{k}\langle u_i, y\rangle s_i(A)u_i \right\|$$

$$= \left(\sum_{i=1}^{k} |\langle u_i, y\rangle|^2 s_i(A)^2 \right)^{1/2} \geq s_k(A).$$

Hence $s_k(A) = \alpha_k$ and the infimum α_k is attained by $P = P_{k-1}$.

Although the second expression (6.6) is included in Theorem 1.27, we give the proof for convenience. When $A \geq 0$, we have

$$s_k(A) = s_k(A^{1/2})^2 = \min\{\|A^{1/2}(I - P)\|^2 : P \text{ is a projection, rank } P = k - 1\}.$$

Since $\|A^{1/2}(I - P)\|^2 = \max_{x \in \mathcal{M}^\perp, \|x\|=1} \langle x, Ax \rangle$ with $\mathcal{M} := \operatorname{ran} P$, the latter expression follows.

(5) Let β_k be the right-hand side of (6.7). Let $X := AP_{k-1}$, where P_{k-1} is as in the above proof of (1). Then we have $\operatorname{rank} X \leq \operatorname{rank} P_{k-1} = k - 1$ so that $\beta_k \leq \|A(I - P_{k-1})\| = s_k(A)$. Conversely, assume that $X \in B(\mathcal{H})$ has rank $< k$. Since $\operatorname{rank} X = \operatorname{rank} |X| = \operatorname{rank} X^*$, the projection P onto $\operatorname{ran} X^*$ has rank $< k$. Then $X(I - P) = 0$ and by (6.5) we have

$$s_k(A) \leq \|A(I - P)\| = \|(A - X)(I - P)\| \leq \|A - X\|,$$

implying that $s_k(A) \leq \beta_k$. Hence $s_k(A) = \beta_k$ and the infimum β_k is attained by AP_{k-1}.

(6) is an immediate consequence of (6.6). It is immediate from (6.5) that $s_n(XA) \leq \|X\| s_n(A)$. Also $s_n(AY) = s_n(Y^*A^*) \leq \|Y\| s_n(A)$ by (3). Hence (7) holds.

Next we show (8)–(10). By (6.7) there exist $X, Y \in B(\mathcal{H})$ with $\operatorname{rank} X < k$, $\operatorname{rank} Y < m$ such that $\|A - X\| = s_k(A)$ and $\|B - Y\| = s_m(B)$. Since $\operatorname{rank}(X+Y) \leq \operatorname{rank} X + \operatorname{rank} Y < k + m - 1$, we have

$$s_{k+m-1}(A + B) \leq \|(A + B) - (X + Y)\| < s_k(A) + s_m(B),$$

implying (8). For $Z := XB + (A - X)Y$ we get

$$\operatorname{rank} Z \leq \operatorname{rank} X + \operatorname{rank} Y < k + m - 1,$$

$$\|AB - Z\| = \|(A - X)(B - Y)\| \leq s_k(A)s_m(B).$$

These imply (9). Letting $m = 1$ and replacing B by $B - A$ in (8) we get

$$s_k(B) \leq s_k(A) + \|B - A\|,$$

which shows (10).

(11) When $A \geq 0$ has the Schmidt decomposition $A = \sum_{i=1}^n s_i(A)|u_i\rangle\langle u_i|$, we have $f(A) = \sum_{i=1}^n f(s_i(A))|u_i\rangle\langle u_i|$. Since $f(s_1(A)) \geq \cdots \geq f(s_n(A)) \geq 0$, $s_k(f(A)) = f(s_k(A))$ follows. $\square$

The next result is called the **Weyl majorization theorem**, whose proof reveals the usefulness of the antisymmetric tensor technique.

Theorem 6.8 *Let $A \in \mathbb{M}_n$ and $\lambda_1(A), \cdots, \lambda_n(A)$ be the eigenvalues of A arranged as $|\lambda_1(A)| \geq \cdots \geq |\lambda_n(A)|$, counting algebraic multiplicities. Then*

$$\prod_{i=1}^k |\lambda_i(A)| \leq \prod_{i=1}^k s_i(A) \quad (1 \leq k \leq n).$$

Proof: If λ is an eigenvalue of A with algebraic multiplicity m, then there exists a set $\{y_1, \ldots, y_m\}$ of independent vectors such that

$$Ay_j - \lambda y_j \in \text{span}\{y_1, \ldots, y_{j-1}\} \quad (1 \leq j \leq m).$$

Hence one can choose independent vectors $x_1, \ldots, x_n$ so that $Ax_i = \lambda_i(A)x_i + z_i$ with $z_i \in \text{span}\{x_1, \ldots, x_{i-1}\}$ for $1 \leq i \leq n$. Then it is readily checked that

$$A^{\wedge k}(x_1 \wedge \cdots \wedge x_k) = Ax_1 \wedge \cdots \wedge Ax_k = \left(\prod_{i=1}^{k} \lambda_i(A)\right) x_1 \wedge \cdots \wedge x_k$$

and $x_1 \wedge \cdots \wedge x_k \neq 0$, implying that $\prod_{i=1}^{k} \lambda_i(A)$ is an eigenvalue of $A^{\wedge k}$. Hence Lemma 1.62 yields that

$$\left|\prod_{i=1}^{k} \lambda_i(A)\right| \leq \|A^{\wedge k}\| = \prod_{i=1}^{k} s_i(A).$$

$\square$

Note that another formulation of the previous theorem is

$$(|\lambda_1(A)|, \ldots, |\lambda_n(A)|) \prec_{w(\log)} s(A).$$

The following majorization results are the celebrated **Lidskii–Wielandt theorem** for the eigenvalues of self-adjoint matrices as well as for the singular values of general matrices.

Theorem 6.9 *If $A, B \in \mathbb{M}_n^{sa}$, then*

$$\lambda(A) - \lambda(B) \prec \lambda(A - B),$$

or equivalently

$$(\lambda_i(A) + \lambda_{n-i+1}(B)) \prec \lambda(A + B).$$

Proof: What we need to prove is that for any choice of $1 \leq i_1 < i_2 < \cdots < i_k \leq n$ we have

$$\sum_{j=1}^{k} (\lambda_{i_j}(A) - \lambda_{i_j}(B)) \leq \sum_{j=1}^{k} \lambda_j(A - B). \tag{6.9}$$

Choose the Schmidt decomposition of $A - B$ as

$$A - B = \sum_{i=1}^{n} \lambda_i(A - B)|u_i\rangle\langle u_i|$$

for an orthonormal basis $\{u_1, \ldots, u_n\}$ of $\mathbb{C}^n$. We may assume without loss of generality that $\lambda_k(A - B) = 0$. In fact, we may replace B by $B + \lambda_k(A - B)I$, which reduces both sides of (6.9) by $k\lambda_k(A - B)$. In this situation, the Jordan decomposition $A - B = (A - B)_+ - (A - B)_-$ is given by

$$(A - B)_+ = \sum_{i=1}^{k} \lambda_i(A - B)|u_i\rangle\langle u_i|, \qquad (A - B)_- = -\sum_{i=k+1}^{n} \lambda_i(A - B)|u_i\rangle\langle u_i|.$$

Since $A = B + (A - B)_+ - (A - B)_- \leq B + (A - B)_+$, it follows from Theorem 1.27 that

$$\lambda_i(A) \leq \lambda_i(B + (A - B)_+), \qquad 1 \leq i \leq n.$$

Since $B \leq B + (A - B)_+$, we also have

$$\lambda_i(B) \leq \lambda_i(B + (A - B)_+), \qquad 1 \leq i \leq n.$$

Hence

$$\sum_{j=1}^{k}(\lambda_{i_j}(A) - \lambda_{i_j}(B)) \leq \sum_{j=1}^{k}(\lambda_{i_j}(B + (A - B)_+) - \lambda_{i_j}(B))$$

$$\leq \sum_{i=1}^{n}(\lambda_i(B + (A - B)_+) - \lambda_i(B))$$

$$= \mathrm{Tr}\,(B + (A - B)_+) - \mathrm{Tr}\,B$$

$$= \mathrm{Tr}\,(A - B)_+ = \sum_{j=1}^{k} \lambda_j(A - B),$$

proving (6.9). Moreover,

$$\sum_{i=1}^{n}(\lambda_i(A) - \lambda_i(B)) = \mathrm{Tr}\,(A - B) = \sum_{i=1}^{n} \lambda_i(A - B).$$

The latter expression is obvious since $\lambda_i(B) = -\lambda_{n-i+1}(-B)$ for $1 \leq i \leq n$. $\qquad\square$

Theorem 6.10 *For all* $A, B \in \mathbb{M}_n$

$$|s(A) - s(B)| \prec_w s(A - B)$$

holds, that is,

$$\sum_{j=1}^{k} |s_{i_j}(A) - s_{i_j}(B)| \le \sum_{j=1}^{k} s_j(A - B)$$

for any choice of $1 \le i_1 < i_2 < \cdots < i_k \le n.$

Proof: Define

$$\mathbf{A} := \begin{bmatrix} 0 & A^* \\ A & 0 \end{bmatrix}, \qquad \mathbf{B} := \begin{bmatrix} 0 & B^* \\ B & 0 \end{bmatrix}.$$

Since

$$\mathbf{A}^*\mathbf{A} = \begin{bmatrix} A^*A & 0 \\ 0 & AA^* \end{bmatrix}, \qquad |\mathbf{A}| = \begin{bmatrix} |A| & 0 \\ 0 & |A^*| \end{bmatrix},$$

it follows from Theorem 6.7 (3) that

$$s(\mathbf{A}) = (s_1(A), s_1(A), s_2(A), s_2(A), \dots, s_n(A), s_n(A)).$$

On the other hand, since

$$\begin{bmatrix} I & 0 \\ 0 & -I \end{bmatrix} \mathbf{A} \begin{bmatrix} I & 0 \\ 0 & -I \end{bmatrix} = -\mathbf{A},$$

we have $\lambda_i(\mathbf{A}) = \lambda_i(-\mathbf{A}) = -\lambda_{2n-i+1}(\mathbf{A})$ for $n \le i \le 2n$. Hence one can write

$$\lambda(\mathbf{A}) = (\lambda_1, \dots, \lambda_n, -\lambda_n, \dots, -\lambda_1),$$

where $\lambda_1 \ge \cdots \ge \lambda_n \ge 0$. Since

$$s(\mathbf{A}) = \lambda(|\mathbf{A}|) = (\lambda_1, \lambda_1, \lambda_2, \lambda_2, \dots, \lambda_n, \lambda_n),$$

we have $\lambda_i = s_i(A)$ for $1 \le i \le n$ and hence

$$\lambda(\mathbf{A}) = (s_1(A), \dots, s_n(A), -s_n(A), \dots, -s_1(A)).$$

Similarly,

$$\lambda(\mathbf{B}) = (s_1(B), \dots, s_n(B), -s_n(B), \dots, -s_1(B)),$$
$$\lambda(\mathbf{A} - \mathbf{B}) = (s_1(A - B), \dots, s_n(A - B), -s_n(A - B), \dots, -s_1(A - B)).$$

Theorem 6.9 implies that

$$\lambda(\mathbf{A}) - \lambda(\mathbf{B}) \prec \lambda(\mathbf{A} - \mathbf{B}).$$

Now we note that the components of $\lambda(\mathbf{A}) - \lambda(\mathbf{B})$ are

$$|s_1(A) - s_1(B)|, \ldots, |s_n(A) - s_n(B)|, -|s_1(A) - s_1(B)|, \ldots, -|s_n(A) - s_n(B)|.$$

Therefore, for any choice of $1 \le i_1 < i_2 < \cdots < i_k \le n$ with $1 \le k \le n$, we have

$$\sum_{j=1}^{k} |s_{i_j}(A) - s_{i_j}(B)| \le \sum_{i=1}^{k} \lambda_i(\mathbf{A} - \mathbf{B}) = \sum_{j=1}^{k} s_j(A - B),$$

the proof is complete. □

The following results due to **Ky Fan** are weaker versions of the Lidskii–Wielandt theorem; indeed, they are consequences of the above theorems.

Corollary 6.11 *If $A, B \in \mathbb{M}_n^{sa}$, then*

$$\lambda(A + B) \prec \lambda(A) + \lambda(B).$$

Proof: Apply Theorem 6.9 to $A + B$ and B. Then

$$\sum_{i=1}^{k} \left(\lambda_i(A + B) - \lambda_i(B) \right) \le \sum_{i=1}^{k} \lambda_i(A),$$

so that

$$\sum_{i=1}^{k} \lambda_i(A + B) \le \sum_{i=1}^{k} \left(\lambda_i(A) + \lambda_i(B) \right).$$

Moreover, $\sum_{i=1}^{n} \lambda_i(A + B) = \text{Tr}\,(A + B) = \sum_{i=1}^{n} \left(\lambda_i(A) + \lambda_i(B) \right)$. □

Corollary 6.12 *If $A, B \in \mathbb{M}_n$, then*

$$s(A + B) \prec_w s(A) + s(B).$$

Proof: Similarly, by Theorem 6.10,

$$\sum_{i=1}^{k} |s_i(A + B) - s_i(B)| \le \sum_{i=1}^{k} s_i(A)$$

so that

$$\sum_{i=1}^{k} s_i(A + B) \le \sum_{i=1}^{k} \left(s_i(A) + s_i(B) \right).$$ □

Another important majorization for singular values of matrices is the **Gel'fand–Naimark theorem**:

Theorem 6.13 *For all $A, B \in \mathbb{M}_n$*

$$(s_i(A)s_{n-i+1}(B)) \prec_{(\log)} s(AB) \tag{6.10}$$

holds, or equivalently

$$\prod_{j=1}^{k} s_{i_j}(AB) \leq \prod_{j=1}^{k} (s_j(A)s_{i_j}(B)) \tag{6.11}$$

for every $1 \leq i_1 < i_2 < \cdots < i_k \leq n$ with equality for $k = n$.

Proof: First assume that A and B are invertible matrices and let $A = U_1 \mathrm{Diag}(s_1, \ldots, s_n)U_2$ be the singular value decomposition (see (6.8)) with the singular values $s_1 \geq \cdots \geq s_n > 0$ of A and unitaries U_1, U_2. Write $D := \mathrm{Diag}(s_1, \ldots, s_n)$. Then $s(AB) = s(U_1 D U_2 B) = s(D U_2 B)$ and $s(B) = s(U_2 B)$, so we may replace A, B by $D, U_2 B$, respectively. Hence we may assume that $A = D = \mathrm{Diag}(s_1, \ldots, s_n)$. Moreover, to prove (6.11), it suffices to assume that $s_k = 1$. In fact, when A is replaced by $s_k^{-1}A$, both sides of (6.11) are multiplied by the same s_k^{-k}. Define $\tilde{A} := \mathrm{Diag}(s_1, \ldots, s_k, 1, \ldots, 1)$; then $\tilde{A}^2 \geq A^2$ and $\tilde{A}^2 \geq I$. We notice from Theorem 6.7 that we have

$$s_i(AB) = s_i((B^* A^2 B)^{1/2}) = s_i(B^* A^2 B)^{1/2}$$
$$\leq s_i(B^* \tilde{A}^2 B)^{1/2} = s_i(\tilde{A}B)$$

for every $i = 1, \ldots, n$ and

$$s_i(\tilde{A}B) = s_i(B^* \tilde{A}^2 B)^{1/2} \geq s_i(B^* B)^{1/2} = s_i(B).$$

Therefore, for any choice of $1 \leq i_1 < \cdots < i_k \leq n$, we have

$$\prod_{j=1}^{k} \frac{s_{i_j}(AB)}{s_{i_j}(B)} \leq \prod_{j=1}^{k} \frac{s_{i_j}(\tilde{A}B)}{s_{i_j}(B)} \leq \prod_{i=1}^{n} \frac{s_i(\tilde{A}B)}{s_i(B)} = \frac{\det |\tilde{A}B|}{\det |B|}$$

$$= \frac{\sqrt{\det(B^* \tilde{A}^2 B)}}{\sqrt{\det(B^* B)}} = \frac{\det \tilde{A} \cdot |\det B|}{|\det B|} = \det \tilde{A} = \prod_{j=1}^{k} s_j(A),$$

proving (6.11). By replacing A and B by AB and B^{-1}, respectively, (6.11) is rephrased as

$$\prod_{j=1}^{k} s_{i_j}(A) \le \prod_{j=1}^{k} \left(s_j(AB)s_{i_j}(B^{-1}) \right).$$

Since $s_i(B^{-1}) = s_{n-i+1}(B)^{-1}$ for $1 \le i \le n$ as readily verified, the above inequality means that

$$\prod_{j=1}^{k} \left(s_{i_j}(A)s_{n-i_j+1}(B) \right) \le \prod_{j=1}^{k} s_j(AB).$$

Hence (6.11) implies (6.10) and vice versa (as long as A, B are invertible).

For general $A, B \in \mathbb{M}_n$ choose a sequence of complex numbers $\alpha_l \in \mathbb{C} \setminus (\sigma(A) \cup \sigma(B))$ such that $\alpha_l \to 0$. Since $A_l := A - \alpha_l I$ and $B_l := B - \alpha_l I$ are invertible, (6.10) and (6.11) hold for those. Then $s_i(A_l) \to s_i(A)$, $s_i(B_l) \to s_i(B)$ and $s_i(A_l B_l) \to s_i(AB)$ as $l \to \infty$ for $1 \le i \le n$. Hence (6.10) and (6.11) hold for general A, B. $\square$

An immediate corollary of this theorem is the following majorization result due to **Horn**.

Corollary 6.14 *For all matrices A and B,*

$$s(AB) \prec_{(\log)} s(A)s(B),$$

where $s(A)s(B) = (s_i(A)s_i(B))$.

Proof: A special case of (6.11) is

$$\prod_{i=1}^{k} s_i(AB) \le \prod_{i=1}^{k} \left(s_i(A)s_i(B) \right)$$

for every $k = 1, \dots, n$. Moreover,

$$\prod_{i=1}^{n} s_i(AB) = \det |AB| = \det |A| \cdot \det |B| = \prod_{i=1}^{n} \left(s_i(A)s_i(B) \right). \qquad \square$$

6.3 Symmetric Norms

A norm $\Phi : \mathbb{R}^n \to \mathbb{R}^+$ is said to be **symmetric** if

$$\Phi(a_1, a_2, \dots, a_n) = \Phi(\varepsilon_1 a_{\pi(1)}, \varepsilon_2 a_{\pi(2)}, \dots, \varepsilon_n a_{\pi(n)}) \qquad (6.12)$$

for every $(a_1, \dots, a_n) \in \mathbb{R}^n$, for any permutation π on $\{1, \dots, n\}$ and $\varepsilon_i = \pm 1$. The normalization is $\Phi(1, 0, \dots, 0) = 1$. Condition (6.12) is equivalently written as

$$\Phi(a) = \Phi(a_1^*, a_2^*, \dots, a_n^*)$$

for $a = (a_1, \ldots, a_n) \in \mathbb{R}^n$, where $(a_1^*, \ldots, a_n^*)$ is the decreasing rearrangement of $(|a_1|, \ldots, |a_n|)$. A symmetric norm is often called a **symmetric gauge function**.

Typical examples of symmetric gauge functions on $\mathbb{R}^n$ are the ℓ_p-norms Φ_p defined by

$$\Phi_p(a) := \begin{cases} \left(\sum_{i=1}^n |a_i|^p \right)^{1/p} & \text{if } 1 \le p < \infty, \\ \max\{|a_i| : 1 \le i \le n\} & \text{if } p = \infty. \end{cases} \tag{6.13}$$

The next lemma characterizes the minimal and maximal normalized symmetric norms.

Lemma 6.15 *Let Φ be a normalized symmetric norm on $\mathbb{R}^n$. If $a = (a_i)$, $b = (b_i) \in \mathbb{R}^n$ and $|a_i| \le |b_i|$ for $1 \le i \le n$, then $\Phi(a) \le \Phi(b)$. Moreover,*

$$\max_{1 \le i \le n} |a_i| \le \Phi(a) \le \sum_{i=1}^n |a_i| \quad (a = (a_i) \in \mathbb{R}^n),$$

which means $\Phi_\infty \le \Phi \le \Phi_1$.

Proof: In view of (6.12) we have

$$\Phi(\alpha a_1, a_2, \ldots, a_n) \le \Phi(a_1, a_2, \ldots, a_n) \quad \text{for } 0 \le \alpha \le 1.$$

This is proved as follows:

$$\begin{aligned} &\Phi(\alpha a_1, a_2, \ldots, a_n) \\ &= \Phi\left(\frac{1+\alpha}{2} a_1 + \frac{1-\alpha}{2}(-a_1), \frac{1+\alpha}{2} a_2 + \frac{1-\alpha}{2} a_2, \ldots, \frac{1+\alpha}{2} a_n + \frac{1-\alpha}{2} a_n \right) \\ &\le \frac{1+\alpha}{2} \Phi(a_1, a_2, \ldots, a_n) + \frac{1-\alpha}{2} \Phi(-a_1, a_2, \ldots, a_n) \\ &= \Phi(a_1, a_2, \ldots, a_n). \end{aligned}$$

(6.12) and the previous inequality imply that

$$|a_i| = \Phi(a_i, 0, \ldots, 0) \le \Phi(a).$$

This means $\Phi_\infty \leq \Phi$. From

$$\Phi(a) \leq \sum_{i=1}^{n} \Phi(a_i, 0, \ldots, 0) = \sum_{i=1}^{n} |a_i|$$

we have $\Phi \leq \Phi_1$. $\qquad\qquad\square$

Lemma 6.16 *If $a = (a_i)$, $b = (b_i) \in \mathbb{R}^n$ and $(|a_1|, \ldots, |a_n|) \prec_w (|b_1|, \ldots, |b_n|)$, then $\Phi(a) \leq \Phi(b)$.*

Proof: Theorem 6.3 implies that there exists a $c \in \mathbb{R}^n$ such that

$$(|a_1|, \ldots, |a_n|) \leq c \prec (|b_1|, \ldots, |b_n|).$$

Theorem 6.1 says that c is a convex combination of coordinate permutations of $(|b_1|, \ldots, |b_n|)$. Lemma 6.15 and (6.12) imply that $\Phi(a) \leq \Phi(c) \leq \Phi(b)$. $\qquad\square$

Let $\mathcal{H}$ be an n-dimensional Hilbert space. A norm $||| \cdot |||$ on $B(\mathcal{H})$ is said to be **unitarily invariant** if

$$|||UAV||| = |||A|||$$

for all $A \in B(\mathcal{H})$ and all unitaries $U, V \in B(\mathcal{H})$. A unitarily invariant norm on $B(\mathcal{H})$ is also called a **symmetric norm**. The following fundamental theorem is due to **von Neumann**.

Theorem 6.17 *There is a bijective correspondence between symmetric gauge functions Φ on $\mathbb{R}^n$ and unitarily invariant norms $||| \cdot |||$ on $B(\mathcal{H})$ determined by the formula*

$$|||A||| = \Phi(s(A)) \qquad (A \in B(\mathcal{H})). \tag{6.14}$$

Proof: Assume that Φ is a symmetric gauge function on $\mathbb{R}^n$. Define $||| \cdot |||$ on $B(\mathcal{H})$ by the formula (6.14). Let $A, B \in B(\mathcal{H})$. Since $s(A + B) \prec_w s(A) + s(B)$ by Corollary 6.12, it follows from Lemma 6.16 that

$$\begin{aligned}
|||A + B||| = \Phi(s(A + B)) &\leq \Phi(s(A) + s(B)) \\
&\leq \Phi(s(A)) + \Phi(s(B)) = |||A||| + |||B|||.
\end{aligned}$$

Furthermore, it is clear that $|||A||| = 0$ if and only if $s(A) = 0$ or $A = 0$. For $\alpha \in \mathbb{C}$ we have

$$|||\alpha A||| = \Phi(|\alpha|s(A)) = |\alpha| \, |||A|||$$

by Theorem 6.7. Hence $||| \cdot |||$ is a norm on $B(\mathcal{H})$, which is unitarily invariant since $s(UAV) = s(A)$ for all unitaries U, V.

Conversely, assume that $||| \cdot |||$ is a unitarily invariant norm on $B(\mathcal{H})$. Choose an orthonormal basis $\{e_1, \ldots, e_n\}$ of $\mathcal{H}$ and define $\Phi : \mathbb{R}^n \to \mathbb{R}$ by

$$\Phi(a) := \left\|\left\| \sum_{i=1}^{n} a_i |e_i\rangle\langle e_i| \right\|\right\| \qquad (a = (a_i) \in \mathbb{R}^n).$$

Then it is immediate that Φ is a norm on $\mathbb{R}^n$. For any permutation π on $\{1, \dots, n\}$ and $\varepsilon_i = \pm 1$, one can define unitaries U, V on $\mathcal{H}$ by $U e_{\pi(i)} = \varepsilon_i e_i$ and $V e_{\pi(i)} = e_i$, $1 \le i \le n$, so that

$$\Phi(a) = \left\|\left\| U \left(\sum_{i=1}^{n} a_{\pi(i)} |e_{\pi(i)}\rangle\langle e_{\pi(i)}| \right) V^* \right\|\right\| = \left\|\left\| \sum_{i=1}^{n} a_{\pi(i)} |U e_{\pi(i)}\rangle\langle V e_{\pi(i)}| \right\|\right\|$$

$$= \left\|\left\| \sum_{i=1}^{n} \varepsilon_i a_{\pi(i)} |e_i\rangle\langle e_i| \right\|\right\| = \Phi(\varepsilon_1 a_{\pi(1)}, \varepsilon_2 a_{\pi(2)}, \dots, \varepsilon_n a_{\pi(n)}).$$

Hence Φ is a symmetric gauge function. For any $A \in B(\mathcal{H})$ let $A = U|A|$ be the polar decomposition of A and $|A| = \sum_{i=1}^{n} s_i(A)|u_i\rangle\langle u_i|$ be the Schmidt decomposition of $|A|$ for an orthonormal basis $\{u_1, \dots, u_n\}$. We have a unitary V defined by $V e_i = u_i$, $1 \le i \le n$. Since

$$A = U|A| = UV \left(\sum_{i=1}^{n} s_i(A)|e_i\rangle\langle e_i| \right) V^*,$$

we have

$$\Phi(s(A)) = \left\|\left\| \sum_{i=1}^{n} s_i(A)|e_i\rangle\langle e_i| \right\|\right\| = \left\|\left\| UV \left(\sum_{i=1}^{n} s_i(A)|e_i\rangle\langle e_i| \right) V^* \right\|\right\| = |\|A\||,$$

and so (6.14) holds. Therefore, the theorem is proved. $\qquad\square$

The next theorem summarizes properties of unitarily invariant (or symmetric) norms on $B(\mathcal{H})$.

Theorem 6.18 *Let* $\||\cdot\||$ *be a unitarily invariant norm on* $B(\mathcal{H})$ *corresponding to a symmetric gauge function* Φ *on* $\mathbb{R}^n$ *and* $A, B, X, Y \in B(\mathcal{H})$. *Then:*

(1) $\||A\|| = \||A^*\||$.
(2) $\||XAY\|| \le \|X\| \, \|Y\| \, \||A\||$.
(3) *If* $s(A) \prec_w s(B)$, *then* $\||A\|| \le \||B\||$.
(4) *Under the normalization we have* $\|A\| \le \||A\|| \le \|A\|_1$.

Proof: By the definition (6.14), (1) follows from Theorem 6.7. By Theorem 6.7 and Lemma 6.15 we have (2) as

$$\||XAY\|| = \Phi(s(XAY)) \le \Phi(\|X\| \, \|Y\| s(A)) = \|X\| \, \|Y\| \, \||A\||.$$

Moreover, (3) and (4) follow from Lemmas 6.16 and 6.15, respectively. $\qquad\square$

For instance, for $1 \leq p \leq \infty$, we have the unitarily invariant norm $\| \cdot \|_p$ on $B(\mathcal{H})$ corresponding to the ℓ_p-norm Φ_p in (6.13), that is, for $A \in B(\mathcal{H})$,

$$\|A\|_p := \Phi_p(s(A)) = \begin{cases} \left(\sum_{i=1}^n s_i(A)^p \right)^{1/p} = (\mathrm{Tr}\, |A|^p)^{1/p} & \text{if } 1 \leq p < \infty, \\ s_1(A) = \|A\| & \text{if } p = \infty. \end{cases}$$

The norm $\| \cdot \|_p$ is called the **Schatten–von Neumann p-norm**. In particular, $\|A\|_1 = \mathrm{Tr}\, |A|$ is the **trace-norm**, $\|A\|_2 = (\mathrm{Tr}\, A^*A)^{1/2}$ is the **Hilbert–Schmidt norm** and $\|A\|_\infty = \|A\|$ is the **operator norm**. (For $0 < p < 1$, we may define $\| \cdot \|_p$ by the same expression as above, but this is not a norm, and is called a quasi-norm.)

Another important class of unitarily invariant norms for $n \times n$ matrices are the **Ky Fan norms** $\| \cdot \|_{(k)}$ defined by

$$\|A\|_{(k)} := \sum_{i=1}^k s_i(A) \qquad \text{for } k = 1, \dots, n.$$

Obviously, $\| \cdot \|_{(1)}$ is the operator norm and $\| \cdot \|_{(n)}$ is the trace-norm. In the next theorem we give two variational expressions for the Ky Fan norms, which are sometimes quite useful since the Ky Fan norms are essential in majorization and norm inequalities for matrices.

The right-hand side of the second expression in the next theorem is known as the **K-functional** in real interpolation theory.

Theorem 6.19 *Let $\mathcal{H}$ be an n-dimensional space. For $A \in B(\mathcal{H})$ and $k = 1, \dots, n$, we have:*

(1) $\|A\|_{(k)} = \max\{\|AP\|_1 : P \text{ is a projection, } \mathrm{rank}\, P = k\}$,
(2) $\|A\|_{(k)} = \min\{\|X\|_1 + k\|Y\| : A = X + Y\}$.

Proof: (1) For any projection P of rank k, we have

$$\|AP\|_1 = \sum_{i=1}^n s_i(AP) = \sum_{i=1}^k s_i(AP) \leq \sum_{i=1}^k s_i(A)$$

by Theorem 6.7. For the converse, take the polar decomposition $A = U|A|$ with a unitary U and the spectral decomposition $|A| = \sum_{i=1}^n s_i(A)P_i$ with mutually orthogonal projections P_i of rank 1. Let $P := \sum_{i=1}^k P_i$. Then

$$\|AP\|_1 = \|U|A|P\|_1 = \left\| \sum_{i=1}^k s_i(A)P_i \right\|_1 = \sum_{i=1}^k s_i(A) = \|A\|_{(k)}.$$

(2) For any decomposition $A = X + Y$, since $s_i(A) \leq s_i(X) + \|Y\|$ by Theorem 6.7 (10), we have

$$\|A\|_{(k)} \le \sum_{i=1}^{k} s_i(X) + k\|Y\| \le \|X\|_1 + k\|Y\|$$

for any decomposition $A = X + Y$. Conversely, with the same notations as in the proof of (1), define

$$X := U \sum_{i=1}^{k} (s_i(A) - s_k(A)) P_i,$$

$$Y := U \Big(s_k(A) \sum_{i=1}^{k} P_i + \sum_{i=k+1}^{n} s_i(A) P_i \Big).$$

Then $X + Y = A$ and

$$\|X\|_1 = \sum_{i=1}^{k} s_i(A) - k s_k(A), \quad \|Y\| = s_k(A).$$

Hence $\|X\|_1 + k\|Y\| = \sum_{i=1}^{k} s_i(A)$.   □

The following is a modification of the above expression in (1):

$$\|A\|_{(k)} = \max\{|\mathrm{Tr}\,(UAP)| : U \text{ a unitary}, P \text{ a projection}, \mathrm{rank}\,P = k\}.$$

Here we prove the **Hölder inequality** for matrices to illustrate the usefulness of the majorization technique.

Theorem 6.20 *Let* $0 < p, p_1, p_2 \le \infty$ *and* $1/p = 1/p_1 + 1/p_2$. *Then*

$$\|AB\|_p \le \|A\|_{p_1} \|B\|_{p_2}, \quad A, B \in B(\mathcal{H}).$$

Proof: When $p_1 = \infty$ or $p_2 = \infty$, the result is obvious. Assume that $0 < p_1, p_2 < \infty$. Since Corollary 6.14 implies that

$$(s_i(AB)^p) \prec_{(\log)} (s_i(A)^p s_i(B)^p),$$

it follows from Theorem 6.5 that

$$(s_i(AB)^p) \prec_w (s_i(A)^p s_i(B)^p).$$

Since $(p_1/p)^{-1} + (p_2/p)^{-1} = 1$, the usual Hölder inequality for vectors shows that

$$\|AB\|_p = \Big(\sum_{i=1}^{n} s_i(AB)^p \Big)^{1/p} \le \Big(\sum_{i=1}^{n} s_i(A)^p s_i(B)^p \Big)^{1/p}$$

$$\leq \Big(\sum_{i=1}^{n} s_i(A)^{p_1} \Big)^{1/p_1} \Big(\sum_{i=1}^{n} s_i(B)^{p_2} \Big)^{1/p_2} \leq \|A\|_{p_1} \|B\|_{p_2}. \qquad \square$$

Corresponding to each symmetric gauge function Φ, define $\Phi' : \mathbb{R}^n \to \mathbb{R}$ by

$$\Phi'(b) := \sup \Big\{ \sum_{i=1}^{n} a_i b_i : a = (a_i) \in \mathbb{R}^n, \ \Phi(a) \leq 1 \Big\}$$

for $b = (b_i) \in \mathbb{R}^n$.

Then Φ' is a symmetric gauge function again, which is said to be **dual** to Φ. For example, when $1 \leq p \leq \infty$ and $1/p + 1/q = 1$, the ℓ_p-norm Φ_p is dual to the ℓ_q-norm Φ_q.

The following is another generalized Hölder inequality, which can be proved similarly to Theorem 6.20.

Lemma 6.21 *Let Φ, Φ_1 and Φ_2 be symmetric gauge functions with the corresponding unitarily invariant norms $||| \cdot |||$, $||| \cdot |||_1$ and $||| \cdot |||_2$ on $B(\mathcal{H})$, respectively. If*

$$\Phi(ab) \leq \Phi_1(a)\Phi_2(b), \qquad a, b \in \mathbb{R}^n,$$

then

$$|||AB||| \leq |||A|||_1 |||B|||_2, \qquad A, B \in B(\mathcal{H}).$$

In particular, if $||| \cdot |||'$ is the unitarily invariant norm corresponding to Φ' dual to Φ, then

$$\|AB\|_1 \leq |||A||| \, |||B|||', \qquad A, B \in B(\mathcal{H}).$$

Proof: By Corollary 6.14, Theorem 6.5, and Lemma 6.16, we have

$$\Phi(s(AB)) \leq \Phi(s(A)s(B)) \leq \Phi_1(s(A))\Phi_2(s(B)) \leq |||A|||_1 |||B|||_2,$$

proving the first assertion. For the second part, note by definition of Φ' that $\Phi_1(ab) \leq \Phi(a)\Phi'(b)$ for $a, b \in \mathbb{R}^n$. $\qquad \square$

Theorem 6.22 *Let Φ and Φ' be dual symmetric gauge functions on $\mathbb{R}^n$ with the corresponding norms $||| \cdot |||$ and $||| \cdot |||'$ on $B(\mathcal{H})$, respectively. Then $||| \cdot |||$ and $||| \cdot |||'$ are dual with respect to the duality $(A, B) \mapsto \mathrm{Tr}\, AB$ for $A, B \in B(\mathcal{H})$, that is,*

$$|||B|||' = \sup\{|\mathrm{Tr}\, AB| : A \in B(\mathcal{H}), \ |||A||| \leq 1\}, \qquad B \in B(\mathcal{H}). \qquad (6.15)$$

Proof: First note that any linear functional on $B(\mathcal{H})$ is represented as $A \in B(\mathcal{H}) \mapsto \mathrm{Tr}\, AB$ for some $B \in B(\mathcal{H})$. We write $|||B|||^\circ$ for the right-hand side of (6.15). From Lemma 6.21 we have

$$|\mathrm{Tr}\, AB| \leq \|AB\|_1 \leq |||A||| \, |||B|||'$$

so that $|||B|||^\circ \leq |||B|||'$ for all $B \in B(\mathcal{H})$. On the other hand, let $B = V|B|$ be the polar decomposition and $|B| = \sum_{i=1}^{n} s_i(B)|v_i\rangle\langle v_i|$ be the Schmidt decomposition of $|B|$. For any $a = (a_i) \in \mathbb{R}^n$ with $\Phi(a) \leq 1$, let $A := \left(\sum_{i=1}^{n} a_i|v_i\rangle\langle v_i|\right)V^*$. Then $s(A) = s\left(\sum_{i=1}^{n} a_i|v_i\rangle\langle v_i|\right) = (a_1^*, \ldots, a_n^*)$, the decreasing rearrangement of $(|a_1|, \ldots, |a_n|)$, and hence $|||A||| = \Phi(s(A)) = \Phi(a) \leq 1$. Moreover,

$$\text{Tr } AB = \text{Tr }\left(\sum_{i=1}^{n} a_i|v_i\rangle\langle v_i|\right)\left(\sum_{i=1}^{n} s_i(B)|v_i\rangle\langle v_i|\right)$$

$$= \text{Tr }\left(\sum_{i=1}^{n} a_i s_i(B)|v_i\rangle\langle v_i|\right) = \sum_{i=1}^{n} a_i s_i(B)$$

so that

$$\sum_{i=1}^{n} a_i s_i(B) \leq |\text{Tr } AB| \leq |||A||| \, |||B|||^\circ \leq |||B|||^\circ.$$

This implies that $|||B|||' = \Phi'(s(B)) \leq |||B|||^\circ$. $\square$

As special cases we have $\|\cdot\|_p' = \|\cdot\|_q$ when $1 \leq p \leq \infty$ and $1/p + 1/q = 1$.

The close relation between the (log-)majorization and the unitarily invariant norm inequalities is summarized in the following proposition.

Theorem 6.23 *Consider the following conditions for $A, B \in B(\mathcal{H})$:*

(i) $s(A) \prec_{w(\log)} s(B)$;
(ii) $|||f(|A|)||| \leq |||f(|B|)|||$ *for every unitarily invariant norm* $||| \cdot |||$ *and every continuous increasing function* $f : \mathbb{R}^+ \to \mathbb{R}^+$ *such that* $f(e^x)$ *is convex;*
(iii) $s(A) \prec_w s(B)$;
(iv) $\|A\|_{(k)} \leq \|B\|_{(k)}$ *for every* $k = 1, \ldots, n$;
(v) $|||A||| \leq |||B|||$ *for every unitarily invariant norm* $||| \cdot |||$;
(vi) $|||f(|A|)||| \leq |||f(|B|)|||$ *for every unitarily invariant norm* $||| \cdot |||$ *and every continuous increasing convex function* $f : \mathbb{R}^+ \to \mathbb{R}^+$.

Then

$$\text{(i)} \iff \text{(ii)} \implies \text{(iii)} \iff \text{(iv)} \iff \text{(v)} \iff \text{(vi)}.$$

Proof: (i) $\Rightarrow$ (ii). Let f be as in (ii). By Theorems 6.5 and 6.7(11) we have

$$s(f(|A|)) = f(s(A)) \prec_w f(s(B)) = s(f(|B|)). \tag{6.16}$$

This implies by Theorem 6.18(3) that $|||f(|A|)||| \leq |||f(|B|)|||$ for any unitarily invariant norm.

(ii) $\Rightarrow$ (i). Let $||| \cdot |||$ be the Ky Fan norm $\|\cdot\|_{(k)}$, and $f(x) = \log(1 + \varepsilon^{-1}x)$ for $\varepsilon > 0$. Then f satisfies the condition in (ii). Since

$$s_i(f(|A|)) = f(s_i(A)) = \log(\varepsilon + s_i(A)) - \log \varepsilon,$$

the inequality $\|f(|A|)\|_{(k)} \leq \|f(|B|)\|_{(k)}$ means that

$$\prod_{i=1}^{k}(\varepsilon + s_i(A)) \leq \prod_{i=1}^{k}(\varepsilon + s_i(B)).$$

Letting $\varepsilon \searrow 0$ gives $\prod_{i=1}^{k} s_i(A) \leq \prod_{i=1}^{k} s_i(B)$ and hence (i) follows.

(i) $\Rightarrow$ (iii) follows from Theorem 6.5. (iii) $\Leftrightarrow$ (iv) is trivial by definition of $\| \cdot \|_{(k)}$ and (vi) $\Rightarrow$ (v) $\Rightarrow$ (iv) is clear. Finally assume (iii) and let f be as in (vi). Theorem 6.7 yields (6.16) again, so that (vi) follows. Hence (iii) $\Rightarrow$ (vi) holds. □

By Theorems 6.9, 6.10 and 6.23 we have:

Corollary 6.24 *For $A, B \in M_n$ and a unitarily invariant norm $||| \cdot |||$, the inequality*

$$|||\text{Diag}(s_1(A) - s_1(B), \ldots, s_n(A) - s_n(B))||| \leq |||A - B|||$$

holds. If A and B are self-adjoint, then

$$|||\text{Diag}(\lambda_1(A) - \lambda_1(B), \ldots, \lambda_n(A) - \lambda_n(B))||| \leq |||A - B|||.$$

The following statements are particular cases for self-adjoint matrices:

$$\left(\sum_{i=1}^{n} |\lambda_i(A) - \lambda_i(B)|^p\right)^{1/p} \leq \|A - B\|_p \quad (1 \leq p < \infty).$$

The following is called **Weyl's inequality**:

$$\max_{1 \leq i \leq n} |\lambda_i(A) - \lambda_i(B)| \leq \|A - B\|.$$

There are similar inequalities in the general case, where λ_i is replaced by s_i.

In the rest of this section we prove symmetric norm inequalities (or eigenvalue majorizations) involving convex/concave functions and expansions. Recall that an operato Z is called an expansion if $Z^*Z \geq I$.

Theorem 6.25 *Let $f : \mathbb{R}^+ \to \mathbb{R}^+$ be a concave function. If $0 \leq A \in M_n$ and $Z \in M_n$ is an expansion, then*

$$|||f(Z^*AZ)||| \leq |||Z^*f(A)Z|||$$

for every unitarily invariant norm $||| \cdot |||$, or equivalently,

$$\lambda(f(Z^*AZ)) \prec_w \lambda(Z^*f(A)Z).$$

Proof: Note that f is automatically non-decreasing. By Theorem 6.22 it suffices to prove the inequality for the Ky Fan k-norms $\| \cdot \|_{(k)}$, $1 \leq k \leq n$. Letting $f_0(x) := f(x) - f(0)$ we have

$$f(Z^*AZ) = f(0)I + f_0(Z^*AZ),$$
$$Z^*f(A)Z = f(0)Z^*Z + Z^*f_0(A)Z \geq f(0)I + Z^*f_0(A)Z,$$

which show that we may assume that $f(0) = 0$. Then there is a spectral projection E of rank k for Z^*AZ such that

$$\|f(Z^*AZ)\|_{(k)} = \sum_{j=1}^{k} f(\lambda_j(Z^*AZ)) = \operatorname{Tr} f(Z^*AZ)E.$$

When we show that

$$\operatorname{Tr} f(Z^*AZ)E \leq \operatorname{Tr} Z^*f(A)ZE, \tag{6.17}$$

it follows that

$$\|f(Z^*AZ)\|_{(k)} \leq \operatorname{Tr} Z^*f(A)ZE \leq \|Z^*f(A)Z\|_{(k)}$$

by Theorem 6.19. For (6.17) we may show that

$$\operatorname{Tr} g(Z^*AZ)E \geq \operatorname{Tr} Z^*g(A)ZE \tag{6.18}$$

for every convex function on $\mathbb{R}^+$ with $g(0) = 0$. Such a function g can be approximated by functions of the form

$$\alpha x + \sum_{i=1}^{m} \alpha_i(x - \beta_i)_+ \tag{6.19}$$

with $\alpha \in \mathbb{R}$ and $\alpha_i, \beta_i > 0$, where $(x - \beta)_+ := \max\{0, x - \beta\}$. Consequently, it suffices to show (6.18) for $g_\beta(x) := (x - \beta)_+$ with $\beta > 0$. From the lemma below we have a unitary U such that

$$g_\beta(Z^*AZ) \geq U^*Z^*g_\beta(A)ZU.$$

We hence have

$$\operatorname{Tr} g_\beta(Z^*AZ)E = \sum_{j=1}^{k} \lambda_j(g_\beta(Z^*AZ)) \geq \sum_{j=1}^{k} \lambda_j(U^*Z^*g_\beta(A)ZU)$$

$$= \sum_{j=1}^{k} \lambda_j (Z^* g_\beta(A)Z) \geq \text{Tr } Z^* g_\beta(A)ZE,$$

that is (6.18) for $g = g_\beta$.   □

Lemma 6.26 *Let $A \in \mathbb{M}_n^+$, $Z \in \mathbb{M}$ be an expansion, and $\beta > 0$. Then there exists a unitary U such that*

$$(Z^* AZ - \beta I)_+ \geq U^* Z^* (A - \beta I)_+ ZU.$$

Proof: Let P be the support projection of $(A - \beta I)_+$ and set $A_\beta := PA$. Let Q be the support projection of $Z^* A_\beta Z$. Since $Z^* AZ \geq Z^* A_\beta Z$ and $(x - \beta)_+$ is a non-decreasing function, for $1 \leq j \leq n$ we have

$$\lambda_j((Z^* AZ - \beta I)_+) = (\lambda_j(Z^* AZ) - \beta)_+$$
$$\geq (\lambda_j(Z^* A_\beta Z) - \beta)_+$$
$$= \lambda_j((Z^* A_\beta Z - \beta I)_+).$$

So there exists a unitary U such that

$$(Z^* AZ - \beta I)_+ \geq U^* (Z^* A_\beta Z - \beta I)_+ U.$$

It is obvious that Q is the support projection of $Z^* PZ$. Also, note that $Z^* PZ$ is unitarily equivalent to $PZZ^* P$. Since $Z^* Z \geq I$, it follows that $ZZ^* \geq I$ and so $PZZ^* P \geq P$. Therefore, we have $Q \leq Z^* PZ$. Since $Z^* A_\beta Z \geq \beta Z^* PZ \geq \beta Q$, we see that

$$(Z^* A_\beta Z - \beta I)_+ = Z^* A_\beta Z - \beta Q \geq Z^* A_\beta Z - \beta Z^* PZ$$
$$= Z^* (A_\beta - \beta P)Z = Z^* (A - \beta I)_+ Z,$$

which gives the conclusion.   □

When f is convex with $f(0) = 0$, the inequality in Theorem 6.25 is reversed.

Theorem 6.27 *Let $f : \mathbb{R}^+ \to \mathbb{R}^+$ be a convex function with $f(0) = 0$. If $0 \leq A \in \mathbb{M}_n$ and $Z \in \mathbb{M}_n$ is an expansion, then*

$$|||f(Z^* AZ)||| \geq |||Z^* f(A)Z|||$$

for every unitarily invariant norm $||| \cdot |||$.

Proof: By approximation we may assume that f is of the form (6.19) with $\alpha \geq 0$ and $\alpha_i, \beta_i > 0$. By Lemma 6.26 we have

$$Z^* f(A) Z = \alpha Z^* A Z + \sum_i \alpha_i Z^* (A - \beta_i I)_+ Z$$

$$\leq \alpha Z^* A Z + \sum_i \alpha_i U_i (Z^* A Z - \beta_i I)_+ U_i^*$$

for some unitaries U_i, $1 \leq i \leq m$. We now consider the Ky Fan k-norms $\| \cdot \|_{(k)}$. For each $k = 1, \ldots, n$ there is a projection E of rank k so that

$$\left\| \alpha Z^* A Z + \sum_i \alpha_i U_i (Z^* A Z - \beta_i I)_+ U_i^* \right\|_{(k)}$$

$$= \text{Tr} \left\{ \alpha Z^* A Z + \sum_i \alpha_i U_i (Z^* A Z - \beta_i I)_+ U_i^* \right\} E$$

$$= \alpha \text{Tr} \, Z^* A Z E + \sum_i \alpha_i \text{Tr} \, (Z^* A Z - \beta_i I)_+ U_i^* E U_i$$

$$\leq \alpha \| Z^* A Z \|_{(k)} + \sum_i \alpha_i \| (Z^* A Z - \beta_i I)_+ \|_{(k)}$$

$$= \sum_{j=1}^{k} \left\{ \alpha \lambda_j (Z^* A Z) + \sum_i \alpha_i (\lambda_j (Z^* A Z) - \beta_i)_+ \right\}$$

$$= \sum_{j=1}^{k} f(\lambda_j (Z^* A Z)) = \| f(Z^* A Z) \|_{(k)},$$

and hence $\| Z^* f(A) Z \|_{(k)} \leq \| f(Z^* A Z) \|_{(k)}$. This implies the conclusion. $\square$

For the trace function the non-negativity assumption of f is not necessary so that we have:

Theorem 6.28 *Let $0 \leq A \in \mathbb{M}_n$ and $Z \in \mathbb{M}_n$ be an expansion. If f is a concave function on $\mathbb{R}^+$ with $f(0) \geq 0$, then*

$$\text{Tr} \, f(Z^* A Z) \leq \text{Tr} \, Z^* f(A) Z.$$

If f is a convex function on $\mathbb{R}^+$ with $f(0) \leq 0$, then

$$\text{Tr} \, f(Z^* A Z) \geq \text{Tr} \, Z^* f(A) Z.$$

Proof: The two assertions are obviously equivalent. To prove the second, by approximation we may assume that f is of the form (6.19) with $\alpha \in \mathbb{R}$ and $\alpha_i, \beta_i > 0$. Then, by Lemma 6.26,

$$\operatorname{Tr} f(Z^*AZ) = \operatorname{Tr}\left\{\alpha Z^*AZ + \sum_i \alpha_i (Z^*AZ - \beta_i I)_+\right\}$$

$$\geq \operatorname{Tr}\left\{\alpha Z^*AZ + \sum_i \alpha_i Z^*(A - \beta_i I)_+ Z\right\} = \operatorname{Tr} Z^* f(A)Z$$

and the statement is proved.   □

6.4 More Majorizations for Matrices

In the first part of this section, we prove a subadditivity property for certain symmetric norm functions. Let $f : \mathbb{R}^+ \to \mathbb{R}^+$ be a concave function. Then f is increasing and it is easy to show that $f(a + b) \leq f(a) + f(b)$ for positive numbers a and b. The **Rotfel'd inequality**

$$\operatorname{Tr} f(A + B) \leq \operatorname{Tr}(f(A) + f(B)) \qquad (A, B \in \mathbb{M}_n^+)$$

is a matrix extension. Another extension is

$$|||f(A + B)||| \leq |||f(A) + f(B)||| \tag{6.20}$$

for all $0 \leq A, B \in \mathbb{M}_n$ and for any unitarily invariant norm $||| \cdot |||$, which will be proved in Theorem 6.33 below.

Lemma 6.29 *Let $g : \mathbb{R}^+ \to \mathbb{R}^+$ be a continuous function. If g is decreasing and $xg(x)$ is increasing, then*

$$\lambda((A + B)g(A + B)) \prec_w \lambda\big(A^{1/2}g(A + B)A^{1/2} + B^{1/2}g(A + B)B^{1/2}\big)$$

for all $A, B \in \mathbb{M}_n^+$.

Proof: Let $\lambda(A+B) = (\lambda_1, \ldots, \lambda_n)$ be the eigenvalue vector arranged in decreasing order and $u_1, \ldots, u_n$ be the corresponding eigenvectors forming an orthonormal basis of $\mathbb{C}^n$. For $1 \leq k \leq n$ let P_k be the orthogonal projection onto the subspace spanned by $u_1, \ldots, u_k$. Since $xg(x)$ is increasing, it follows that

$$\lambda((A + B)g(A + B)) = (\lambda_1 g(\lambda_1), \ldots, \lambda_n g(\lambda_n)).$$

Hence, what we need to prove is

$$\operatorname{Tr}(A + B)g(A + B)P_k \leq \operatorname{Tr}\left(A^{1/2}g(A + B)A^{1/2} + B^{1/2}g(A + B)B^{1/2}\right)P_k,$$

since the left-hand side is equal to $\sum_{i=1}^{k} \lambda_i g(\lambda_i)$ and the right-hand side is less than or equal to $\sum_{i=1}^{k} \lambda_i \left(A^{1/2} g(A+B) A^{1/2} + B^{1/2} g(A+B) B^{1/2} \right)$. The above inequality immediately follows by summing the following two inequalities:

$$\operatorname{Tr} g(A+B)^{1/2} A g(A+B)^{1/2} P_k \leq \operatorname{Tr} A^{1/2} g(A+B) A^{1/2} P_k, \qquad (6.21)$$

$$\operatorname{Tr} g(A+B)^{1/2} B g(A+B)^{1/2} P_k \leq \operatorname{Tr} B^{1/2} g(A+B) B^{1/2} P_k. \qquad (6.22)$$

To prove (6.21), we write P_k, $H := g(A+B)$ and $A^{1/2}$ as

$$P_k = \begin{bmatrix} I_{\mathcal{K}} & 0 \\ 0 & 0 \end{bmatrix}, \quad H = \begin{bmatrix} H_1 & 0 \\ 0 & H_2 \end{bmatrix}, \quad A^{1/2} = \begin{bmatrix} A_{11} & A_{12} \\ A_{12}^* & A_{22} \end{bmatrix}$$

in the form of 2×2 block matrices corresponding to the orthogonal decomposition $\mathbb{C}^n = \mathcal{K} \oplus \mathcal{K}^\perp$ with $\mathcal{K} := P_k \mathbb{C}^n$. Then

$$P_k g(A+B)^{1/2} A g(A+B)^{1/2} P_k = \begin{bmatrix} H_1^{1/2} A_{11}^2 H_1^{1/2} + H_1^{1/2} A_{12} A_{12}^* H_1^{1/2} & 0 \\ 0 & 0 \end{bmatrix},$$

$$P_k A^{1/2} g(A+B) A^{1/2} P_k = \begin{bmatrix} A_{11} H_1 A_{11} + A_{12} H_2 A_{12}^* & 0 \\ 0 & 0 \end{bmatrix}.$$

Since g is decreasing, we notice that

$$H_1 \leq g(\lambda_k) I_{\mathcal{K}}, \qquad H_2 \geq g(\lambda_k) I_{\mathcal{K}^\perp}.$$

Therefore, we have

$$\operatorname{Tr} H_1^{1/2} A_{12} A_{12}^* H_1^{1/2} = \operatorname{Tr} A_{12}^* H_1 A_{12} \leq g(\lambda_k) \operatorname{Tr} A_{12}^* A_{12}$$
$$= g(\lambda_k) \operatorname{Tr} A_{12} A_{12}^* \leq \operatorname{Tr} A_{12} H_2 A_{12}^*$$

so that

$$\operatorname{Tr} \left(H_1^{1/2} A_{11}^2 H_1^{1/2} + H_1^{1/2} A_{12} A_{12}^* H_1^{1/2} \right) \leq \operatorname{Tr} \left(A_{11} H_1 A_{11} + A_{12} H_2 A_{12}^* \right),$$

which proves (6.21). (6.22) is proved similarly. □

In the next result matrix concavity is assumed.

Theorem 6.30 *Let* $f : \mathbb{R}^+ \to \mathbb{R}^+$ *be a continuous matrix monotone (equivalently, matrix concave) function. Then* (6.20) *holds for all* $0 \leq A, B \in \mathbb{M}_n$ *and for any unitarily invariant norm* $||| \cdot |||$.

Proof: By continuity we may assume that A, B are invertible. Let $g(x) := f(x)/x$; then g satisfies the assumptions of Lemma 6.29. Hence the lemma implies that

$$|||f(A+B)||| \leq |||A^{1/2}(A+B)^{-1/2}f(A+B)(A+B)^{-1/2}A^{1/2}$$
$$+B^{1/2}(A+B)^{-1/2}f(A+B)(A+B)^{-1/2}B^{1/2}|||. \quad (6.23)$$

Since $C := A^{1/2}(A+B)^{-1/2}$ is a contraction, Theorem 4.23 implies from the matrix concavity that

$$A^{1/2}(A+B)^{-1/2}f(A+B)(A+B)^{-1/2}A^{1/2}$$
$$= Cf(A+B)C^* \leq f(C(A+B)C^*) = f(A),$$

and similarly

$$B^{1/2}(A+B)^{-1/2}f(A+B)(A+B)^{-1/2}B^{1/2} \leq f(B).$$

Therefore, the right-hand side of (6.23) is less than or equal to $|||f(A)+f(B)|||$. $\square$

A particular case of the next theorem is $|||(A+B)^m||| \geq |||A^m+B^m|||$ for $m \in \mathbb{N}$, which was proved by Bhatia and Kittaneh in [23].

Theorem 6.31 *Let* $g : \mathbb{R}^+ \to \mathbb{R}^+$ *be an increasing bijective function whose inverse function is operator monotone. Then*

$$|||g(A+B)||| \geq |||g(A)+g(B)||| \quad (6.24)$$

for all $0 \leq A, B \in \mathbb{M}_n$ *and* $||| \cdot |||$.

Proof: Let f be the inverse function of g. For every $0 \leq A, B \in \mathbb{M}_n$, Theorem 6.30 implies that
$$f(\lambda(A+B)) \prec_w \lambda(f(A)+f(B)).$$

Now, replace A and B by $g(A)$ and $g(B)$, respectively. Then we have

$$f(\lambda(g(A)+g(B))) \prec_w \lambda(A+B).$$

Since f is concave and hence g is convex (and increasing), we have by Example 6.4

$$\lambda(g(A)+g(B)) \prec_w g(\lambda(A+B)) = \lambda(g(A+B)),$$

which means by Theorem 6.23 that $|||g(A)+g(B)||| \leq |||g(A+B)|||$. $\square$

The above theorem can be extended to the next theorem due to Kosem [57], which is the first main result of this section. The simpler proof below is from [30].

Theorem 6.32 *Let* $g : \mathbb{R}^+ \to \mathbb{R}^+$ *be a continuous convex function with* $g(0) = 0$. *Then* (6.24) *holds for all* A, B *and* $||| \cdot |||$ *as above.*

Proof: First, note that a convex function $g \geq 0$ on $\mathbb{R}^+$ with $g(0) = 0$ is non-decreasing. Let Γ denote the set of all non-negative functions g on $\mathbb{R}^+$ for which the conclusion of the theorem holds. It is obvious that Γ is closed under pointwise convergence and multiplication by non-negative scalars. When $f, g \in \Gamma$, for the Ky Fan norms $\| \cdot \|_{(k)}$, $1 \leq k \leq n$, and for $0 \leq A, B \in \mathbb{M}_n$ we have

$$
\begin{aligned}
\|(f + g)(A + B)\|_{(k)} &= \|f(A + B)\|_{(k)} + \|g(A + B)\|_{(k)} \\
&\geq \|f(A) + f(B)\|_{(k)} + \|g(A) + g(B)\|_{(k)} \\
&\geq \|(f + g)(A) + (f + g)(B)\|_{(k)},
\end{aligned}
$$

where the above equality is guaranteed by the fact that f and g are non-decreasing and the latter inequality is the triangle inequality. Hence $f + g \in \Gamma$ by Theorem 6.23 so that Γ is a convex cone. Notice that any convex function $g \geq 0$ on $\mathbb{R}^+$ with $g(0) = 0$ is the pointwise limit of an increasing sequence of functions of the form $\sum_{l=1}^{m} c_l \gamma_{a_l}(x)$ with $c_l, a_l > 0$, where γ_a is the angle function at $a > 0$ given as $\gamma_a(x) := \max\{x - a, 0\}$. Hence it suffices to show that $\gamma_a \in \Gamma$ for all $a > 0$. To do this, for $a, r > 0$ we define

$$
h_{a,r}(x) := \frac{1}{2}\left(\sqrt{(x - a)^2 + r} + x - \sqrt{a^2 + r}\right), \qquad x \geq 0,
$$

which is an increasing bijective function on $\mathbb{R}^+$ and whose inverse is

$$
x - \frac{r/2}{2x + \sqrt{a^2 + r} - a} + \frac{\sqrt{a^2 + r} + a}{2}. \tag{6.25}
$$

Since (6.25) is operator monotone on $\mathbb{R}^+$, we have $h_{a,r} \in \Gamma$ by Theorem 6.31. Therefore, $\gamma_a \in \Gamma$ since $h_{a,r} \to \gamma_a$ as $r \searrow 0$. $\square$

The next subadditivity inequality extending Theorem 6.30 was proved by **Bourin** and **Uchiyama** in [30], and is the second main result.

Theorem 6.33 *Let $f : \mathbb{R}^+ \to \mathbb{R}^+$ be a continuous concave function. Then (6.20) holds for all A, B and $||| \cdot |||$ as above.*

Proof: Let λ_i and u_i, $1 \leq i \leq n$, be taken as in the proof of Lemma 6.29, and P_k, $1 \leq k \leq n$, also be as defined there. We may prove the weak majorization

$$
\sum_{i=1}^{k} f(\lambda_i) \leq \sum_{i=1}^{k} \lambda_i(f(A) + f(B)) \qquad (1 \leq k \leq n).
$$

To do this, it suffices to show that

$$
\mathrm{Tr}\, f(A + B)P_k \leq \mathrm{Tr}\,(f(A) + f(B))P_k. \tag{6.26}
$$

Indeed, since a concave f is necessarily increasing, the left-hand side of (6.26) is $\sum_{i=1}^{k} f(\lambda_i)$ and the right-hand side is less than or equal to $\sum_{i=1}^{k} \lambda_i(f(A) + f(B))$. Here, note by Exercise 12 that f is the pointwise limit of a sequence of functions of the form $\alpha + \beta x - g(x)$ where $\alpha \geq 0$, $\beta > 0$, and $g \geq 0$ is a continuous convex function on $\mathbb{R}^+$ with $g(0) = 0$. Hence, to prove (6.26), it suffices to show that

$$\operatorname{Tr} g(A + B)P_k \geq \operatorname{Tr}(g(A) + g(B))P_k$$

for any continuous convex function $g \geq 0$ on $\mathbb{R}^+$ with $g(0) = 0$. In fact, this is seen as follows:

$$\operatorname{Tr} g(A + B)P_k = \|g(A + B)\|_{(k)} \geq \|g(A) + g(B)\|_{(k)} \geq \operatorname{Tr}(g(A) + g(B))P_k,$$

where the above equality is due to the fact that g is increasing and the first inequality follows from Theorem 6.32. $\qquad\square$

The subadditivity inequality of Theorem 6.32 was further extended by J.-C. Bourin in such a way that if f is a positive continuous concave function on $\mathbb{R}^+$ then

$$|||f(|A + B|)||| \leq |||f(|A|) + f(|B|)|||$$

for all normal matrices $A, B \in \mathbb{M}_n$ and for any unitarily invariant norm $||| \cdot |||$. In particular,

$$|||f(|Z|)||| \leq |||f(|A|) + f(|B|)|||$$

when $Z = A + iB$ is the Descartes decomposition of Z.

In the second part of this section, we prove the inequality between norms of $f(|A - B|)$ and $f(A) - f(B)$ (or the weak majorization for their singular values) when f is a positive operator monotone function on $\mathbb{R}^+$ and $A, B \in \mathbb{M}_n^+$. We first prepare some simple facts for the next theorem.

Lemma 6.34 *For self-adjoint $X, Y \in \mathbb{M}_n$, let $X = X_+ - X_-$ and $Y = Y_+ - Y_-$ be the Jordan decompositions.*

(1) *If $X \leq Y$ then $s_i(X_+) \leq s_i(Y_+)$ for all i.*
(2) *If $s(X_+) \prec_w s(Y_+)$ and $s(X_-) \prec_w s(Y_-)$, then $s(X) \prec_w s(Y)$.*

Proof: (1) Let Q be the support projection of X_+. Since

$$X_+ = QXQ \leq QYQ \leq QY_+Q,$$

we have $s_i(X_+) \leq s_i(QY_+Q) \leq s_i(Y_+)$ by Theorem 6.7(7).

(2) It is rather easy to see that $s(X)$ is the decreasing rearrangement of the combination of $s(X_+)$ and $s(X_-)$. Hence for each $k \in \mathbb{N}$ we can choose $0 \leq m \leq k$ so that

$$\sum_{i=1}^{k} s_i(X) = \sum_{i=1}^{m} s_i(X_+) + \sum_{i=1}^{k-m} s_i(X_-).$$

Hence

$$\sum_{i=1}^{k} s_i(X) \leq \sum_{i=1}^{m} s_i(Y_+) + \sum_{i=1}^{k-m} s_i(Y_-) \leq \sum_{i=1}^{k} s_i(Y),$$

as desired.   □

Theorem 6.35 *Let* $f : \mathbb{R}^+ \to \mathbb{R}^+$ *be a matrix monotone function. Then*

$$|||f(A) - f(B)||| \leq |||f(|A - B|)|||$$

for all $0 \leq A, B \in \mathbb{M}_n$ *and for any unitarily invariant norm* $||| \cdot |||$. *Equivalently,*

$$s(f(A) - f(B)) \prec_w s(f(|A - B|)) \tag{6.27}$$

holds.

Proof: First assume that $A \geq B \geq 0$ and let $C := A - B \geq 0$. In view of Theorem 6.23, it suffices to prove that

$$\|f(B + C) - f(B)\|_{(k)} \leq \|f(C)\|_{(k)} \quad (1 \leq k \leq n). \tag{6.28}$$

For each $\lambda \in (0, \infty)$ let

$$h_\lambda(x) := \frac{x}{x + \lambda} = 1 - \frac{\lambda}{x + \lambda},$$

which is increasing on $\mathbb{R}^+$ with $h_\lambda(0) = 0$. According to the integral representation (4.19) for f with $a, b \geq 0$ and a positive measure μ on $(0, \infty)$, we have

$$s_i(f(C)) = f(s_i(C))$$
$$= a + bs_i(C) + \int_{(0,\infty)} \frac{\lambda s_i(C)}{s_i(C) + \lambda} \, d\mu(\lambda)$$
$$= a + bs_i(C) + \int_{(0,\infty)} \lambda s_i(h_\lambda(C)) \, d\mu(\lambda),$$

so that

$$\|f(C)\|_{(k)} \geq b\|C\|_{(k)} + \int_{(0,\infty)} \lambda \|h_\lambda(C)\|_{(k)} \, d\mu(\lambda). \tag{6.29}$$

On the other hand, since

$$f(B + C) = aI + b(B + C) + \int_{(0,\infty)} \lambda h_\lambda(B + C) \, d\mu(\lambda)$$

as well as the analogous expression for $f(B)$, we have

$$f(B + C) - f(B) = bC + \int_{(0,\infty)} \lambda(h_\lambda(B + C) - h_\lambda(B)) \, d\mu(\lambda),$$

so that

$$\|f(B + C) - f(B)\|_{(k)} \leq b\|C\|_{(k)} + \int_{(0,\infty)} \lambda\|h_\lambda(B + C) - h_\lambda(B)\|_{(k)} \, d\mu(\lambda).$$

By this inequality and (6.29), it suffices for (6.28) to show that

$$\|h_\lambda(B + C) - h_\lambda(B)\|_{(k)} \leq \|h_\lambda(C)\|_{(k)} \qquad (\lambda \in (0, \infty), \ 1 \leq k \leq n).$$

As $h_\lambda(x) = h_1(x/\lambda)$, it is enough to show this inequality for the case $\lambda = 1$ since we may replace B and C with $\lambda^{-1}B$ and $\lambda^{-1}C$, respectively. Thus, what remains to be proven is the following:

$$\|(B + I)^{-1} - (B + C + I)^{-1}\|_{(k)} \leq \|I - (C + I)^{-1}\|_{(k)} \quad (1 \leq k \leq n). \quad (6.30)$$

Since

$$(B+I)^{-1}-(B+C+I)^{-1} = (B+I)^{-1/2}h_1((B+I)^{-1/2}C(B+I)^{-1/2})(B+I)^{-1/2}$$

and $\|(B + I)^{-1/2}\| \leq 1$, we obtain

$$\begin{aligned}
s_i((B + I)^{-1} - (B + C + I)^{-1}) &\leq s_i(h_1((B + I)^{-1/2}C(B + I)^{-1/2})) \\
&= h_1(s_i((B + I)^{-1/2}C(B + I)^{-1/2})) \\
&\leq h_1(s_i(C)) = s_i(I - (C + I)^{-1})
\end{aligned}$$

by repeated use of Theorem 6.7 (7). Therefore, (6.30) is proved.

Next, let us prove the assertion in the general case $A, B \geq 0$. Since $0 \leq A \leq B + (A - B)_+$, it follows that

$$f(A) - f(B) \leq f(B + (A - B)_+) - f(B),$$

which implies by Lemma 6.34 (1) that

$$\|(f(A) - f(B))_+\|_{(k)} \leq \|f(B + (A - B)_+) - f(B)\|_{(k)}.$$

Applying (6.28) to $B + (A - B)_+$ and B, we have

$$\|f(B + (A - B)_+) - f(B)\|_{(k)} \le \|f((A - B)_+)\|_{(k)}.$$

Therefore,

$$s((f(A) - f(B))_+) \prec_w s(f((A - B)_+)). \tag{6.31}$$

Exchanging the role of A, B gives

$$s((f(A) - f(B))_-) \prec_w s(f((A - B)_-)). \tag{6.32}$$

Here, we may assume that $f(0) = 0$ since f can be replaced by $f - f(0)$. Then we immediately see that

$$f((A - B)_+)f((A - B)_-) = 0, \quad f((A - B)_+) + f((A - B)_-) = f(|A - B|).$$

Hence $s(f(A) - f(B)) \prec_w s(f(|A - B|))$ follows from (6.31) and (6.32) thanks to Lemma 6.34 (2). $\qquad\square$

When $f(x) = x^\theta$ with $0 < \theta < 1$, the weak majorization (6.27) gives the norm inequality formerly proved by Birman, Koplienko and Solomyak:

$$\|A^\theta - B^\theta\|_{p/\theta} \le \|A - B\|_p^\theta$$

for all $A, B \in \mathbb{M}_n^+$ and $\theta \le p \le \infty$. The case where $\theta = 1/2$ and $p = 1$ is known as the **Powers–Størmer inequality**.

The following is an immediate corollary of Theorem 6.35, whose proof is similar to that of Theorem 6.31.

Corollary 6.36 *Let $g : \mathbb{R}^+ \to \mathbb{R}^+$ be an increasing bijective function whose inverse function is operator monotone. Then*

$$|||g(A) - g(B)||| \ge |||g(|A - B|)|||$$

for all A, B and $||| \cdot |||$ as above.

In [13], Audenaert and Aujla pointed out that Theorem 6.35 is not true in the case where $f : \mathbb{R}^+ \to \mathbb{R}^+$ is a general continuous concave function and that Corollary 6.36 is not true in the case where $g : \mathbb{R}^+ \to \mathbb{R}^+$ is a general continuous convex function.

In the last part of this section we prove log-majorizations results, which give inequalities strengthening or complementing the Golden–Thompson inequality. The following log-majorization is due to Huzihiro Araki.

Theorem 6.37 *For all $A, B \in \mathbb{M}_n^+$,*

$$s((A^{1/2}BA^{1/2})^r) \prec_{(\log)} s(A^{r/2}B^r A^{r/2}) \quad (r \ge 1), \tag{6.33}$$

or equivalently

$$s((A^{p/2}B^p A^{p/2})^{1/p}) \prec_{(\log)} s((A^{q/2}B^q A^{q/2})^{1/q}) \qquad (0 < p \le q). \qquad (6.34)$$

Proof: We can pass to the limit from $A + \varepsilon I$ and $B + \varepsilon I$ as $\varepsilon \searrow 0$ by Theorem 6.7(10). So we may assume that A and B are invertible.

First we show that

$$\|(A^{1/2}B A^{1/2})^r\| \le \|A^{r/2}B^r A^{r/2}\| \qquad (r \ge 1). \qquad (6.35)$$

It is enough to check that $A^{r/2}B^r A^{r/2} \le I$ implies $A^{1/2}B A^{1/2} \le I$ which is equivalent to a monotonicity: $B^r \le A^{-r}$ implies $B \le A^{-1}$.

We have

$$((A^{1/2}B A^{1/2})^r)^{\wedge k} = ((A^{\wedge k})^{1/2}(B^{\wedge k})(A^{\wedge k})^{1/2})^r,$$
$$(A^{r/2}B^r A^{r/2})^{\wedge k} = (A^{\wedge k})^{r/2}(B^{\wedge k})^r (A^{\wedge k})^{r/2},$$

and in place of A, B in (6.35) we put $A^{\wedge k}, B^{\wedge k}$:

$$\|((A^{1/2}B A^{1/2})^r)^{\wedge k}\| \le \|(A^{r/2}B^r A^{r/2})^{\wedge k}\|.$$

This means, thanks to Lemma 1.62, that

$$\prod_{i=1}^{k} s_i((A^{1/2}B A^{1/2})^r) \le \prod_{i=1}^{k} s_i(A^{r/2}B^r A^{r/2}).$$

Moreover,

$$\prod_{i=1}^{n} s_i((A^{1/2}B A^{1/2})^r) = (\det A \cdot \det B)^r = \prod_{i=1}^{n} s_i(A^{r/2}B^r A^{r/2}).$$

Hence (6.33) is proved. If we replace A, B by A^p, B^p and take $r = q/p$, then

$$s((A^{p/2}B^p A^{p/2})^{q/p}) \prec_{(\log)} s(A^{q/2}B^q A^{q/2}),$$

which implies (6.34) by Theorem 6.7(11). □

Let $0 \le A, B \in \mathbb{M}_m$, $s, t \in \mathbb{R}^+$ and $t \ge 1$. Then the theorem implies

$$\mathrm{Tr}\,(A^{1/2}B A^{1/2})^{st} \le \mathrm{Tr}\,(A^{t/2}B A^{t/2})^s \qquad (6.36)$$

which is called the **Araki–Lieb–Thirring inequality**. The case $s = 1$ and integer t was the **Lieb–Thirring inequality**.

Theorems 6.27 and 6.37 yield:

Corollary 6.38 *Let* $0 \leq A, B \in \mathbb{M}_n$ *and* $|||\cdot|||$ *be any unitarily invariant norm. If f is a continuous increasing function on $\mathbb{R}^+$ such that $f(0) \geq 0$ and $f(e^t)$ is convex, then*

$$|||f((A^{1/2}BA^{1/2})^r)||| \leq |||f(A^{r/2}B^rA^{r/2})||| \quad (r \geq 1).$$

In particular,

$$|||(A^{1/2}BA^{1/2})^r||| \leq |||A^{r/2}B^rA^{r/2}||| \quad (r \geq 1).$$

The next corollary strengthens the **Golden–Thompson inequality** to the form of a log-majorization.

Corollary 6.39 *For all self-adjoint $H, K \in \mathbb{M}_n$,*

$$s(e^{H+K}) \prec_{(\log)} s((e^{rH/2}e^{rK}e^{rH/2})^{1/r}) \quad (r > 0).$$

Hence, for every unitarily invariant norm $|||\cdot|||$,

$$|||e^{H+K}||| \leq |||(e^{rH/2}e^{rK}e^{rH/2})^{1/r}||| \quad (r > 0),$$

and the above right-hand side decreases to $|||e^{H+K}|||$ *as* $r \searrow 0$. *In particular,*

$$|||e^{H+K}||| \leq |||e^{H/2}e^Ke^{H/2}||| \leq |||e^He^K|||. \tag{6.37}$$

Proof: The log-majorization follows by letting $p \searrow 0$ in (6.34) thanks to the above lemma. The second assertion follows from the first and Theorem 6.23. Thanks to Theorem 6.7 (3) and Theorem 6.37 we have

$$|||e^He^K||| = |||\,|e^Ke^H|\,||| = |||(e^He^{2K}e^H)^{1/2}||| \geq |||e^{H/2}e^Ke^{H/2}|||,$$

which is the second inequality of (6.37).   □

The specialization of the inequality (6.37) to the trace-norm $||\cdot||_1$ is the **Golden–Thompson** trace inequality $\operatorname{Tr} e^{H+K} \leq \operatorname{Tr} e^He^K$. It was shown in [79] that $\operatorname{Tr} e^{H+K} \leq \operatorname{Tr}(e^{H/n}e^{K/n})^n$ for every $n \in \mathbb{N}$. The extension (6.37) was given in [61, 80] and, for the operator norm, is known as **Segal's inequality** (see [77, p. 260]).

Theorem 6.40 *If $A, B, X \in \mathbb{M}_n$ and*

$$\begin{bmatrix} A & X \\ X & B \end{bmatrix} \geq 0,$$

then we have

$$\lambda\left(\begin{bmatrix} A & X \\ X & B \end{bmatrix}\right) \prec \lambda\left(\begin{bmatrix} A+B & 0 \\ 0 & 0 \end{bmatrix}\right).$$

Proof: By Example 2.6 and the Ky Fan majorization (Corollary 6.11), we have

$$\lambda\left(\begin{bmatrix} A & X \\ X & B \end{bmatrix}\right) \prec \lambda\left(\begin{bmatrix} \frac{A+B}{2} & 0 \\ 0 & 0 \end{bmatrix}\right) + \lambda\left(\begin{bmatrix} 0 & 0 \\ 0 & \frac{A+B}{2} \end{bmatrix}\right) = \lambda\left(\begin{bmatrix} A+B & 0 \\ 0 & 0 \end{bmatrix}\right).$$

This is the result.   □

The following statement is a special case of the previous theorem.

Example 6.41 For all $X, Y \in \mathbb{M}_n$ such that X^*Y is Hermitian, we have

$$\lambda(XX^* + YY^*) \prec \lambda(X^*X + Y^*Y).$$

Since

$$\begin{bmatrix} XX^* + YY^* & 0 \\ 0 & 0 \end{bmatrix} = \begin{bmatrix} X & Y \\ 0 & 0 \end{bmatrix}\begin{bmatrix} X^* & 0 \\ Y^* & 0 \end{bmatrix}$$

is unitarily conjugate to

$$\begin{bmatrix} X^* & 0 \\ Y^* & 0 \end{bmatrix}\begin{bmatrix} X & Y \\ 0 & 0 \end{bmatrix} = \begin{bmatrix} X^*X & X^*Y \\ Y^*X & Y^*Y \end{bmatrix}$$

and X^*Y is Hermitian by assumption, the above corollary implies that

$$\lambda\left(\begin{bmatrix} XX^* + YY^* & 0 \\ 0 & 0 \end{bmatrix}\right) \prec \lambda\left(\begin{bmatrix} X^*X + Y^*Y & 0 \\ 0 & 0 \end{bmatrix}\right).$$

So the statement follows.   □

Next we study log-majorizations and norm inequalities. These involve the **weighted geometric means**

$$A \#_\alpha B = A^{1/2}(A^{-1/2}BA^{-1/2})^\alpha A^{1/2},$$

where $0 \leq \alpha \leq 1$. The log-majorization in the next theorem is due to Ando and Hiai [8] and is considered as complementary to Theorem 6.37.

Theorem 6.42 *For all $A, B \in \mathbb{M}_n^+$,*

$$s(A^r \#_\alpha B^r) \prec_{(\log)} s((A \#_\alpha B)^r) \quad (r \geq 1), \tag{6.38}$$

or equivalently

$$s((A^p \#_\alpha B^p)^{1/p}) \prec_{(\log)} s((A^q \#_\alpha B^q)^{1/q}) \quad (p \geq q > 0). \tag{6.39}$$

Proof: First assume that both A and B are invertible. Note that

$$\det(A^r \#_\alpha B^r) = (\det A)^{r(1-\alpha)}(\det B)^{r\alpha} = \det(A \#_\alpha B)^r.$$

For every $k = 1, \ldots, n$, it is easily verified from the properties of the antisymmetric tensor powers that

$$(A^r \#_\alpha B^r)^{\wedge k} = (A^{\wedge k})^r \#_\alpha (B^{\wedge k})^r,$$
$$((A \#_\alpha B)^r)^{\wedge k} = ((A^{\wedge k}) \#_\alpha (B^{\wedge k}))^r.$$

So it suffices to show that

$$\|A^r \#_\alpha B^r\| \le \|(A \#_\alpha B)^r\| \quad (r \ge 1), \tag{6.40}$$

because (6.38) follows from Lemma 1.62 by taking $A^{\wedge k}$, $B^{\wedge k}$ instead of A, B in (6.40). To show (6.40), we prove that $A \#_\alpha B \le I$ implies $A^r \#_\alpha B^r \le I$. When $1 \le r \le 2$, let us write $r = 2 - \varepsilon$ with $0 \le \varepsilon \le 1$. Let $C := A^{-1/2} B A^{-1/2}$. Suppose that $A \#_\alpha B \le I$. Then $C^\alpha \le A^{-1}$ and

$$A \le C^{-\alpha}, \tag{6.41}$$

so that, since $0 \le \varepsilon \le 1$,

$$A^{1-\varepsilon} \le C^{-\alpha(1-\varepsilon)}. \tag{6.42}$$

Now we have

$$
\begin{aligned}
A^r \#_\alpha B^r &= A^{1-\frac{\varepsilon}{2}} \{A^{-1+\frac{\varepsilon}{2}} B \cdot B^{-\varepsilon} \cdot B A^{-1+\frac{\varepsilon}{2}}\}^\alpha A^{1-\frac{\varepsilon}{2}} \\
&= A^{1-\frac{\varepsilon}{2}} \{A^{-\frac{1-\varepsilon}{2}} C A^{1/2} (A^{-1/2} C^{-1} A^{-1/2})^\varepsilon A^{1/2} C A^{-\frac{1-\varepsilon}{2}}\}^\alpha A^{1-\frac{\varepsilon}{2}} \\
&= A^{1/2} \{A^{1-\varepsilon} \#_\alpha [C(A \#_\varepsilon C^{-1})C]\} A^{1/2} \\
&\le A^{1/2} \{C^{-\alpha(1-\varepsilon)} \#_\alpha [C(C^{-\alpha} \#_\varepsilon C^{-1})C]\} A^{1/2}
\end{aligned}
$$

by using (6.41), (6.42), and the joint monotonicity of power means. Since

$$C^{-\alpha(1-\varepsilon)} \#_\alpha [C(C^{-\alpha} \#_\varepsilon C^{-1})C] = C^{-\alpha(1-\varepsilon)(1-\alpha)}[C(C^{-\alpha(1-\varepsilon)} C^{-\varepsilon})C]^\alpha = C^\alpha,$$

we have

$$A^r \#_\alpha B^r \le A^{1/2} C^\alpha A^{1/2} = A \#_\alpha B \le I.$$

Therefore (6.38) is proved when $1 \le r \le 2$. When $r > 2$, write $r = 2^m s$ with $m \in \mathbb{N}$ and $1 \le s \le 2$. Repeating the above argument we have

$$s(A^r \#_\alpha B^r) \prec_{w(\log)} s(A^{2^{m-1}s} \#_\alpha B^{2^{m-1}s})^2$$

$$\vdots$$

$$\prec_{w(\log)} s(A^s \#_\alpha B^s)^{2^m}$$

$$\prec_{w(\log)} s(A \#_\alpha B)^r.$$

For general $A, B \in B(\mathcal{H})^+$ let $A_\varepsilon := A + \varepsilon I$ and $B_\varepsilon := B + \varepsilon I$ for $\varepsilon > 0$. Since

$$A^r \#_\alpha B^r = \lim_{\varepsilon \searrow 0} A_\varepsilon^r \#_\alpha B_\varepsilon^r \quad \text{and} \quad (A \#_\alpha B)^r = \lim_{\varepsilon \searrow 0} (A_\varepsilon \#_\alpha B_\varepsilon)^r,$$

we have (6.38) by the above case and Theorem 6.7 (10). Finally, (6.39) readily follows from (6.38) as in the last part of the proof of Theorem 6.37. □

By Theorems 6.42 and 6.23 we have:

Corollary 6.43 *Let $0 \le A, B \in \mathbb{M}_n$ and $||| \cdot |||$ be any unitarily invariant norm. If f is a continuous increasing function on $\mathbb{R}^+$ such that $f(0) \ge 0$ and $f(e^t)$ is convex, then*

$$||| f(A^r \#_\alpha B^r) ||| \le ||| f((A \#_\alpha B)^r) ||| \quad (r \ge 1).$$

In particular,

$$||| A^r \#_\alpha B^r ||| \le ||| (A \#_\alpha B)^r ||| \quad (r \ge 1).$$

Corollary 6.44 *For all self-adjoint $H, K \in \mathbb{M}_n$,*

$$s((e^{rH} \#_\alpha e^{rK})^{1/r}) \prec_{w(\log)} s(e^{(1-\alpha)H + \alpha K}) \quad (r > 0).$$

Hence, for every unitarily invariant norm $||| \cdot |||$,

$$||| (e^{rH} \#_\alpha e^{rK})^{1/r} ||| \le ||| e^{(1-\alpha)H + \alpha K} ||| \quad (r > 0),$$

and the above left-hand side increases to $||| e^{(1-\alpha)H + \alpha K} |||$ as $r \searrow 0$.

Specializing to trace inequality we have

$$\text{Tr} \, (e^{rH} \#_\alpha e^{rK})^{1/r} \le \text{Tr} \, e^{(1-\alpha)H + \alpha K} \quad (r > 0),$$

which was first proved in [47]. The following logarithmic trace inequalities are also known for all $0 \le A, B \in B(\mathcal{H})$ and every $r > 0$:

$$\frac{1}{r} \text{Tr} \, A \log B^{r/2} A^r B^{r/2} \le \text{Tr} \, A(\log A + \log B) \le \frac{1}{r} \text{Tr} \, A \log A^{r/2} B^r A^{r/2}, \quad (6.43)$$

$$\frac{1}{r}\text{Tr }A\log(A^r \# B^r)^2 \leq \text{Tr }A(\log A + \log B). \tag{6.44}$$

The **exponential function** has a generalization:

$$\exp_p(X) = (I + pX)^{\frac{1}{p}}, \tag{6.45}$$

where $X = X^* \in \mathbb{M}_n$ and $p \in (0, 1]$. (If $p \to 0$, then the limit is $\exp X$.) There is a corresponding extension of the Golden–Thompson trace inequality.

Theorem 6.45 *For* $0 \leq X, Y \in \mathbb{M}_n$ *and* $p \in (0, 1]$ *the following inequalities hold:*

$$\text{Tr }\exp_p(X + Y) \leq \text{Tr }\exp_p(X + Y + pY^{1/2}XY^{1/2})$$
$$\leq \text{Tr }\exp_p(X + Y + pXY) \leq \text{Tr }\exp_p(X)\exp_p(Y).$$

Proof: Let $X_1 := pX$, $Y_1 := pY$ and $q := 1/p$. Then

$$\begin{aligned}
\text{Tr }\exp_p(X + Y) &\leq \text{Tr }\exp_p(X + Y + pY^{1/2}XY^{1/2}) \\
&= \text{Tr }[(I + X_1 + Y_1 + Y_1^{1/2}X_1Y_1^{1/2})^q] \\
&\leq \text{Tr }[(I + X_1 + Y_1 + X_1Y_1)^q] \\
&= \text{Tr }[((I + X_1)(I + Y_1))^q].
\end{aligned}$$

The first inequality is immediate from the monotonicity of the function $(1 + px)^{1/p}$ and the second follows from Lemma 6.46 below. Next we take

$$\text{Tr }[((I + X_1)(I + Y_1))^q] \leq \text{Tr }[(I + X_1)^q(I + y_1)^q] = \text{Tr }[\exp_p(X)\exp_p(Y)],$$

which is, by the Araki–Lieb–Thirring inequality, (6.36).   □

Lemma 6.46 *For* $0 \leq X, Y \in \mathbb{M}_n$ *we have the following:*

$$\text{Tr }[(I + X + Y + Y^{1/2}XY^{1/2})^p] \leq \text{Tr }[(I + X + Y + XY)^p] \quad if\ p \geq 1,$$

$$\text{Tr }[(I + X + Y + Y^{1/2}XY^{1/2})^p] \geq \text{Tr }[(I + X + Y + XY)^p] \quad if\ 0 \leq p \leq 1.$$

Proof: For every $A, B \in \mathbb{M}_n^{sa}$, let $X = A$ and $Z = (BA)^k$ for any $k \in \mathbb{N}$. Since $X^*Z = A(BA)^k$ is Hermitian, we have

$$\lambda(A^2 + (BA)^k(AB)^k) \prec \lambda(A^2 + (AB)^k(BA)^k). \tag{6.46}$$

When $k = 1$, by Theorem 6.1 this majorization yields the trace inequalities:

$$\text{Tr}\,[(A^2 + BA^2B)^p] \leq \text{Tr}\,[(A^2 + AB^2A)^p] \quad \text{if } p \geq 1,$$
$$\text{Tr}\,[(A^2 + BA^2B)^p] \geq \text{Tr}\,[(A^2 + AB^2A)^p] \quad \text{if } 0 \leq p \leq 1.$$

Moreover, for $0 \leq X, Y \in \mathbb{M}_n$, let $A = (I + X)^{1/2}$ and $B = Y^{1/2}$. Notice that

$$\text{Tr}\,[(A^2 + BA^2B)^p] = \text{Tr}\,[(I + X + Y + Y^{1/2}XY^{1/2})^p]$$

and

$$\begin{aligned}
\text{Tr}\,[(A^2 + BA^2B)^p] &= \text{Tr}\,[((I + X)^{1/2}(I + Y)(I + X)^{1/2})^p] \\
&= \text{Tr}\,[((I + X)(I + Y))^p] = \text{Tr}\,[(I + X + Y + XY)^p],
\end{aligned}$$

where $(I + X)(I + Y)$ has eigenvalues in $(0, \infty)$ so that $((I + X)(I + Y))^p$ is defined via the analytic functional calculus (3.17). The statement then follows. $\square$

The inequalities of Theorem 6.45 can be extended to the symmetric norm inequality, as shown below, together with the complementary geometric mean inequality.

Theorem 6.47 *Let $||| \cdot |||$ be a symmetric norm on $\mathbb{M}_n$ and $p \in (0, 1]$. For all $0 \leq X, Y \in \mathbb{M}_n$ we have*

$$||| \exp_p(2X) \# \exp_p(2Y) ||| \leq ||| \exp_p(X + Y) |||$$
$$\leq ||| \exp_p(X)^{1/2} \exp_p(Y) \exp_p(X)^{1/2} |||$$
$$\leq ||| \exp_p(X) \exp_p(Y) |||.$$

Proof: We have

$$\begin{aligned}
\lambda(\exp_p(2X) \# \exp_p(2Y)) &= \lambda((I + 2pX)^{1/p} \# (I + 2pY)^{1/p}) \\
&\prec_{(\log)} (((I + 2pX) \# (I + 2pY))^{1/p}) \\
&\leq \lambda(\exp_p(X + Y)),
\end{aligned}$$

where the log-majorization is due to (6.38) and the inequality is due to the arithmetic-geometric mean inequality:

$$(I + 2pX) \# (I + 2pY) \leq \frac{(I + 2pX) + (I + 2pY)}{2} = I + p(X + Y).$$

On the other hand, let $A := (I + pX)^{1/2}$ and $B := (pY)^{1/2}$. We can use (6.46) and Theorem 6.37:

$$\lambda(\exp_p(X+Y)) \le \lambda((A^2 + BA^2B)^{1/p})$$
$$\prec \lambda((A^2 + AB^2A)^{1/p})$$
$$= \lambda(((I+pX)^{1/2}(I+pY)(I+pX)^{1/2})^{1/p})$$
$$\prec_{(\log)} \lambda((I+pX)^{1/2p}(I+pY)^{1/p}(I+pX)^{1/2p})$$
$$= \lambda(\exp_p(X)^{1/2}\exp_p(Y)\exp_p(X)^{1/2})$$
$$\prec_{(\log)} \lambda((\exp_p(X)\exp_p(Y)^2\exp_p(X))^{1/2})$$
$$= \lambda(|\exp_p(X)\exp_p(Y)|).$$

The above majorizations give the stated norm inequalities.   □

6.5 Notes and Remarks

The quote at the beginning of the chapter is from the paper John **von Neumann**, Some matrix inequalities and metrization of matric-space, Tomsk. Univ. Rev. **1**(1937), 286–300. (The paper is also in the book *John von Neumann Collected Works.*) Theorem 6.17 and the duality of the ℓ_p norm also appeared in this chapter.

Example 6.2 is from the paper M. A. Nielsen and J. Kempe, Separable states are more disordered globally than locally, Phys. Rev. Lett. **86**(2001), 5184–5187. The most comprehensive text on majorization theory for vectors and matrices is Marshall and Olkin's monograph [66]. (There is a recently reprinted version: A. W. Marshall, I. Olkin and B. C. Arnold, Inequalities: Theory of Majorization and Its Applications, Second ed., Springer, New York, 2011.) The content presented here is mostly based on Fumio Hiai [43]. Two survey articles [5, 6] of Tsuyoshi **Ando** are the best sources on majorizations for the eigenvalues and the singular values of matrices.

The first complete proof of the Lidskii–Wielandt theorem (Theorem 6.9) was obtained by Helmut Wielandt in 1955, who proved a complicated minimax representation by induction. The proofs of Theorems 6.9 and 6.13 presented here are surprisingly elementary and short (compared with previously known proofs), which are from the paper C.-K. **Li** and R. **Mathias**, The Lidskii–Mirsky–Wielandt theorem—additive and multiplicative versions, Numer. Math. **81**(1999), 377–413.

Here is a brief remark on the famous Horn conjecture that was affirmatively solved just before 2000. The conjecture concerns three real vectors $a = (a_1, \dots, a_n)$, $b = (b_1, \dots, b_n)$, and $c = (c_1, \dots, c_n)$. If there are two $n \times n$ Hermitian matrices A and B such that $a = \lambda(A)$, $b = \lambda(B)$, and $c = \lambda(A+B)$, that is, a, b, c are the eigenvalues of $A, B, A+B$, then the three vectors obey many inequalities of the form

$$\sum_{k \in K} c_k \le \sum_{i \in I} a_i + \sum_{j \in J} b_j$$

for certain triples (I, J, K) of subsets of $\{1, \ldots, n\}$, including those coming from the Lidskii–Wielandt theorem, together with the obvious equality

$$\sum_{i=1}^{n} c_i = \sum_{i=1}^{n} a_i + \sum_{i=1}^{n} b_i.$$

Horn [52] proposed a procedure which produces such triples (I, J, K) and conjectured that all the inequalities obtained in this way are sufficient to characterize a, b, c that are the eigenvalues of Hermitian matrices A, B, $A+B$. This long-standing Horn conjecture was solved by a combination of two papers, one by Klyachko [55] and the other by Knuston and Tao [56].

The Lieb–Thirring inequality was proved in 1976 by Elliott H. **Lieb** and Walter **Thirring** in a physical proceeding. It is interesting that Bellmann proved the particular case $\text{Tr}\,(AB)^2 \leq \text{Tr}\,A^2 B^2$ in 1980 and he conjectured $\text{Tr}\,(AB)^n \leq \text{Tr}\,A^n B^n$. The extension was proved by Huzihiro **Araki**, On an inequality of Lieb and Thirring, Lett. Math. Phys. **19**(1990), 167–170.

Theorem 6.25 is from J.-C. **Bourin** [29]. Theorem 6.27 from [28] also appeared in the paper of Aujla and Silva [15] with the inequality reversed for a contraction instead of an expansion. The subadditivity inequality in Theorem 6.30 was first obtained by T. **Ando** and X. **Zhan**, Norm inequalities related to operator monotone functions, Math. Ann. **315**(1999), 771–780. The proof of Theorem 6.30 presented here is simpler and is due to M. **Uchiyama** [82]. Theorem 6.35 is due to Ando [4].

In the papers [8, 47] there are more details on the logarithmic trace inequalities (6.43) and (6.44). Theorem 6.45 is in the paper S. **Furuichi** and M. Lin, A matrix trace inequality and its application, Linear Algebra Appl. **433**(2010), 1324–1328.

6.6 Exercises

1. Let S be a doubly substochastic $n \times n$ matrix. Show that there exists a doubly stochastic $n \times n$ matrix D such that $S_{ij} \leq D_{ij}$ for all $1 \leq i, j \leq n$.
2. Let Δ_n denote the set of all probability vectors in $\mathbb{R}^n$, i.e.,

$$\Delta_n := \{ p = (p_1, \ldots, p_n) : p_i \geq 0, \ \sum_{i=1}^{n} p_i = 1 \}.$$

Prove that

$$(1/n, 1/n, \ldots, 1/n) \prec p \prec (1, 0, \ldots, 0) \qquad (p \in \Delta_n).$$

The **Shannon entropy** of $p \in \Delta_n$ is $H(p) := -\sum_{i=1}^{n} p_i \log p_i$. Show that $H(q) \leq H(p) \leq \log n$ for all $p \prec q$ in Δ_n and $H(p) = \log n$ if and only if $p = (1/n, \ldots, 1/n)$.

3. Let $A \in \mathbb{M}_n^{sa}$. Prove the equality

$$\sum_{i=1}^{k} \lambda_i(A) = \max\{\operatorname{Tr} AP : P \text{ is a projection, rank } P = k\}$$

for $1 \le k \le n$.

4. Let $A, B \in \mathbb{M}_n^{sa}$. Show that $A \le B$ implies $\lambda_k(A) \le \lambda_k(B)$ for $1 \le k \le n$.

5. Show that the statement of Theorem 6.13 is equivalent to the inequality

$$\prod_{j=1}^{k} \left(s_{n+1-j}(A)s_{i_j}(B) \right) \le \prod_{j=1}^{k} s_{i_j}(AB)$$

for any choice of $1 \le i_1 < \cdots < i_k \le n$.

6. Give an example showing that, for the generalized inverse, $(AB)^\dagger = B^\dagger A^\dagger$ is not always true.

7. Describe the generalized inverse of a row matrix.

8. What is the generalized inverse of an orthogonal projection?

9. Let $A \in B(\mathcal{H})$ have polar decomposition $A = U|A|$. Prove that

$$|\langle x, Ax \rangle| \le \frac{\langle x, |A|x \rangle + \langle x, U|A|U^*x \rangle}{2}$$

for $x \in \mathcal{H}$.

10. Show that $|\operatorname{Tr} A| \le \|A\|_1$ for $A \in B(\mathcal{H})$.

11. Let $0 < p, p_1, p_2 \le \infty$ and $1/p = 1/p_1 + 1/p_2$. Prove the Hölder inequality for the vectors $a, b \in \mathbb{R}^n$:

$$\Phi_p(ab) \le \Phi_{p_1}(a)\Phi_{p_2}(b),$$

where $ab = (a_i b_i)$.

12. Show that a continuous concave function $f : \mathbb{R}^+ \to \mathbb{R}^+$ is the pointwise limit of a sequence of functions of the form

$$\alpha + \beta x - \sum_{\ell=1}^{m} c_\ell \gamma_{a_\ell}(x),$$

where $\alpha \ge 0$, $\beta, c_\ell, a_\ell > 0$ and γ_a is as given in the proof of Theorem 6.32.

13. Prove for self-adjoint matrices H, K the Lie–Trotter formula:

$$\lim_{r \to 0} (e^{rH/2} e^{rK} e^{rH/2})^{1/r} = e^{H+K}.$$

14. Prove for self-adjoint matrices H, K that

$$\lim_{r \to 0} (e^{rH} \#_\alpha e^{rK})^{1/r} = e^{(1-\alpha)H + \alpha K}.$$

15. Let f be a real function on $[a, b]$ with $a \le 0 \le b$. Prove the converse of Corollary 4.27, that is, if

$$\mathrm{Tr}\, f(Z^*AZ) \le \mathrm{Tr}\, Z^* f(A)Z$$

for every $A \in \mathbb{M}_2^{sa}$ with $\sigma(A) \subset [a, b]$ and every contraction $Z \in \mathbb{M}_2$, then f is convex on $[a, b]$ and $f(0) \le 0$.

16. Prove Theorem 4.28 in a direct way similar to the proof of Theorem 4.26.

17. Provide an example of a pair A, B of 2×2 Hermitian matrices such that

$$\lambda_1(|A + B|) < \lambda_1(|A| + |B|) \quad \text{and} \quad \lambda_2(|A + B|) > \lambda_2(|A| + |B|).$$

From this, show that Theorems 4.26 and 4.28 are not true for a simple convex function $f(x) = |x|$.

Chapter 7
Some Applications

Matrices are important in many areas of both pure and applied mathematics. In particular, they play essential roles in quantum probability and quantum information. A discrete classical probability is a vector $(p_1, p_2, \ldots, p_n)$ of $p_i \geq 0$ with $\sum_{i=1}^{n} p_i = 1$. Its counterpart in quantum theory is a matrix $D \in \mathbb{M}_n(\mathbb{C})$ such that $D \geq 0$ and $\operatorname{Tr} D = 1$; such matrices are called density matrices. Thus matrix analysis is central to quantum probability/statistics and quantum information. We remark that classical theory is included in quantum theory as a special case, where the relevant matrices are restricted to diagonal matrices. On the other hand, the concepts of classical probability theory can be reformulated with matrices, for instance, the covariance matrices typical in Gaussian probabilities and Fisher information matrices in the Cramér–Rao inequality.

This chapter is devoted to some aspects of the applications of matrices. One of the most important concepts in probability theory is the Markov property. This concept is discussed in the first section in the setting of Gaussian probabilities. The structure of covariance matrices for Gaussian probabilities with the Markov property is clarified in connection with the Boltzmann entropy. Its quantum analogue in the setting of CCR-algebras $\mathrm{CCR}(\mathcal{H})$ is the subject of Sect. 7.3. The counterpart of the notion of Gaussian probabilities is that of Gaussian or quasi-free states ω_A induced by positive operators A (similar to covariance matrices) on the underlying Hilbert space $\mathcal{H}$. In the setting of the triplet CCR-algebra

$$\mathrm{CCR}(\mathcal{H}_1 \oplus \mathcal{H}_2 \oplus \mathcal{H}_3) = \mathrm{CCR}(\mathcal{H}_1) \otimes \mathrm{CCR}(\mathcal{H}_2) \otimes \mathrm{CCR}(\mathcal{H}_3),$$

the special structure of A on $\mathcal{H}_1 \oplus \mathcal{H}_2 \oplus \mathcal{H}_3$ and equality in the strong subadditivity of the von Neumann entropy of ω_A emerge as equivalent conditions for the Markov property of ω_A.

The most useful entropy in both classical and quantum probabilities is the relative entropy $S(D_1 \| D_2) := \operatorname{Tr} D_1 (\log D_1 - \log D_2)$ for density matrices D_1, D_2, which was already discussed in Sects. 3.2 and 4.5. (It is also known as the Kullback–Leibler divergence in the classical case.) The notion was extended to the quasi-entropy:

F. Hiai and D. Petz, *Introduction to Matrix Analysis and Applications*,
Universitext, DOI: 10.1007/978-3-319-04150-6_7,
© Hindustan Book Agency 2014

$$S_f^A(D_1 \| D_2) := \langle AD_2^{1/2}, f(\Delta(D_1/D_2))(AD_2^{1/2})\rangle$$

associated with a certain function $f : \mathbb{R}^+ \to \mathbb{R}$ and a reference matrix A, where $\Delta(D_1/D_2)X := D_1 X D_2^{-1} = \mathbb{L}_{D_1}\mathbb{R}_{D_2}^{-1}(X)$. (Recall that $M_f(\mathbb{L}_A, \mathbb{R}_B) = f(\mathbb{L}_A\mathbb{R}_B^{-1})\mathbb{R}_B$ was used for the matrix mean transformation in Sect. 5.4.) The original relative entropy $S(D_1 \| D_2)$ is recovered by taking $f(x) = x \log x$ and $A = I$. The monotonicity and the joint convexity properties are two major properties of the quasi-entropies, which are the subject of Sect. 7.2. Another important topic in this section is the monotone Riemannian metrics on the manifold of invertible positive density matrices.

In a quantum system with a state D, several measurements may be performed to recover D, which forms the subject of the quantum state tomography. Here, a measurement is given by a POVM (positive operator-valued measure) $\{F(x) : x \in \mathcal{X}\}$, i.e., a finite set of positive matrices $F(x) \in \mathbb{M}_n(\mathbb{C})$ such that $\sum_{x \in \mathcal{X}} F(x) = I$. In Sect. 7.4 we study a few results concerning how to construct optimal quantum measurements.

The last section is concerned with the quantum version of the Cramér–Rao inequality, which is a certain matrix inequality between a kind of generalized variance and the quantum Fisher information. This subject belongs to quantum estimation theory and is also related to the monotone Riemannian metrics.

7.1 Gaussian Markov Property

In probability theory the matrices typically have real entries, but the content of this section can be modified for the complex case.

Given a positive definite real matrix $M \in \mathbb{M}_n(\mathbb{R})$ a **Gaussian probability** density is defined on $\mathbb{R}^n$ as

$$p(x) := \sqrt{\frac{\det M}{(2\pi)^n}} \exp\left(-\tfrac{1}{2}\langle x, Mx\rangle\right) \qquad (x \in \mathbb{R}^n).$$

Obviously $p(x) > 0$ and the integral

$$\int_{\mathbb{R}^n} p(x)\,dx = 1$$

follows due to the constant factor. Since

$$\int_{\mathbb{R}^n} \langle x, Bx\rangle p(x)\,dx = \operatorname{Tr} BM^{-1},$$

the particular case $B = E(ij)$ gives

$$\int_{\mathbb{R}^n} x_i x_j \, p(x) \, dx = \int_{\mathbb{R}^n} \langle x, E(ij)x \rangle p(x) \, dx = \operatorname{Tr} E(ij)M^{-1} = (M^{-1})_{ij}.$$

Thus the inverse of the matrix M is the covariance matrix.

The **Boltzmann entropy** is

$$S(p) = -\int_{\mathbb{R}^n} p(x) \log p(x) \, dx = \frac{n}{2} \log(2\pi e) - \frac{1}{2} \operatorname{Tr} \log M. \qquad (7.1)$$

(Instead of $\operatorname{Tr} \log M$, the formulation $\log \det M$ is often used.)

If $\mathbb{R}^n = \mathbb{R}^k \times \mathbb{R}^\ell$, then the probability density $p(x)$ has a reduction $p_1(y)$ on $\mathbb{R}^k$:

$$p_1(y) := \sqrt{\frac{\det M_1}{(2\pi)^k}} \exp\left(-\tfrac{1}{2} \langle y, M_1 y \rangle \right) \qquad (y \in \mathbb{R}^k).$$

To describe how M and M_1 are related we take the block matrix form

$$M = \begin{bmatrix} M_{11} & M_{12} \\ M_{12}^* & M_{22} \end{bmatrix},$$

where $M_{11} \in \mathbb{M}_k(\mathbb{R})$. Then we have

$$p_1(y) = \sqrt{\frac{\det M}{(2\pi)^m \det M_{22}}} \exp\left(-\tfrac{1}{2} \langle y, (M_{11} - M_{12} M_{22}^{-1} M_{12}^*) y \rangle \right),$$

see Example 2.7. Therefore $M_1 = M_{11} - M_{12} M_{22}^{-1} M_{12}^* = M/M_{22}$, which is called the **Schur complement** of M_{22} in M. We have $\det M_1 \cdot \det M_{22} = \det M$.

Let $p_2(z)$ be the reduction of $p(x)$ to $\mathbb{R}^\ell$ and denote the Gaussian matrix by M_2. In this case $M_2 = M_{22} - M_{12}^* M_{11}^{-1} M_{12} = M/M_{11}$. The following equivalent conditions hold:

(1) $S(p) \le S(p_1) + S(p_2)$;
(2) $-\operatorname{Tr} \log M \le -\operatorname{Tr} \log M_1 - \operatorname{Tr} \log M_2$;
(3) $\operatorname{Tr} \log M \le \operatorname{Tr} \log M_{11} + \operatorname{Tr} \log M_{22}$.

(1) is known as the subadditivity of the Boltzmann entropy. The equivalence of (1) and (2) follows directly from formula (7.1). (2) can be rewritten as

$$-\log \det M \le -(\log \det M - \log \det M_{22}) - (\log \det M - \log \det M_{11})$$

and we have (3). The equality condition is $M_{12} = 0$. If

$$M^{-1} = S = \begin{bmatrix} S_{11} & S_{12} \\ S_{12}^* & S_{22} \end{bmatrix},$$

then $M_{12} = 0$ is obviously equivalent to $S_{12} = 0$. It is interesting to note that (2) is equivalent to the inequality

(2*) $\text{Tr} \log S \leq \text{Tr} \log S_{11} + \text{Tr} \log S_{22}$.

The three-fold factorization $\mathbb{R}^n = \mathbb{R}^k \times \mathbb{R}^\ell \times \mathbb{R}^m$ is more interesting and includes essential properties. The Gaussian matrix of the probability density p is

$$M = \begin{bmatrix} M_{11} & M_{12} & M_{13} \\ M_{12}^* & M_{22} & M_{23} \\ M_{13}^* & M_{23}^* & M_{33} \end{bmatrix}, \tag{7.2}$$

where $M_{11} \in \mathbb{M}_k(\mathbb{R})$, $M_{22} \in \mathbb{M}_\ell(\mathbb{R})$, $M_{33} \in \mathbb{M}_m(\mathbb{R})$. Denote the reduced probability densities of p by $p_1, p_2, p_3, p_{12}, p_{23}$. The strong subadditivity of the Boltzmann entropy

$$S(p) + S(p_2) \leq S(p_{12}) + S(p_{23}) \tag{7.3}$$

is equivalent to the inequality

$$\text{Tr} \log S + \text{Tr} \log S_{22} \leq \text{Tr} \log \begin{bmatrix} S_{11} & S_{12} \\ S_{12}^* & S_{22} \end{bmatrix} + \text{Tr} \log \begin{bmatrix} S_{22} & S_{23} \\ S_{23}^* & S_{33} \end{bmatrix}, \tag{7.4}$$

where

$$M^{-1} = S = \begin{bmatrix} S_{11} & S_{12} & S_{13} \\ S_{12}^* & S_{22} & S_{23} \\ S_{13}^* & S_{23}^* & S_{33} \end{bmatrix}.$$

The **Markov property** in probability theory is typically defined as

$$\frac{p(x_1, x_2, x_3)}{p_{12}(x_1, x_2)} = \frac{p_{23}(x_2, x_3)}{p_2(x_2)} \quad (x_1 \in \mathbb{R}^k, x_2 \in \mathbb{R}^\ell, x_3 \in \mathbb{R}^m).$$

Taking the logarithm and integrating with respect to dp, we obtain

$$-S(p) + S(p_{12}) = -S(p_{23}) + S(p_2) \tag{7.5}$$

and this is the equality case in (7.3) and in (7.4). The equality case of (7.4) is described in Theorem 4.50, so we have the following:

Theorem 7.1 *The Gaussian probability density described by the block matrix (7.2) has the Markov property if and only if $S_{13} = S_{12} S_{22}^{-1} S_{23}$ for the inverse.*

Another condition comes from the inverse property of a 3×3 block matrix.

Theorem 7.2 *Let* $S = [S_{ij}]_{i,j=1}^3$ *be an invertible block matrix and assume that* S_{22} *and* $[S_{ij}]_{i,j=2}^3$ *are invertible. Then the* $(1,3)$ *entry of the inverse* $S^{-1} = [M_{ij}]_{i,j=1}^3$ *is given by the following formula:*

$$\left(S_{11} - [S_{12}, S_{13}] \begin{bmatrix} S_{22}, & S_{23} \\ S_{32} & S_{33} \end{bmatrix}^{-1} \begin{bmatrix} S_{12} \\ S_{13} \end{bmatrix} \right)^{-1}$$

$$\times (S_{12} S_{22}^{-1} S_{23} - S_{13})(S_{33} - S_{32} S_{22}^{-1} S_{23})^{-1}.$$

Hence $M_{13} = 0$ *if and only if* $S_{13} = S_{12} S_{22}^{-1} S_{23}$.

It follows that the Gaussian block matrix (7.2) has the Markov property if and only if $M_{13} = 0$.

7.2 Entropies and Monotonicity

Entropy and relative entropy are important notions in information theory. The quantum versions are in matrix theory. Recall that $0 \le D \in \mathbb{M}_n$ is a **density matrix** if $\text{Tr } D = 1$. This means that the eigenvalues $(\lambda_1, \lambda_2, \ldots, \lambda_n)$ form a probabilistic set: $\lambda_i \ge 0$, $\sum_i \lambda_i = 1$. The von Neumann entropy $S(D) = -\text{Tr } D \log D$ of the density matrix D is the Shannon entropy of the probabilistic set, $-\sum_i \lambda_i \log \lambda_i$.

The **partial trace** $\text{Tr}_1 : \mathbb{M}_n \otimes \mathbb{M}_m \to \mathbb{M}_m$ is a linear mapping which is defined by the formula $\text{Tr}_1(A \otimes B) = (\text{Tr } A)B$ on elementary tensors. It is called the partial trace, since only the trace of the first tensor factor is taken. $\text{Tr}_2 : \mathbb{M}_n \otimes \mathbb{M}_m \to \mathbb{M}_n$ is similarly defined.

The first example includes the strong subadditivity of the von Neumann entropy and a condition of the equality is also included. (Other conditions will appear in Theorem 7.6.)

Example 7.3 Here, we shall need the concept of a three-fold tensor product and reduced densities. Let D_{123} be a density matrix in $\mathbb{M}_k \otimes \mathbb{M}_\ell \otimes \mathbb{M}_m$. The reduced density matrices are defined by the partial traces:

$$D_{12} := \text{Tr}_3 D_{123} \in \mathbb{M}_k \otimes \mathbb{M}_\ell, \quad D_2 := \text{Tr}_{13} D_{123} \in \mathbb{M}_\ell, \quad D_{23} := \text{Tr}_1 D_{123} \in \mathbb{M}_k.$$

The **strong subadditivity** of S is the inequality

$$S(D_{123}) + S(D_2) \le S(D_{12}) + S(D_{23}), \tag{7.6}$$

which is equivalent to

$$\text{Tr } D_{123} (\log D_{123} - (\log D_{12} - \log D_2 + \log D_{23})) \ge 0.$$

The operator
$$\exp(\log D_{12} - \log D_2 + \log D_{23})$$

is positive and can be written as λD for a density matrix D. Actually,

$$\lambda = \mathrm{Tr}\ \exp(\log D_{12} - \log D_2 + \log D_{23}).$$

We have

$$S(D_{12}) + S(D_{23}) - S(D_{123}) - S(D_2)$$
$$= \mathrm{Tr}\ D_{123}\ (\log D_{123} - (\log D_{12} - \log D_2 + \log D_{23}))$$
$$= S(D_{123}\|\lambda D) = S(D_{123}\|D) - \log \lambda.$$

Here $S(X\|Y) := \mathrm{Tr}\ X(\log X - \log Y)$ is the **relative entropy**. If X and Y are density matrices, then $S(X\|Y) \geq 0$, see the Streater inequality (3.13).

Therefore, $\lambda \leq 1$ implies the positivity of the left-hand side (and the strong subadditivity). By Theorem 4.55, we have

$$\mathrm{Tr}\ \exp(\log D_{12} - \log D_2 + \log D_{23}) \leq \int_0^\infty \mathrm{Tr}\ D_{12}(tI + D_2)^{-1} D_{23}(tI + D_2)^{-1}\, dt.$$

Applying the partial traces we have

$$\mathrm{Tr}\ D_{12}(tI + D_2)^{-1} D_{23}(tI + D_2)^{-1} = \mathrm{Tr}\ D_2(tI + D_2)^{-1} D_2(tI + D_2)^{-1}$$

which can be integrated out. Hence

$$\int_0^\infty \mathrm{Tr}\ D_{12}(tI + D_2)^{-1} D_{23}(tI + D_2)^{-1}\, dt = \mathrm{Tr}\ D_2 = 1$$

obtaining $\lambda \leq 1$, and the strong subadditivity is proved.

If equality holds in (7.6), then $\exp(\log D_{12} - \log D_2 + \log D_{23})$ is a density matrix and

$$S(D_{123}\| \exp(\log D_{12} - \log D_2 + \log D_{23})) = 0$$

implies

$$\log D_{123} = \log D_{12} - \log D_2 + \log D_{23}.$$

This is the necessary and sufficient condition for equality. $\qquad\square$

For a density matrix D one can define the q-entropy as

$$S_q(D) = \frac{1 - \mathrm{Tr}\, D^q}{q - 1} = \frac{\mathrm{Tr}\,(D^q - D)}{1 - q} \qquad (q > 1).$$

This is also called the **quantum Tsallis entropy**. The limit $q \to 1$ is the von Neumann entropy.

The next theorem is the subadditivity of the q-entropy. The result has an elementary proof, but it was not known for several years.

Theorem 7.4 *When the density matrix $D \in \mathbb{M}_n \otimes \mathbb{M}_m$ has the partial densities $D_1 := \mathrm{Tr}_2 D$ and $D_2 := \mathrm{Tr}_1 D$, the subadditivity inequality $S_q(D) \leq S_q(D_1) + S_q(D_2)$, or equivalently*

$$\mathrm{Tr}\, D_1^q + \mathrm{Tr}\, D_2^q = \|D_1\|_q^q + \|D_2\|_q^q \leq 1 + \|D\|_q^q = 1 + \mathrm{Tr}\, D^q$$

holds for $q \geq 1$.

Proof: It is enough to show the case $q > 1$. First we use the q-norms and we prove

$$1 + \|D\|_q \geq \|D_1\|_q + \|D_2\|_q. \tag{7.7}$$

Lemma 7.5 below will be used.

If $1/q + 1/q' = 1$, then for $A \geq 0$ we have

$$\|A\|_q := \max\{\mathrm{Tr}\, AB : B \geq 0, \|B\|_{q'} \leq 1\}.$$

It follows that

$$\|D_1\|_q = \mathrm{Tr}\, X D_1 \quad \text{and} \quad \|D_2\|_q = \mathrm{Tr}\, Y D_2$$

with some $X \geq 0$ and $Y \geq 0$ such that $\|X\|_{q'} \leq 1$ and $\|Y\|_{q'} \leq 1$. It follows from Lemma 7.5 that

$$\|(X \otimes I_m + I_n \otimes Y - I_n \otimes I_m)_+\|_{q'} \leq 1$$

and we have $Z \geq 0$ such that

$$Z \geq X \otimes I_m + I_n \otimes Y - I_n \otimes I_m$$

and $\|Z\|_{q'} = 1$. It follows that

$$\mathrm{Tr}\,(ZD) + 1 \geq \mathrm{Tr}\,(X \otimes I_m + I_n \otimes Y)D = \mathrm{Tr}\, X D_1 + \mathrm{Tr}\, Y D_2.$$

Since

$$\|D\|_q \geq \mathrm{Tr}\,(ZD),$$

we have the inequality (7.7).

We examine the maximum of the function $f(x, y) = x^q + y^q$ in the domain

$$M := \{(x, y) : 0 \le x \le 1, 0 \le y \le 1, x + y \le 1 + \|D\|_q\}.$$

Since f is convex, it is sufficient to check the extreme points $(0, 0)$, $(1, 0)$, $(1, \|D\|_q)$, $(\|D\|_q, 1)$, $(0, 1)$. It follows that $f(x, y) \le 1 + \|D\|_q^q$. The inequality (7.7) gives that $(\|D_1\|_q, \|D_2\|_q) \in M$ and this gives $f(\|D_1\|_q, \|D_2\|_q) \le 1 + \|D\|_q^q$, which is the statement. □

Lemma 7.5 *For $q \ge 1$ and for the positive matrices $0 \le X \in \mathbb{M}_n$ and $0 \le Y \in \mathbb{M}_m$ assume that $\|X\|_q, \|Y\|_q \le 1$. Then the quantity*

$$\|(X \otimes I_m + I_n \otimes Y - I_n \otimes I_m)_+\|_q \le 1 \qquad (7.8)$$

holds.

Proof: Let $\{x_i : 1 \le i \le n\}$ and $\{y_j : 1 \le j \le m\}$ be the eigenvalues of X and Y, respectively. Then

$$\sum_{i=1}^n x_i^q \le 1, \qquad \sum_{j=1}^m y_j^q \le 1$$

and

$$\|(X \otimes I_m + I_n \otimes Y - I_n \otimes I_m)_+\|_q^q = \sum_{i,j}((x_i + y_j - 1)_+)^q.$$

The function $a \mapsto (a + b - 1)_+$ is convex for any real value of b:

$$\left(\frac{a_1 + a_2}{2} + b - 1\right)_+ \le \frac{1}{2}(a_1 + b - 1)_+ + \frac{1}{2}(a_2 + b - 1)_+ .$$

It follows that the vector-valued function

$$a \mapsto ((a + y_j - 1)_+ : j)$$

is convex as well. Since the ℓ^q norm for positive real vectors is convex and monotonically increasing, we conclude that

$$f(a) := \left(\sum_j ((a + y_j - 1)_+)^q\right)^{1/q}$$

is a convex function. Since $f(0) = 0$ and $f(1) = 1$, we have the inequality $f(a) \le a$ for $0 \le a \le 1$. Actually, we need this for x_i. Since $0 \le x_i \le 1$, $f(x_i) \le x_i$ follows and

$$\sum_i \sum_j ((x_i + y_j - 1)_+)^q = \sum_i f(x_i)^q \le \sum_i x_i^q \le 1.$$

So (7.8) is proved.   □

The next theorem is stated in the setting of Example 7.3.

Theorem 7.6 *The following conditions are equivalent:*

  (i) $S(D_{123}) + S(D_2) = S(D_{12}) + S(D_{23})$;
 (ii) $D_{123}^{it} D_{23}^{-it} = D_{12}^{it} D_2^{-it}$ *for every real t*;
(iii) $D_{123}^{1/2} D_{23}^{-1/2} = D_{12}^{1/2} D_2^{-1/2}$;
 (iv) $\log D_{123} - \log D_{23} = \log D_{12} - \log D_2$;
  (v) *there are positive matrices* $X \in \mathbb{M}_k \otimes \mathbb{M}_\ell$ *and* $Y \in \mathbb{M}_\ell \otimes \mathbb{M}_m$ *such that*
      $D_{123} = (X \otimes I_m)(I_k \otimes Y)$.

In the mathematical formalism of quantum mechanics, instead of n-tuples of numbers one works with $n \times n$ complex matrices. They form an algebra and this allows an algebraic approach.

For positive definite matrices $D_1, D_2 \in \mathbb{M}_n$, for $A \in \mathbb{M}_n$ and a function $f :$ $\mathbb{R}^+ \to \mathbb{R}$, the **quasi-entropy** is defined as

$$S_f^A(D_1 \| D_2) := \langle AD_2^{1/2}, f(\Delta(D_1/D_2))(AD_2^{1/2}) \rangle$$
$$= \operatorname{Tr} D_2^{1/2} A^* f(\Delta(D_1/D_2))(AD_2^{1/2}), \qquad (7.9)$$

where $\langle B, C \rangle := \operatorname{Tr} B^* C$ is the so-called **Hilbert–Schmidt inner product** and $\Delta(D_1/D_2) : \mathbb{M}_n \to \mathbb{M}_n$ is a linear mapping acting on matrices as follows:

$$\Delta(D_1/D_2)A := D_1 A D_2^{-1}.$$

This concept was introduced by Petz in [70, 72]. An alternative terminology is the **quantum f-divergence**.

For a positive definite matrix $D \in \mathbb{M}_n$ the left and the right multiplication operators acting on matrices are defined by

$$\mathbb{L}_D(X) := DX, \qquad \mathbb{R}_D(X) := XD \qquad (X \in \mathbb{M}_n). \qquad (7.10)$$

If we set

$$\mathbb{J}_{D_1, D_2}^f := f(\mathbb{L}_{D_1} \mathbb{R}_{D_2}^{-1}) \mathbb{R}_{D_2},$$

then the quasi-entropy has the form

$$S_f^A(D_1 \| D_2) = \langle A, \mathbb{J}_{D_1, D_2}^f A \rangle. \qquad (7.11)$$

It is clear from the definition that

$$S_f^A(\lambda D_1 \| \lambda D_2) = \lambda S_f^A(D_1 \| D_2)$$

for a positive number λ.

Let $\alpha : \mathbb{M}_n \to \mathbb{M}_m$ be a mapping between two matrix algebras. The dual α^* : $\mathbb{M}_m \to \mathbb{M}_n$ with respect to the Hilbert–Schmidt inner product is positive if and only if α is positive. Moreover, α is unital if and only if α^* is trace preserving. $\alpha : \mathbb{M}_n \to \mathbb{M}_m$ is called a **Schwarz mapping** if

$$\alpha(B^*B) \geq \alpha(B^*)\alpha(B) \tag{7.12}$$

for every $B \in \mathbb{M}_n$.

The quasi-entropies are monotone and jointly convex.

Theorem 7.7 *Assume that $f : \mathbb{R}^+ \to \mathbb{R}$ is a matrix monotone function with $f(0) \geq 0$ and $\alpha : \mathbb{M}_n \to \mathbb{M}_m$ is a unital Schwarz mapping. Then*

$$S_f^A(\alpha^*(D_1) \| \alpha^*(D_2)) \geq S_f^{\alpha(A)}(D_1 \| D_2) \tag{7.13}$$

holds for $A \in \mathbb{M}_n$ and for invertible density matrices D_1 and D_2 in the matrix algebra $\mathbb{M}_m$.

Proof: The proof is based on inequalities for matrix monotone and matrix concave functions. First note that

$$S_{f+c}^A(\alpha^*(D_1) \| \alpha^*(D_2)) = S_f^A(\alpha^*(D_1) \| \alpha^*(D_2)) + c \operatorname{Tr} D_1 \alpha(A^*A)$$

and

$$S_{f+c}^{\alpha(A)}(D_1 \| D_2) = S_f^{\alpha(A)}(D_1 \| D_2) + c \operatorname{Tr} D_1 (\alpha(A)^* \alpha(A))$$

for a positive constant c. By the Schwarz inequality (7.12), we may assume that $f(0) = 0$.

Let $\Delta := \Delta(D_1/D_2)$ and $\Delta_0 := \Delta(\alpha^*(D_1)/\alpha^*(D_2))$. The operator

$$V X \alpha^*(D_2)^{1/2} = \alpha(X) D_2^{1/2} \qquad (X \in \mathcal{M}_0)$$

is a contraction:

$$\begin{aligned} \|\alpha(X) D_2^{1/2}\|^2 &= \operatorname{Tr} D_2(\alpha(X)^* \alpha(X)) \\ &\leq \operatorname{Tr} D_2(\alpha(X^*X) = \operatorname{Tr} \alpha^*(D_2) X^*X = \|X \alpha^*(D_2)^{1/2}\|^2 \end{aligned}$$

since the Schwarz inequality is applicable to α. A similar simple computation gives that

$$V^* \Delta V \leq \Delta_0 .$$

Since f is matrix monotone, we have $f(\Delta_0) \geq f(V^*\Delta V)$. Recall that f is matrix concave. Therefore $f(V^*\Delta V) \geq V^*f(\Delta)V$ and we conclude

$$f(\Delta_0) \geq V^*f(\Delta)V.$$

Application to the vector $A\alpha^*(D_2)^{1/2}$ gives the statement.   □

It is remarkable that for a multiplicative α (i.e., α is a $*$-homomorphism) we do not need the condition $f(0) \geq 0$. Moreover, since $V^*\Delta V = \Delta_0$, we do not need the matrix monotonicity of the function f. In this case matrix concavity is the only condition needed to obtain the result analogous to Theorem 7.7. If we apply the monotonicity (7.13) (with $-f$ in place of f) to the embedding $\alpha(X) = X \oplus X$ of $\mathbb{M}_n$ into $\mathbb{M}_n \oplus \mathbb{M}_n \subset \mathbb{M}_n \otimes \mathbb{M}_2$ and to the densities $D_1 = \lambda E_1 \oplus (1 - \lambda)F_1$, $D_2 = \lambda E_2 \oplus (1 - \lambda)F_2$, then we obtain the joint convexity of the quasi-entropy:

Theorem 7.8 *If $f : \mathbb{R}^+ \to \mathbb{R}$ is a matrix convex function, then $S_f^A(D_1\|D_2)$ is jointly convex in the variables D_1 and D_2.*

If we consider the quasi-entropy in the terminology of means, then we obtain another proof. The joint convexity of the mean is the inequality

$$f(\mathbb{L}_{(A_1+A_2)/2}\mathbb{R}_{(B_1+B_2)/2}^{-1})\mathbb{R}_{(B_1+B_2)/2} \leq \tfrac{1}{2}f(\mathbb{L}_{A_1}\mathbb{R}_{B_1}^{-1})\mathbb{R}_{B_1} + \tfrac{1}{2}f(\mathbb{L}_{A_2}\mathbb{R}_{B_2}^{-1})\mathbb{R}_{B_2},$$

which can be simplified as

$$f(\mathbb{L}_{A_1+A_2}\mathbb{R}_{B_1+B_2}^{-1}) \leq \mathbb{R}_{B_1+B_2}^{-1/2}\mathbb{R}_{B_1}^{1/2}f(\mathbb{L}_{A_1}\mathbb{R}_{B_1}^{-1})\mathbb{R}_{B_1}^{1/2}\mathbb{R}_{B_1+B_2}^{-1/2}$$

$$+\mathbb{R}_{B_1+B_2}^{-1/2}\mathbb{R}_{B_2}^{1/2}f(\mathbb{L}_{A_2}\mathbb{R}_{B_2}^{-1})\mathbb{R}_{B_2}^{1/2}\mathbb{R}_{B_1+B_2}^{-1/2}$$

$$= Cf(\mathbb{L}_{A_1}\mathbb{R}_{B_1}^{-1})C^* + Df(\mathbb{L}_{A_2}\mathbb{R}_{B_2}^{-1})D^*.$$

Here $CC^* + DD^* = I$ and

$$C(\mathbb{L}_{A_1}\mathbb{R}_{B_1}^{-1})C^* + D(\mathbb{L}_{A_2}\mathbb{R}_{B_2}^{-1})D^* = \mathbb{L}_{A_1+A_2}\mathbb{R}_{B_1+B_2}^{-1}.$$

So the joint convexity of the quasi-entropy has the form

$$f(CXC^* + DYD^*) \leq Cf(X)C^* + Df(Y)D^*$$

which is true for a matrix convex function f, see Theorem 4.22.

Example 7.9 The concept of quasi-entropies includes some important special cases. If $f(t) = t^\alpha$, then

$$S_f^A(D_1\|D_2) = \mathrm{Tr}\, A^*D_1^\alpha AD_2^{1-\alpha}.$$

If $0 < \alpha < 1$, then f is matrix monotone. The joint concavity in (D_1, D_2) is the famous **Lieb's concavity theorem** [63].

In the case where $A = I$ we have a kind of relative entropy. For $f(x) = x \log x$ we have the **Umegaki relative entropy**

$$S(D_1 \| D_2) = \operatorname{Tr} D_1 (\log D_1 - \log D_2).$$

(If we want a matrix monotone function, then we can take $f(x) = \log x$ and then we have $S(D_2 \| D_1)$.) The Umegaki relative entropy is the most important example.

Let

$$f_\alpha(x) = \frac{1}{\alpha(1 - \alpha)} (1 - x^\alpha).$$

This function is matrix monotone decreasing for $\alpha \in (-1, 1)$. (For $\alpha = 0$, the limit is taken, which is equal to $- \log x$.) Then the **relative entropies of degree** α are produced:

$$S_\alpha(D_1 \| D_2) := \frac{1}{\alpha(1 - \alpha)} \operatorname{Tr} (I - D_1^\alpha D_2^{-\alpha}) D_2.$$

These quantities are essential in the quantum case.   □

Let $\mathcal{M}_n$ be the set of positive definite density matrices in $\mathbb{M}_n$. This is a differentiable manifold and the set of tangent vectors is $\{A = A^* \in \mathbb{M}_n \; : \; \operatorname{Tr} A = 0\}$. A **Riemannian metric** is a family of real inner products $\gamma_D(A, B)$ on the tangent vectors. For a function $f : (0, \infty) \to (0, \infty)$ with $x f(x^{-1}) = f(x)$, a possible definition is similar to (7.11): for $A, B \in \mathbb{M}_n^{sa}$ with $\operatorname{Tr} A = \operatorname{Tr} B = 0$,

$$\gamma_D^f(A, B) := \operatorname{Tr} A (\mathbb{J}_D^f)^{-1}(B). \tag{7.14}$$

(Here $\mathbb{J}_D^f = \mathbb{J}_{D,D}^f$.) The condition $x f(x^{-1}) = f(x)$ implies that $(\mathbb{J}_D^f)^{-1}(B) \in \mathbb{M}_n^{sa}$ if $B \in \mathbb{M}_n^{sa}$. Hence (7.14) actually defines a real inner product.

By a **monotone metric** we mean a family γ_D of Riemannian metrics on all manifolds $\mathcal{M}_n$ such that

$$\gamma_{\beta(D)}(\beta(A), \beta(A)) \le \gamma_D(A, A) \tag{7.15}$$

for every completely positive trace-preserving mapping $\beta : \mathbb{M}_n \to \mathbb{M}_m$ and every $A \in \mathbb{M}_n^{sa}$ with $\operatorname{Tr} A = 0$. If f is matrix monotone, then γ_D^f satisfies this monotonicity, see [71].

Let $\beta : \mathbb{M}_n \otimes \mathbb{M}_2 \to \mathbb{M}_n$ be defined as

$$\begin{bmatrix} B_{11} & B_{12} \\ B_{21} & B_{22} \end{bmatrix} \mapsto B_{11} + B_{22}.$$

This is completely positive and trace-preserving, making it a so-called partial trace. For

$$D = \begin{bmatrix} \lambda D_1 & 0 \\ 0 & (1-\lambda)D_2 \end{bmatrix}, \qquad A = \begin{bmatrix} \lambda A_1 & 0 \\ 0 & (1-\lambda)A_2 \end{bmatrix}$$

the inequality (7.15) gives

$$\gamma_{\lambda D_1 + (1-\lambda)D_2}(\lambda A_1 + (1-\lambda)A_2, \lambda A_1 + (1-\lambda)A_2)$$
$$\leq \gamma_{\lambda D_1}(\lambda A_1, \lambda A_1) + \gamma_{(1-\lambda)D_2}((1-\lambda)A_2, (1-\lambda)A_2).$$

Since $\gamma_{tD}(tA, tB) = t\gamma_D(A, B)$, we obtain the joint convexity:

Theorem 7.10 *For a matrix monotone function f, the monotone metric $\gamma_D^f(A, A)$ is a jointly convex function of (D, A) for positive definite D and general $A \in \mathbb{M}_n$.*

Now let $f : (0, \infty) \to (0, \infty)$ be a continuous function; the definition of f at 0 is not necessary here. Define $g, h : (0, \infty) \to (0, \infty)$ by $g(x) := xf(x^{-1})$ and

$$h(x) := \left(\frac{f(x)^{-1} + g(x)^{-1}}{2} \right)^{-1}, \qquad x > 0.$$

Obviously, h is symmetric, i.e., $h(x) = xh(x^{-1})$ for $x > 0$, so we may call h the harmonic symmetrization of f.

The difference between two parameters in $\mathbb{J}_{D_1, D_2}^f$ and one parameter in $\mathbb{J}_{D, D}^f$ is not essential if the matrix size can be changed. We need the next lemma.

Lemma 7.11 *For $D_1, D_2 > 0$ and general X in $\mathbb{M}_n$ let*

$$D := \begin{bmatrix} D_1 & 0 \\ 0 & D_2 \end{bmatrix}, \quad Y := \begin{bmatrix} 0 & X \\ 0 & 0 \end{bmatrix}, \quad A := \begin{bmatrix} 0 & X \\ X^* & 0 \end{bmatrix}.$$

Then

$$\langle Y, (\mathbb{J}_D^f)^{-1}Y \rangle = \langle X, (\mathbb{J}_{D_1, D_2}^f)^{-1}X \rangle, \tag{7.16}$$

$$\langle A, (\mathbb{J}_D^f)^{-1}A \rangle = 2\langle X, (\mathbb{J}_{D_1, D_2}^h)^{-1}X \rangle. \tag{7.17}$$

Proof: First we show that

$$(\mathbb{J}_D^f)^{-1}\begin{bmatrix} X_{11} & X_{12} \\ X_{21} & X_{22} \end{bmatrix} = \begin{bmatrix} (J_{D_1}^f)^{-1}X_{11} & (J_{D_1, D_2}^f)^{-1}X_{12} \\ (J_{D_2, D_1}^f)^{-1}X_{21} & (J_{D_2}^f)^{-1}X_{22} \end{bmatrix}. \tag{7.18}$$

Since continuous functions can be approximated by polynomials, it is enough to check (7.18) for $f(x) = x^k$, which is easy. From (7.18), (7.16) is obvious and

$$\langle A, (\mathbb{J}_D^f)^{-1} A \rangle = \langle X, (\mathbb{J}_{D_1,D_2}^f)^{-1} X \rangle + \langle X^*, (\mathbb{J}_{D_2,D_1}^f)^{-1} X^* \rangle.$$

From the spectral decompositions

$$D_1 = \sum_i \lambda_i P_i \quad \text{and} \quad D_2 = \sum_j \mu_j Q_j$$

we have

$$\mathbb{J}_{D_1,D_2}^f A = \sum_{i,j} m_f(\lambda_i, \mu_j) P_i A Q_j$$

and

$$\langle X, (\mathbb{J}_{D_1,D_2}^g)^{-1} X \rangle = \sum_{i,j} m_g(\lambda_i, \mu_j) \text{Tr } X^* P_i X Q_j$$

$$= \sum_{i,j} m_f(\mu_j, \lambda_i) \text{Tr } X Q_j X^* P_i$$

$$= \langle X^*, (\mathbb{J}_{D_2,D_1}^f)^{-1} X^* \rangle. \tag{7.19}$$

Therefore,

$$\langle A, (\mathbb{J}_D^f)^{-1} A \rangle = \langle X, (\mathbb{J}_{D_1,D_2}^f)^{-1} X \rangle + \langle X, (\mathbb{J}_{D_1,D_2}^g)^{-1} X \rangle = 2 \langle X, (\mathbb{J}_{D_1,D_2}^h)^{-1} X \rangle.$$

$\square$

Theorem 7.12 *In the above setting consider the following conditions:*

(i) *f is matrix monotone;*

(ii) *$(D, A) \mapsto \langle A, (\mathbb{J}_D^f)^{-1} A \rangle$ is jointly convex for positive definite D and general A in $\mathbb{M}_n$ for every n;*

(iii) *$(D_1, D_2, A) \mapsto \langle A, (\mathbb{J}_{D_1,D_2}^f)^{-1} A \rangle$ is jointly convex for positive definite D_1, D_2 and general A in $\mathbb{M}_n$ for every n;*

(iv) *$(D, A) \mapsto \langle A, (\mathbb{J}_D^f)^{-1} A \rangle$ is jointly convex for positive definite D and self-adjoint A in $\mathbb{M}_n$ for every n;*

(v) *h is matrix monotone.*

Then (i) $\Leftrightarrow$ (ii) $\Leftrightarrow$ (iii) $\Rightarrow$ (iv) $\Leftrightarrow$ (v).

Proof: (i) $\Rightarrow$ (ii) is Theorem 7.10 and (ii) $\Rightarrow$ (iii) follows from (7.16). We prove (iii) $\Rightarrow$ (i). For each $\xi \in \mathbb{C}^n$ let $X_\xi := [\xi \, 0 \, \cdots \, 0] \in \mathbb{M}_n$, i.e., the first column of X_ξ is ξ and all other entries of X_ξ are zero. When $D_2 = I$ and $X = X_\xi$, we have for $D > 0$ in $\mathbb{M}_n$

$$\langle X_\xi, (\mathbb{J}_{D,I}^f)^{-1} X_\xi \rangle = \langle X_\xi, f(D)^{-1} X_\xi \rangle = \langle \xi, f(D)^{-1} \xi \rangle.$$

Hence it follows from (iii) that $\langle \xi, f(D)^{-1}\xi \rangle$ is jointly convex in $D > 0$ in $\mathbb{M}_n$ and $\xi \in \mathbb{C}^n$. By a standard convergence argument we see that $(D, \xi) \mapsto \langle \xi, f(D)^{-1}\xi \rangle$ is jointly convex for positive invertible $D \in B(\mathcal{H})$ and $\xi \in \mathcal{H}$, where $B(\mathcal{H})$ is the set of bounded operators on a separable infinite-dimensional Hilbert space $\mathcal{H}$. Now Theorem 3.1 in [9] is used to conclude that $1/f$ is matrix monotone decreasing, so f is matrix monotone.

(ii) $\Rightarrow$ (iv) is trivial. Assume (iv); then it follows from (7.17) that (iii) holds for h instead of f, so (v) holds thanks to (iii) $\Rightarrow$ (i) for h. From (7.19) when $A = A^*$ and $D_1 = D_2 = D$, it follows that

$$\langle A, (\mathbb{J}_D^f)^{-1}A \rangle = \langle A, (\mathbb{J}_D^g)^{-1}A \rangle = \langle A, (\mathbb{J}_D^h)^{-1}A \rangle.$$

Hence (v) implies (iv) by applying (i) $\Rightarrow$ (ii) to h. $\square$

Example 7.13 The χ^2-**divergence**

$$\chi^2(p, q) := \sum_i \frac{(p_i - q_i)^2}{q_i} = \sum_i \left(\frac{p_i}{q_i} - 1 \right)^2 q_i$$

was first introduced by Karl Pearson in 1900 for probability densities p and q. Since

$$\left(\sum_i |p_i - q_i| \right)^2 = \left(\sum_i \left| \frac{p_i}{q_i} - 1 \right| q_i \right)^2 \leq \sum_i \left(\frac{p_i}{q_i} - 1 \right)^2 q_i,$$

we have

$$\|p - q\|_1^2 \leq \chi^2(p, q). \tag{7.20}$$

A quantum generalization was introduced very recently: for density matrices ρ and σ,

$$\chi_\alpha^2(\rho, \sigma) = \mathrm{Tr}\left((\rho - \sigma)\sigma^{-\alpha}(\rho - \sigma)\sigma^{\alpha-1} \right) = \mathrm{Tr}\,\rho\sigma^{-\alpha}\rho\sigma^{\alpha-1} - 1$$
$$= \langle \rho, (\mathbb{J}_\sigma^f)^{-1}\rho \rangle - 1,$$

where $\alpha \in [0, 1]$ and $f(x) = x^\alpha$. If ρ and σ commute, then this formula is independent of α.

The monotonicity of the χ^2-divergence follows from (7.15). The monotonicity and the classical inequality (7.20) imply that

$$\|\rho - \sigma\|_1^2 \leq \chi^2(\rho, \sigma).$$

Indeed, if E is the conditional expectation onto the commutative algebra generated by $\rho - \sigma$, then

$$\|\rho - \sigma\|_1^2 = \|E(\rho) - E(\sigma)\|_1^2 \le \chi^2(E(\rho), E(\sigma)) \le \chi^2(\rho, \sigma).$$

□

7.3 Quantum Markov Triplets

The CCR-algebra used in this section is an infinite-dimensional C*-algebra, but its parametrization will be by a finite-dimensional Hilbert space $\mathcal{H}$. (CCR is the abbreviation of "canonical commutation relation" and the book [69] contains the details.)

Assume that for every $f \in \mathcal{H}$ a unitary operator $W(f)$ is given so that the relations

$$W(f_1)W(f_2) = W(f_1 + f_2) \exp(i\,\sigma(f_1, f_2)),$$

$$W(-f) = W(f)^*$$

hold for $f_1, f_2, f \in \mathcal{H}$ with $\sigma(f_1, f_2) := \mathrm{Im}\langle f_1, f_2\rangle$. The C*-algebra generated by these unitaries is unique and denoted by $\mathrm{CCR}(\mathcal{H})$. Given a positive operator $A \in B(\mathcal{H})$, a functional $\omega_A : \mathrm{CCR}(\mathcal{H}) \to \mathbb{C}$ can be defined as

$$\omega_A(W(f)) := \exp\left(-\|f\|^2/2 - \langle f, Af\rangle\right).$$

This is called a **Gaussian** or **quasi-free state**. In the so-called Fock representation of $\mathrm{CCR}(\mathcal{H})$ the quasi-free state ω_A has the density operator $D_A, D_A \ge 0$ and $\mathrm{Tr}\, D_A = 1$. We do not describe D_A here but we remark that if the λ_i's are the eigenvalues of A, then D_A has the eigenvalues

$$\prod_i \frac{1}{1 + \lambda_i} \left(\frac{\lambda_i}{1 + \lambda_i}\right)^{n_i},$$

where the n_i's are non-negative integers. Therefore the von Neumann entropy is

$$S(\omega_A) := -\mathrm{Tr}\, D_A \log D_A = \mathrm{Tr}\, \kappa(A), \tag{7.21}$$

where $\kappa(t) := -t \log t + (t + 1) \log(t + 1)$ is an interesting special function.

Assume that $\mathcal{H} = \mathcal{H}_1 \oplus \mathcal{H}_2$ and write the positive mapping $A \in B(\mathcal{H})$ in the form of a block matrix:

$$A = \begin{bmatrix} A_{11} & A_{12} \\ A_{21} & A_{22} \end{bmatrix}.$$

If $f \in \mathcal{H}_1$, then

$$\omega_A(W(f \oplus 0)) = \exp\left(-\|f\|^2/2 - \langle f, A_{11}f\rangle\right).$$

Therefore the restriction of the quasi-free state ω_A to $\mathrm{CCR}(\mathcal{H}_1)$ is the quasi-free state $\omega_{A_{11}}$.

Let $\mathcal{H} = \mathcal{H}_1 \oplus \mathcal{H}_2 \oplus \mathcal{H}_3$ be a finite-dimensional Hilbert space and consider the CCR-algebras $\mathrm{CCR}(\mathcal{H}_i)$ $(1 \le i \le 3)$. Then

$$\mathrm{CCR}(\mathcal{H}) = \mathrm{CCR}(\mathcal{H}_1) \otimes \mathrm{CCR}(\mathcal{H}_2) \otimes \mathrm{CCR}(\mathcal{H}_3)$$

holds. Assume that D_{123} is a density operator in $\mathrm{CCR}(\mathcal{H})$ and we denote by D_{12}, D_2, D_{23} its reductions into the subalgebras $\mathrm{CCR}(\mathcal{H}_1) \otimes \mathrm{CCR}(\mathcal{H}_2)$, $\mathrm{CCR}(\mathcal{H}_2)$, $\mathrm{CCR}(\mathcal{H}_2) \otimes \mathrm{CCR}(\mathcal{H}_3)$, respectively. These subalgebras form a **Markov triplet** with respect to the state D_{123} if

$$S(D_{123}) - S(D_{23}) = S(D_{12}) - S(D_2), \qquad (7.22)$$

where S denotes the von Neumann entropy and we assume that both sides are finite in the equation. (Note (7.22) is the quantum analogue of (7.5).)

Now we concentrate on the Markov property of a quasi-free state $\omega_A \equiv \omega_{123}$ with the density operator D_{123}, where A is a positive operator acting on $\mathcal{H} = \mathcal{H}_1 \oplus \mathcal{H}_2 \oplus \mathcal{H}_3$ and it has a block matrix form

$$A = \begin{bmatrix} A_{11} & A_{12} & A_{13} \\ A_{21} & A_{22} & A_{23} \\ A_{31} & A_{32} & A_{33} \end{bmatrix}.$$

Then the restrictions D_{12}, D_{23} and D_2 are also Gaussian states with the positive operators

$$B = \begin{bmatrix} A_{11} & A_{12} & 0 \\ A_{21} & A_{22} & 0 \\ 0 & 0 & I \end{bmatrix}, \quad C = \begin{bmatrix} I & 0 & 0 \\ 0 & A_{22} & A_{23} \\ 0 & A_{32} & A_{33} \end{bmatrix} \quad \text{and} \quad D = \begin{bmatrix} I & 0 & 0 \\ 0 & A_{22} & 0 \\ 0 & 0 & I \end{bmatrix},$$

respectively. Formula (7.21) tells us that the Markov condition (7.22) is equivalent to

$$\mathrm{Tr}\,\kappa(A) + \mathrm{Tr}\,\kappa(D) = \mathrm{Tr}\,\kappa(B) + \mathrm{Tr}\,\kappa(C).$$

(This kind of condition has appeared already in the study of strongly subadditive functions, see Theorem 4.50.)

Denote by P_i the orthogonal projection from $\mathcal{H}$ onto $\mathcal{H}_i$, $1 \le i \le 3$. Of course, $P_1 + P_2 + P_3 = I$ and we also use the notation $P_{12} := P_1 + P_2$ and $P_{23} := P_2 + P_3$.

Theorem 7.14 *Assume that $A \in B(\mathcal{H})$ is a positive invertible operator and the corresponding quasi-free state is denoted by $\omega_A \equiv \omega_{123}$ on $\mathrm{CCR}(\mathcal{H})$. Then the following conditions are equivalent.*

(a) $S(\omega_{123}) + S(\omega_2) = S(\omega_{12}) + S(\omega_{23})$;
(b) $\mathrm{Tr}\,\kappa(A) + \mathrm{Tr}\,\kappa(P_2 A P_2) = \mathrm{Tr}\,\kappa(P_{12} A P_{12}) + \mathrm{Tr}\,\kappa(P_{23} A P_{23})$;

(c) *there is a projection $P \in B(\mathcal{H})$ such that $P_1 \leq P \leq P_1 + P_2$ and $PA = AP$.*

Proof: By the formula (7.21), (a) and (b) are equivalent.

Condition (c) tells us that the matrix A has a special form:

$$
A = \begin{bmatrix} A_{11} & \begin{bmatrix} a & 0 \end{bmatrix} & 0 \\ \begin{bmatrix} a^* \\ 0 \end{bmatrix} & \begin{bmatrix} c & 0 \\ 0 & d \end{bmatrix} & \begin{bmatrix} 0 \\ b \end{bmatrix} \\ 0 & \begin{bmatrix} 0 & b^* \end{bmatrix} & A_{33} \end{bmatrix} = \begin{bmatrix} \begin{bmatrix} A_{11} & a \\ a^* & c \end{bmatrix} & 0 \\ 0 & \begin{bmatrix} d & b \\ b^* & A_{33} \end{bmatrix} \end{bmatrix}, \tag{7.23}
$$

where the parameters a, b, c, d (and 0) are operators. This is a block diagonal matrix:

$$
A = \begin{bmatrix} A_1 & 0 \\ 0 & A_2 \end{bmatrix},
$$

and the projection P is

$$
\begin{bmatrix} I & 0 \\ 0 & 0 \end{bmatrix}
$$

in this setting.

The Hilbert space $\mathcal{H}_2$ is decomposed as $\mathcal{H}_2^L \oplus \mathcal{H}_2^R$, where $\mathcal{H}_2^L$ is the range of the projection $P P_2$. Therefore,

$$
\mathrm{CCR}(\mathcal{H}) = \mathrm{CCR}(\mathcal{H}_1 \oplus \mathcal{H}_2^L) \otimes \mathrm{CCR}(\mathcal{H}_2^R \oplus \mathcal{H}_3)
$$

and ω_{123} becomes a product state $\omega_L \otimes \omega_R$. From this we can easily show the implication $(c) \Rightarrow (a)$.

The essential part is the proof of $(b) \Rightarrow (c)$. Now assume (b), that is,

$$
\mathrm{Tr}\, \kappa(A) + \mathrm{Tr}\, \kappa(A_{22}) = \mathrm{Tr}\, \kappa(B) + \mathrm{Tr}\, \kappa(C).
$$

We notice that the function $\kappa(x) = -x \log x + (x+1) \log(x+1)$ admits the integral representation

$$
\kappa(x) = \int_1^\infty t^{-2} \log(tx + 1)\, dt.
$$

By Theorem 4.50 applied to $tA + I$ we have

$$
\mathrm{Tr}\, \log(tA + I) + \mathrm{Tr}\, \log(tA_{22} + I) \leq \mathrm{Tr}\, \log(tB + I) + \mathrm{Tr}\, \log(tC + I) \tag{7.24}
$$

for every $t > 1$. Hence it follows from (7.24) that equality holds in (7.24) for almost every $t > 1$. By Theorem 4.50 again this implies that

$$tA_{13} = tA_{12}(tA_{22} + I)^{-1}tA_{23}$$

for almost every $t > 1$. The continuity gives that for every $t > 1$ we have

$$A_{13} = A_{12}(A_{22} + t^{-1}I)^{-1}A_{23}.$$

Since $A_{12}(A_{22} + zI)^{-1}A_{23}$ is an analytic function in $\{z \in \mathbb{C} : \operatorname{Re} z > 0\}$, we have

$$A_{13} = A_{12}(A_{22} + sI)^{-1}A_{23} \qquad (s \in \mathbb{R}^+).$$

Letting $s \to \infty$ shows that $A_{13} = 0$. Since $A_{12}s(A_{22} + sI)^{-1}A_{23} \to A_{12}A_{23}$ as $s \to \infty$, we also have $A_{12}A_{23} = 0$. The latter condition means that ran $A_{23} \subset \ker A_{12}$, or equivalently $(\ker A_{12})^{\perp} \subset \ker A_{23}^*$.

The linear combinations of the functions $x \mapsto 1/(s + x)$ form an algebra and by the Stone–Weierstrass theorem $A_{12}g(A_{22})A_{23} = 0$ for any continuous function g.

We want to show that the equality implies the structure (7.23) of the operator A. We have $A_{23} : \mathcal{H}_3 \to \mathcal{H}_2$ and $A_{12} : \mathcal{H}_2 \to \mathcal{H}_1$. To deduce the structure (7.23), we have to find a subspace $H \subset \mathcal{H}_2$ such that

$$A_{22}H \subset H, \quad H^{\perp} \subset \ker A_{12}, \quad H \subset \ker A_{23}^*,$$

or alternatively $K \, (= H^{\perp}) \subset \mathcal{H}_2$ should be an invariant subspace of A_{22} such that

$$\operatorname{ran} A_{23} \subset K \subset \ker A_{12}.$$

Let

$$K := \left\{ \sum_i A_{22}^{n_i} A_{23} x_i : x_i \in \mathcal{H}_3,\, n_i \geq 0 \right\}$$

where the sum is always finite. K is a subspace of $\mathcal{H}_2$. The property ran $A_{23} \subset K$ and the invariance under A_{22} are obvious. Since

$$A_{12}A_{22}^n A_{23}x = 0,$$

$K \subset \ker A_{12}$ also follows. The proof is complete. $\qquad\qquad\square$

In the theorem it was assumed that $\mathcal{H}$ is a finite-dimensional Hilbert space, but the proof also works in infinite dimensions. In the theorem the formula (7.23) shows that A should be a block diagonal matrix. There are nontrivial Markovian Gaussian states which are not a product in the time localization ($\mathcal{H} = \mathcal{H}_1 \oplus \mathcal{H}_2 \oplus \mathcal{H}_3$). However, the first and the third subalgebras are always independent.

The next two theorems give different descriptions (but they are not essentially different).

Theorem 7.15 *For a quasi-free state* ω_A *the Markov property* (7.22) *is equivalent to the condition*

$$A^{it}(I+A)^{-it}D^{-it}(I+D)^{it} = B^{it}(I+B)^{-it}C^{-it}(I+C)^{it}$$

for every real t.

Theorem 7.16 *The block matrix*

$$A = \begin{bmatrix} A_{11} & A_{12} & A_{13} \\ A_{21} & A_{22} & A_{23} \\ A_{31} & A_{32} & A_{33} \end{bmatrix}$$

gives a Gaussian state with the Markov property if and only if

$$A_{13} = A_{12}f(A_{22})A_{23}$$

for any continuous function $f : \mathbb{R} \to \mathbb{R}$.

This shows that the CCR condition is much more restrictive than the classical one.

7.4 Optimal Quantum Measurements

In the matrix formalism the state of a quantum system is a density matrix $0 \le \rho \in \mathbb{M}_d(\mathbb{C})$ with the property $\mathrm{Tr}\,\rho = 1$. A finite set $\{F(x) : x \in \mathbb{X}\}$ of positive matrices is called a **positive operator-valued measure (POVM)** if

$$\sum_{x \in \mathbb{X}} F(x) = I,$$

where $F(x) \ne 0$ can be assumed. Quantum state tomography can recover the state ρ from the probability set $\{\mathrm{Tr}\,\rho F(x) : x \in \mathbb{X}\}$. In this section there are arguments for the optimal POVM set. There are a few rules from quantum theory, but the essential part is the notion of frames in the Hilbert space $\mathbb{M}_d(\mathbb{C})$.

The space $\mathbb{M}_d(\mathbb{C})$ of matrices equipped with the Hilbert–Schmidt inner product $\langle A|B \rangle = \mathrm{Tr}\,A^*B$ is a Hilbert space. We use the **bra-ket** notation for operators: $\langle A|$ is an operator bra and $|B\rangle$ is an operator ket. Then $|A\rangle\langle B|$ is a linear mapping $\mathbb{M}_d(\mathbb{C}) \to \mathbb{M}_d(\mathbb{C})$. For example,

$$|A\rangle\langle B|C = (\mathrm{Tr}\,B^*C)A, \qquad (|A\rangle\langle B|)^* = |B\rangle\langle A|,$$

$$|\mathbf{A}_1 A\rangle\langle \mathbf{A}_2 B| = \mathbf{A}_1|A\rangle\langle B|\mathbf{A}_2^*$$

when $\mathbf{A}_1, \mathbf{A}_2 : \mathbb{M}_d(\mathbb{C}) \to \mathbb{M}_d(\mathbb{C})$.

For an orthonormal basis $\{|E_k\rangle : 1 \le k \le d^2\}$ of $\mathbb{M}_d(\mathbb{C})$, a linear superoperator $\mathbf{S} : \mathbb{M}_d(\mathbb{C}) \to \mathbb{M}_d(\mathbb{C})$ can then be written as $\mathbf{S} = \sum_{j,k} s_{jk} |E_j\rangle\langle E_k|$ and its action is defined as

$$\mathbf{S}|A\rangle = \sum_{j,k} s_{jk} |E_j\rangle\langle E_k|A\rangle = \sum_{j,k} s_{jk} E_j \operatorname{Tr}\left(E_k^* A\right).$$

We denote the identity superoperator by $\mathbf{I}$, and so $\mathbf{I} = \sum_k |E_k\rangle\langle E_k|$.

The Hilbert space $\mathbb{M}_d(\mathbb{C})$ has an orthogonal decomposition

$$\{cI : c \in \mathbb{C}\} \oplus \{A \in \mathbb{M}_d(\mathbb{C}) : \operatorname{Tr} A = 0\}.$$

In the block matrix form under this decomposition,

$$\mathbf{I} = \begin{bmatrix} 1 & 0 \\ 0 & I_{d^2-1} \end{bmatrix} \quad \text{and} \quad |I\rangle\langle I| = \begin{bmatrix} d & 0 \\ 0 & 0 \end{bmatrix}.$$

Let $\mathbb{X}$ be a finite set. An **operator frame** is a family of operators $\{A(x) : x \in \mathbb{X}\}$ for which there exists a constant $a > 0$ such that

$$a\langle C|C\rangle \le \sum_{x \in \mathbb{X}} |\langle A(x)|C\rangle|^2 \tag{7.25}$$

for all $C \in \mathbb{M}_d(\mathbb{C})$. The **frame superoperator** is defined as

$$\mathbf{A} = \sum_{x \in \mathbb{X}} |A(x)\rangle\langle A(x)|.$$

It has the properties

$$\mathbf{A}B = \sum_{x \in \mathbb{X}} |A(x)\rangle\langle A(x)|B\rangle = \sum_{x \in \mathbb{X}} |A(x)\rangle \operatorname{Tr} A(x)^* B,$$

$$\operatorname{Tr} \mathbf{A}^2 = \sum_{x,y \in \mathbb{X}} |\langle A(x)|A(y)\rangle|^2. \tag{7.26}$$

The operator $\mathbf{A}$ is positive (and self-adjoint), since

$$\langle B|\mathbf{A}|B\rangle = \sum_{x \in \mathbb{X}} |\langle A(x)|B\rangle|^2 \ge 0.$$

Since this formula shows that (7.25) is equivalent to

$$a\mathbf{I} \le \mathbf{A},$$

it follows that (7.25) holds if and only if $\mathbf{A}$ has an inverse. The frame is called **tight** if $\mathbf{A} = a\mathbf{I}$.

Let $\tau : \mathbb{X} \to (0, \infty)$. Then $\{A(x) : x \in \mathbb{X}\}$ is an operator frame if and only if $\{\tau(x)A(x) : x \in \mathbb{X}\}$ is an operator frame.

Let $\{A_i \in \mathbb{M}_d(\mathbb{C}) : 1 \leq i \leq k\}$ be a subset of $\mathbb{M}_d(\mathbb{C})$ such that the linear span is $\mathbb{M}_d(\mathbb{C})$. (Then $k \geq d^2$.) This is a simple example of an operator frame. If $k = d^2$, then the operator frame is tight if and only if $\{A_i \in \mathbb{M}_d(\mathbb{C}) : 1 \leq i \leq d^2\}$ is an orthonormal basis up to a constant multiple.

A set $\{A(x) : x \in \mathbb{X}\}$ of positive matrices is **informationally complete (IC)** if for each pair of distinct quantum states $\rho \neq \sigma$ there exists an element $x \in \mathbb{X}$ such that $\operatorname{Tr} A(x)\rho \neq \operatorname{Tr} A(x)\sigma$. When the $A(x)$'s are all of unit rank, A is said to be **rank-one**. It is clear that for numbers $\lambda(x) > 0$ the set $\{A(x) : x \in \mathbb{X}\}$ is IC if and only if $\{\lambda(x)A(x) : x \in \mathbb{X}\}$ is IC.

Theorem 7.17 *Let $\{F(x) : x \in \mathbb{X}\}$ be a POVM. Then F is informationally complete if and only if $\{F(x) : x \in \mathbb{X}\}$ is an operator frame.*

Proof: Let

$$\mathbf{A} = \sum_{x \in \mathbb{X}} |F(x)\rangle\langle F(x)|, \tag{7.27}$$

which is a positive operator.

Suppose that F is informationally complete and take an operator A with self-adjoint decomposition $A_1 + iA_2$ such that

$$\langle A|\mathbf{A}|A\rangle = \sum_{x \in \mathbb{X}} |\operatorname{Tr} F(x)A|^2 = \sum_{x \in \mathbb{X}} |\operatorname{Tr} F(x)A_1|^2 + \sum_{x \in \mathbb{X}} |\operatorname{Tr} F(x)A_2|^2 = 0,$$

then we must have $\operatorname{Tr} F(x)A_1 = \operatorname{Tr} F(x)A_2 = 0$. The operators A_1 and A_2 are traceless:

$$\operatorname{Tr} A_i = \sum_{x \in \mathbb{X}} \operatorname{Tr} F(x)A_i = 0 \qquad (i = 1, 2).$$

Take a positive definite state ρ and a small number $\varepsilon > 0$. Then $\rho + \varepsilon A_i$ can be a state and we have

$$\operatorname{Tr} F(x)(\rho + \varepsilon A_i) = \operatorname{Tr} F(x)\rho \qquad (x \in \mathbb{X}).$$

The informationally complete property gives $A_1 = A_2 = 0$ and so $A = 0$. It follows that $\mathbf{A}$ is invertible and the operator frame property follows.

For the converse, assume that for the distinct quantum states $\rho \neq \sigma$ we have

$$\langle \rho - \sigma|\mathbf{A}|\rho - \sigma\rangle = \sum_{x \in \mathbb{X}} |\operatorname{Tr} F(x)(\rho - \sigma)|^2 > 0.$$

Then there must exist an $x \in \mathbb{X}$ such that

$$\mathrm{Tr}\,(F(x)(\rho - \sigma)) \neq 0,$$

or equivalently, $\mathrm{Tr}\,F(x)\rho \neq \mathrm{Tr}\,F(x)\sigma$, which means that F is informationally complete. □

Suppose that a POVM $\{F(x) : x \in \mathbb{X}\}$ is used for quantum measurement when the state is ρ. The outcome of the measurement is an element $x \in \mathbb{X}$ and its probability is $p(x) = \mathrm{Tr}\,\rho F(x)$. If N measurements are performed on N independent quantum systems (in the same state), then the results are $y_1, \ldots, y_N$. The outcome $x \in \mathbb{X}$ occurs with some multiplicity and the estimate for the probability is

$$\hat{p}(x) = \hat{p}(x; y_1, \ldots, y_N) := \frac{1}{N} \sum_{k=1}^{N} \delta(x, y_k). \tag{7.28}$$

From this information the state estimate has the form

$$\hat{\rho} = \sum_{x \in \mathbb{X}} \hat{p}(x) Q(x),$$

where $\{Q(x) : x \in \mathbb{X}\}$ is a set of matrices. If we require that

$$\rho = \sum_{x \in \mathbb{X}} \mathrm{Tr}\,(\rho F(x)) Q(x)$$

should hold for every state ρ, then $\{Q(x) : x \in \mathbb{X}\}$ should satisfy some conditions. This idea will need the concept of a dual frame.

For a frame $\{A(x) : x \in \mathbb{X}\}$, a **dual frame** $\{B(x) : x \in \mathbb{X}\}$ is a frame such that

$$\sum_{x \in \mathbb{X}} |B(x)\rangle\langle A(x)| = \mathbf{I},$$

or equivalently for all $C \in \mathbb{M}_d(\mathbb{C})$ we have

$$C = \sum_{x \in \mathbb{X}} \langle A(x)|C\rangle B(x) = \sum_{x \in \mathbb{X}} \langle B(x)|C\rangle A(x).$$

The existence of a dual frame is equivalent to the frame inequality (7.25), but we also have a canonical construction: The **canonical dual frame** is defined by the operators

$$|\tilde{A}(x)\rangle := \mathbf{A}^{-1}|A(x)\rangle. \tag{7.29}$$

Recall that the inverse of **A** exists whenever $\{A(x) : x \in \mathbb{X}\}$ is an operator frame. Note that given any operator frame $\{A(x) : x \in \mathbb{X}\}$ we can construct a tight frame as $\{\mathbf{A}^{-1/2}|A(x)\rangle : x \in \mathbb{X}\}$.

Theorem 7.18 *If $\{\tilde{A}(x) : x \in \mathbb{X}\}$ is the canonical dual of an operator frame $\{A(x) : x \in \mathbb{X}\}$ with superoperator **A**, then*

$$\mathbf{A}^{-1} = \sum_{x \in \mathbb{X}} |\tilde{A}(x)\rangle\langle\tilde{A}(x)|$$

and the canonical dual of $\{\tilde{A}(x) : x \in \mathbb{X}\}$ is $\{A(x) : x \in \mathbb{X}\}$. For an arbitrary dual frame $\{B(x) : x \in \mathbb{X}\}$ of $\{A(x) : x \in \mathbb{X}\}$ the inequality

$$\sum_{x \in \mathbb{X}} |B(x)\rangle\langle B(x)| \geq \sum_{x \in \mathbb{X}} |\tilde{A}(x)\rangle\langle\tilde{A}(x)|$$

holds and equality holds only if $B \equiv \tilde{A}$.

Proof: **A** and $\mathbf{A}^{-1}$ are self-adjoint superoperators and we have

$$\sum_{x \in \mathbb{X}} |\tilde{A}(x)\rangle\langle\tilde{A}(x)| = \sum_{x \in \mathbb{X}} |\mathbf{A}^{-1}A(x)\rangle\langle\mathbf{A}^{-1}A(x)|$$

$$= \mathbf{A}^{-1}\Big(\sum_{x \in \mathbb{X}} |A(x)\rangle\langle A(x)|\Big)\mathbf{A}^{-1}$$

$$= \mathbf{A}^{-1}\mathbf{A}\mathbf{A}^{-1} = \mathbf{A}^{-1}.$$

The second statement is $\mathbf{A}|\tilde{A}(x)\rangle = |A(x)\rangle$, which comes immediately from $|\tilde{A}(x)\rangle = \mathbf{A}^{-1}|A(x)\rangle$.

Let B be a dual frame of A and define $D(x) := B(x) - \tilde{A}(x)$. Then

$$\sum_{x \in \mathbb{X}} |\tilde{A}(x)\rangle\langle D(x)| = \sum_{x \in \mathbb{X}} \Big(|\tilde{A}(x)\rangle\langle B(x)| - |\tilde{A}(x)\rangle\langle\tilde{A}(x)|\Big)$$

$$= \mathbf{A}^{-1}\sum_{x \in \mathbb{X}} |A(x)\rangle\langle B(x)| - \mathbf{A}^{-1}\sum_{x \in \mathbb{X}} |A(x)\rangle\langle A(x)|\mathbf{A}^{-1}$$

$$= \mathbf{A}^{-1}\mathbf{I} - \mathbf{A}^{-1}\mathbf{A}\mathbf{A}^{-1} = 0.$$

The adjoint gives

$$\sum_{x \in \mathbb{X}} |D(x)\rangle\langle\tilde{A}(x)| = 0\,,$$

and

$$\sum_{x \in \mathbb{X}} |B(x)\rangle\langle B(x)| = \sum_{x \in \mathbb{X}} |\tilde{A}(x)\rangle\langle \tilde{A}(x)| + \sum_{x \in \mathbb{X}} |\tilde{A}(x)\rangle\langle D(x)|$$

$$+ \sum_{x \in \mathbb{X}} |D(x)\rangle\langle \tilde{A}(x)| + \sum_{x \in \mathbb{X}} |D(x)\rangle\langle D(x)|$$

$$= \sum_{x \in \mathbb{X}} |\tilde{A}(x)\rangle\langle \tilde{A}(x)| + \sum_{x \in \mathbb{X}} |D(x)\rangle\langle D(x)|$$

$$\geq \sum_{x \in \mathbb{X}} |\tilde{A}(x)\rangle\langle \tilde{A}(x)|$$

with equality if and only if $D \equiv 0$. $\qquad\square$

We have the following inequality, which is also known as the frame bound.

Theorem 7.19 *Let* $\{A(x) : x \in \mathbb{X}\}$ *be an operator frame with superoperator* **A**. *Then the inequality*

$$\sum_{x,y \in \mathbb{X}} |\langle A(x)|A(y)\rangle|^2 \geq \frac{(\mathrm{Tr}\,\mathbf{A})^2}{d^2} \qquad (7.30)$$

holds, and we have equality if and only if $\{A(x) : x \in \mathbb{X}\}$ *is a tight operator frame.*

Proof: By (7.26) the left-hand side is $\mathrm{Tr}\,\mathbf{A}^2$, so the inequality holds. It is clear that equality holds if and only if all eigenvalues of **A** are the same, that is, $\mathbf{A} = c\mathbf{I}$. $\qquad\square$

The trace measure τ is defined by $\tau(x) := \mathrm{Tr}\,F(x)$. The useful superoperator is

$$\mathbf{F} = \sum_{x \in \mathbb{X}} |F(x)\rangle\langle F(x)|(\tau(x))^{-1}.$$

Formally this is different from the frame superoperator (7.27). Therefore, we express the POVM F as

$$F(x) = P_0(x)\sqrt{\tau(x)} \qquad (x \in \mathbb{X})$$

where $\{P_0(x) : x \in \mathbb{X}\}$ is called a **positive operator-valued density** (POVD). Then

$$\mathbf{F} = \sum_{x \in \mathbb{X}} |P_0(x)\rangle\langle P_0(x)| = \sum_{x \in \mathbb{X}} |F(x)\rangle\langle F(x)|(\tau(x))^{-1}. \qquad (7.31)$$

F is invertible if and only if **A** in (7.27) is invertible. As a corollary, we see that for an informationally complete POVM F, the POVD P_0 can be considered as a generalized operator frame. The canonical dual frame (in the sense of (7.29)) then defines a **reconstruction operator-valued density**

$$|R_0(x)\rangle := \mathbf{F}^{-1}|P_0(x)\rangle \qquad (x \in \mathbb{X}).$$

We also use the notation $R(x) := R_0(x)\tau(x)^{-1/2}$. The identity

$$\sum_{x \in \mathbb{X}} |R(x)\rangle\langle F(x)| = \sum_{x \in \mathbb{X}} |R_0(x)\rangle\langle P_0(x)| = \sum_{x \in \mathbb{X}} \mathbf{F}^{-1} |P_0(x)\rangle\langle P_0(x)| = \mathbf{F}^{-1}\mathbf{F} = \mathbf{I}$$

(7.32)

then allows state reconstruction in terms of the measurement statistics:

$$\rho = \left(\sum_{x \in \mathbb{X}} |R(x)\rangle\langle F(x)|\right)\rho = \sum_{x \in \mathbb{X}} (\mathrm{Tr}\, F(x)\rho)\, R(x).$$
(7.33)

So this **state-reconstruction formula** is an immediate consequence of the action of (7.32) on ρ.

Theorem 7.20 *We have*

$$\mathbf{F}^{-1} = \sum_{x \in \mathbb{X}} |R(x)\rangle\langle R(x)|\tau(x)$$
(7.34)

and the operators $R(x)$ are self-adjoint and $\mathrm{Tr}\, R(x) = 1$.

Proof: From the mutual canonical dual relation of $\{P_0(x) : x \in \mathbb{X}\}$ and $\{R_0(x) : x \in \mathbb{X}\}$ we have

$$\mathbf{F}^{-1} = \sum_{x \in \mathbb{X}} |R_0(x)\rangle\langle R_0(x)|$$

by Theorem 7.18, and this is (7.34).

The operators $R(x)$ are self-adjoint since $\mathbf{F}$ and thus $\mathbf{F}^{-1}$ map self-adjoint operators to self-adjoint operators. For an arbitrary POVM, the identity operator is always an eigenvector of the POVM superoperator:

$$\mathbf{F}|I\rangle = \sum_{x \in \mathbb{X}} |F(x)\rangle\langle F(x)|I\rangle(\tau(x))^{-1} = \sum_{x \in \mathbb{X}} |F(x)\rangle = |I\rangle.$$
(7.35)

Thus $|I\rangle$ is also an eigenvector of $\mathbf{F}^{-1}$, and we obtain

$$\mathrm{Tr}\, R(x) = \langle I|R(x)\rangle = \tau(x)^{-1/2}\langle I|R_0(x)\rangle = \tau(x)^{-1/2}\langle I|\mathbf{F}^{-1}P_0(x)\rangle$$
$$= \tau(x)^{-1/2}\langle I|P_0(x)\rangle = \tau(x)^{-1}\langle I|F(x)\rangle = \tau(x)^{-1}\tau(x) = 1.$$

$\square$

Note that we need $|\mathbb{X}| \geq d^2$ for F to be informationally complete. If this were not the case then $\mathbf{F}$ could not have full rank. An IC-POVM with $|\mathbb{X}| = d^2$ is called *minimal*. In this case the reconstruction OVD is unique. In general, however, there will be many different choices.

Example 7.21 Let $x_1, x_2, \ldots, x_d$ be an orthonormal basis of $\mathbb{C}^d$. Then $Q_i = |x_i\rangle\langle x_i|$ are projections and $\{Q_i : 1 \le i \le d\}$ is a POVM. However, it is not informationally complete. The subset

$$\mathcal{A} := \left\{ \sum_{i=1}^{d} \lambda_i |x_i\rangle\langle x_i| : \lambda_1, \lambda_2, \ldots, \lambda_d \in \mathbb{C} \right\} \subset \mathbb{M}_d(\mathbb{C})$$

is a maximal abelian *-subalgebra, called a **MASA**.

A good example of an IC-POVM comes from $d + 1$ similar sets:

$$\{Q_k^{(m)} : 1 \le k \le d, \ 1 \le m \le d + 1\}$$

consists of projections of rank one and

$$\text{Tr}\, Q_k^{(m)} Q_l^{(n)} = \begin{cases} \delta_{kl} & \text{if} \quad m = n, \\ 1/d & \text{if} \quad m \neq n. \end{cases}$$

A POVM is defined by

$$\mathbb{X} := \{(k, m) : 1 \le k \le d, \ 1 \le m \le d + 1\}$$

and

$$F(k, m) := \frac{1}{d + 1}\, Q_k^{(m)}, \qquad \tau(k, m) := \frac{1}{d + 1}$$

for $(k, m) \in \mathbb{X}$. (Here τ is constant and this is a uniformity.) We have

$$\sum_{(k,m)\in\mathbb{X}} |F(k, m)\rangle\langle F(k, m)| Q_l^{(n)} = \frac{1}{(d + 1)^2} \left(Q_l^{(n)} + I \right).$$

This implies

$$\mathbf{F} A = \left(\sum_{x\in\mathbb{X}} |F(x)\rangle\langle F(x)| (\tau(x))^{-1} \right) A = \frac{1}{(d + 1)} \left(A + (\text{Tr}\, A)I \right).$$

So $\mathbf{F}$ is rather simple: if $\text{Tr}\, A = 0$, then $\mathbf{F} A = \frac{1}{d+1} A$ and $\mathbf{F} I = I$. (Another formulation is given in (7.36).)

This example is a complete set of **mutually unbiased bases** (**MUBs**) [54, 85]. The definition

$$\mathcal{A}_m := \left\{ \sum_{k=1}^{d} \lambda_k Q_k^{(m)} : \lambda_1, \lambda_2, \ldots, \lambda_d \in \mathbb{C} \right\} \subset \mathbb{M}_d(\mathbb{C})$$

gives $d + 1$ MASAs. These MASAs are **quasi-orthogonal** in the following sense. If $A_i \in \mathcal{A}_i$ and $\mathrm{Tr}\, A_i = 0$ $(1 \le i \le d + 1)$, then $\mathrm{Tr}\, A_i A_j = 0$ for $i \ne j$. The construction of $d + 1$ quasi-orthogonal MASAs is known when d is a prime-power (see also [31]). But $d = 6$ is not a prime-power and it is already a problematic example. $\square$

It is straightforward to confirm that we have the decomposition

$$\mathbf{F} = \frac{1}{d} |I\rangle\langle I| + \sum_{x \in \mathbb{X}} |P(x) - I/d\rangle\langle P(x) - I/d|\tau(x)$$

for any POVM superoperator (7.31), where $P(x) := P_0(x)\tau(x)^{-1/2} = F(x)\tau(x)^{-1}$ and

$$\frac{1}{d} |I\rangle\langle I| = \begin{bmatrix} 1 & 0 \\ 0 & 0 \end{bmatrix}$$

is the projection onto the subspace $\mathbb{C}I$. Setting

$$\mathbf{I}_0 := \begin{bmatrix} 0 & 0 \\ 0 & I_{d^2-1} \end{bmatrix},$$

an IC-POVM $\{F(x) : x \in \mathbb{X}\}$ is **tight** if

$$\sum_{x \in \mathbb{X}} |P(x) - I/d\rangle\langle P(x) - I/d|\tau(x) = a\mathbf{I}_0.$$

Theorem 7.22 *F is a tight rank-one IC-POVM if and only if*

$$\mathbf{F} = \frac{\mathbf{I} + |I\rangle\langle I|}{d + 1} = \begin{bmatrix} 1 & 0 \\ 0 & \frac{1}{d+1} I_{d^2-1} \end{bmatrix}. \tag{7.36}$$

(The latter is in the block matrix form.)

Proof: The constant a can be found by taking the superoperator trace:

$$a = \frac{1}{d^2 - 1} \sum_{x \in \mathbb{X}} \langle P(x) - I/d | P(x) - I/d\rangle \tau(x)$$

$$= \frac{1}{d^2 - 1} \left(\sum_{x \in \mathbb{X}} \langle P(x) | P(x)\rangle \tau(x) - 1 \right).$$

The POVM superoperator of a tight IC-POVM satisfies the identity

$$\mathbf{F} = a\mathbf{I} + \frac{1 - a}{d} |I\rangle\langle I|. \tag{7.37}$$

In the special case of a rank-one POVM a takes its maximum possible value $1/(d+1)$. Since this is in fact only possible for rank-one POVMs, by noting that (7.37) can be taken as an alternative definition in the general case, we obtain the proposition.   □

It follows from (7.36) that

$$\mathbf{F}^{-1} = \begin{bmatrix} 1 & 0 \\ 0 & (d+1)I_{d^2-1} \end{bmatrix} = (d+1)\mathbf{I} - |I\rangle\langle I|.$$

This shows that Example 7.21 contains a tight rank-one IC-POVM. Here is another example.

Example 7.23 An example of an IC-POVM is the **symmetric informationally complete POVM** (SIC POVM). The set $\{Q_k : 1 \le k \le d^2\}$ consists of projections of rank one such that

$$\mathrm{Tr}\, Q_k Q_l = \frac{1}{d+1} \qquad (k \ne l).$$

Then $\mathbb{X} := \{x : 1 \le x \le d^2\}$ and

$$F(x) = \frac{1}{d} Q_x, \qquad \mathbf{F} = \frac{1}{d} \sum_{x \in \mathbb{X}} |Q_x\rangle\langle Q_x|.$$

We have some simple computations: $\mathbf{F}I = I$ and

$$\mathbf{F}(Q_k - I/d) = \frac{1}{d+1}(Q_k - I/d).$$

This implies that if $\mathrm{Tr}\, A = 0$, then

$$\mathbf{F}A = \frac{1}{d+1} A.$$

So the SIC POVM is a tight rank-one IC-POVM.

SIC-POVMs are conjectured to exist in all dimensions [12, 86].   □

The next theorem tells us that the SIC POVM is characterized by the IC POVM property.

Theorem 7.24 *If a set $\{Q_k \in \mathbb{M}_d(\mathbb{C}) : 1 \le k \le d^2\}$ consists of projections of rank one such that*

$$\sum_{k=1}^{d^2} \lambda_k |Q_k\rangle\langle Q_k| = \frac{\mathbf{I} + |I\rangle\langle I|}{d+1} \tag{7.38}$$

with numbers $\lambda_k > 0$, *then*

$$\lambda_i = \frac{1}{d}, \quad \text{Tr } Q_i Q_j = \frac{1}{d+1} \quad (i \neq j).$$

Proof: Note that if both sides of (7.38) are applied to $|I\rangle$, then we get

$$\sum_{i=1}^{d^2} \lambda_i Q_i = I. \tag{7.39}$$

First we show that $\lambda_i = 1/d$. From (7.38) we have

$$\sum_{i=1}^{d^2} \lambda_i \langle A|Q_i\rangle\langle Q_i|A\rangle = \langle A|\frac{\mathbf{I} + |I\rangle\langle I|}{d+1}|A\rangle \tag{7.40}$$

with

$$A := Q_k - \frac{1}{d+1}I.$$

(7.40) becomes

$$\lambda_k \frac{d^2}{(d+1)^2} + \sum_{j \neq k} \lambda_j \left(\text{Tr } Q_j Q_k - \frac{1}{d+1} \right)^2 = \frac{d}{(d+1)^2}. \tag{7.41}$$

The inequality

$$\lambda_k \frac{d^2}{(d+1)^2} \leq \frac{d}{(d+1)^2}$$

gives $\lambda_k \leq 1/d$ for every $1 \leq k \leq d^2$. The trace of (7.39) is

$$\sum_{i=1}^{d^2} \lambda_i = d.$$

Hence it follows that $\lambda_k = 1/d$ for every $1 \leq k \leq d^2$. So from (7.41) we have

$$\sum_{j \neq k} \lambda_j \left(\text{Tr } Q_j Q_k - \frac{1}{d+1} \right)^2 = 0$$

and this gives the result.   □

The state-reconstruction formula for a tight rank-one IC-POVM also takes an elegant form. From (7.33) we have

$$\rho = \sum_{x \in \mathbb{X}} R(x)p(x) = \sum_{x \in \mathbb{X}} \mathbf{F}^{-1}P(x)p(x) = \sum_{x \in \mathbb{X}} ((d+1)P(x) - I)p(x)$$

and obtain

$$\rho = (d+1)\sum_{x \in \mathbb{X}} P(x)p(x) \; - \; I \; .$$

Finally, let us rewrite the frame bound (Theorem 7.19) for the context of quantum measurements.

Theorem 7.25 *Let* $\{F(x) : x \in \mathbb{X}\}$ *be a POVM. Then*

$$\sum_{x,y \in \mathbb{X}} \langle P(x)|P(y)\rangle^2 \tau(x)\tau(y) \;\geq\; 1 + \frac{(\mathrm{Tr}\,\mathbf{F} - 1)^2}{d^2 - 1}, \tag{7.42}$$

with equality if and only if F *is a tight IC-POVM.*

Proof: The frame bound (7.30) has a slightly improved form

$$\mathrm{Tr}\,(\mathbf{A}^2) \geq \big(\mathrm{Tr}\,(\mathbf{A})\big)^2/D,$$

where D is the dimension of the operator space. Setting $\mathbf{A} = \mathbf{F} - \frac{1}{d}|I\rangle\langle I|$ and $D = d^2 - 1$ for $\mathbb{M}_d(\mathbb{C}) \ominus \mathbb{C}I$ then gives (7.42) (using (7.35)). $\square$

Informationally complete quantum measurements are precisely those measurements which can be used for quantum state tomography. We will show that, amongst all IC-POVMs, the tight rank-one IC-POVMs are the most robust against statistical error in the quantum tomographic process. We will also find that, for an arbitrary IC-POVM, the canonical dual frame with respect to the trace measure is the optimal dual frame for state reconstruction. These results are shown only for the case of linear quantum state tomography.

Consider a state-reconstruction formula of the form

$$\rho = \sum_{x \in \mathbb{X}} p(x)Q(x) = \sum_{x \in \mathbb{X}} (\mathrm{Tr}\,F(x)\rho)Q(x), \tag{7.43}$$

where $Q(x) : \mathbb{X} \to \mathbb{M}_d(\mathbb{C})$ is an operator-valued density. If this formula is to remain valid for all ρ, then we must have

$$\sum_{x \in \mathbb{X}} |Q(x)\rangle\langle F(x)| = \mathbf{I} = \sum_{x \in \mathbb{X}} |Q_0(x)\rangle\langle P_0(x)|, \tag{7.44}$$

where $Q_0(x) = \tau(x)^{1/2}Q(x)$ and $P_0(x) = \tau(x)^{-1/2}F(x)$. Equation (7.44) forces $\{Q(x) : x \in \mathbb{X}\}$ to be a dual frame of $\{F(x) : x \in \mathbb{X}\}$. Similarly $\{Q_0(x) : x \in \mathbb{X}\}$ is a dual frame of $\{P_0(x) : x \in \mathbb{X}\}$. Our first goal is to find the optimal dual frame.

Suppose that we take N independent random samples, $y_1, \ldots, y_N$, and the outcome x occurs with some unknown probability $p(x)$. Our estimate for this probability is (7.28) which of course obeys the expectation $\mathrm{E}[\hat{p}(x)] = p(x)$. An elementary calculation shows that the expected covariance for two samples is

$$\mathrm{E}\big[(p(x) - \hat{p}(x))(p(y) - \hat{p}(y))\big] = \frac{1}{N}\Big(p(x)\delta(x, y) - p(x)p(y)\Big). \quad (7.45)$$

Now suppose that the $p(x)$'s are the outcome probabilities for an informationally complete quantum measurement of the state ρ, $p(x) = \mathrm{Tr}\, F(x)\rho$. The estimate of ρ is

$$\hat{\rho} = \hat{\rho}(y_1, \ldots, y_N) := \sum_{x \in \mathbb{X}} \hat{p}(x; y_1, \ldots, y_N)Q(x),$$

and the error can be measured by the squared Hilbert–Schmidt distance:

$$\|\rho - \hat{\rho}\|_2^2 = \langle \rho - \hat{\rho}, \rho - \hat{\rho}\rangle = \sum_{x,y \in \mathbb{X}} (p(x) - \hat{p}(x))(p(y) - \hat{p}(y))\langle Q(x), Q(y)\rangle,$$

which has the expectation $\mathrm{E}\big[\|\rho - \hat{\rho}\|_2^2\big]$. We want to minimize this quantity, not for an arbitrary ρ, but for some average. (Integration will be on the set of unitary matrices with respect to the Haar measure.)

Theorem 7.26 *Let $\{F(x) : x \in \mathbb{X}\}$ be an informationally complete POVM which has a dual frame $\{Q(x) : x \in \mathbb{X}\}$ as an operator-valued density. The quantum system has a state σ and $y_1, \ldots, y_N$ are random samples of the measurements. Then*

$$\hat{p}(x) := \frac{1}{N}\sum_{k=1}^{N} \delta(x, y_k), \qquad \hat{\rho} := \sum_{x \in \mathbb{X}} \hat{p}(x)Q(x).$$

Finally let $\rho = \rho(\sigma, U) := U\sigma U^*$ be parametrized by a unitary U. Then for the average squared distance

$$\int_{\mathcal{U}} \mathrm{E}\big[\|\rho - \hat{\rho}\|_2^2\big]\, d\mu(U) \geq \frac{1}{N}\left(\frac{1}{d}\mathrm{Tr}\,(\mathbf{F}^{-1}) - \mathrm{Tr}\,(\sigma^2)\right) \quad (7.46)$$

$$\geq \frac{1}{N}\Big(d(d + 1) - 1 - \mathrm{Tr}\,(\sigma^2)\Big). \quad (7.47)$$

Equality in the inequality (7.46) occurs if and only if Q is the reconstruction operator-valued density (defined as $|R(x)\rangle = \mathbf{F}^{-1}|P(x)\rangle$) and equality in the inequality (7.47) occurs if and only if F is a tight rank-one IC-POVM.

Proof: For a fixed IC-POVM F we have

$$\mathrm{E}\big[\|\rho - \hat{\rho}\|_2^2\big] = \frac{1}{N} \sum_{x,y \in X} \big(p(x)\delta(x, y) - p(x)p(y)\big)\langle Q(x), Q(y)\rangle$$

$$= \frac{1}{N}\left(\sum_{x \in X} p(x)\langle Q(x), Q(x)\rangle - \langle \sum_{x \in X} p(x)Q(x) \sum_{y \in X} p(y)Q(y)\rangle\right)$$

$$= \frac{1}{N}\big(\Delta_p(Q) - \mathrm{Tr}\,(\rho^2)\big),$$

where the formulas (7.45) and (7.43) are used and moreover

$$\Delta_p(Q) := \sum_{x \in X} p(x)\,\langle Q(x), Q(x)\rangle.$$

Since we have no control over $\mathrm{Tr}\,\rho^2$, we want to minimize $\Delta_p(Q)$. The IC-POVM which minimizes $\Delta_p(Q)$ will in general depend on the quantum state under examination. We thus set $\rho = \rho(\sigma, U) := U\sigma U^*$, and now remove this dependence by taking the Haar average $\mu(U)$ over all $U \in \mathrm{U}(d)$. Note that

$$\int_{\mathrm{U}(d)} U P U^* \, d\mu(U)$$

is the same constant C for any projection of rank 1. If $\sum_{i=1}^d P_i = I$, then

$$dC = \sum_{i=1}^d \int_{\mathrm{U}(d)} U P U^* \, d\mu(U) = I$$

and we have $C = I/d$. Therefore for $A = \sum_{i=1}^d \lambda_i P_i$ we have

$$\int_{\mathrm{U}(d)} U A U^* \, d\mu(U) = \sum_{i=1}^d \lambda_i C = \frac{I}{d}\mathrm{Tr}\,A.$$

This fact implies

$$\int_{\mathrm{U}(d)} \Delta_p(Q)\, d\mu(U) = \sum_{x \in X} \mathrm{Tr}\left(F(x)\int_{\mathrm{U}(d)} U\sigma U^* \, d\mu(U)\right)\langle Q(x), Q(x)\rangle$$

$$= \frac{1}{d}\sum_{x \in X} \mathrm{Tr}\,F(x)\,\mathrm{Tr}\,\sigma\langle Q(x), Q(x)\rangle$$

$$= \frac{1}{d}\sum_{x \in X} \tau(x)\,\langle Q(x), Q(x)\rangle =: \frac{1}{d}\Delta_\tau(Q),$$

where $\tau(x) := \operatorname{Tr} F(x)$. We will now minimize $\Delta_\tau(Q)$ over all choices for Q, while keeping the IC-POVM F fixed. Our only constraint is that $\{Q(x) : x \in \mathbb{X}\}$ remains a dual frame to $\{F(x) : x \in \mathbb{X}\}$ (see (7.44)), so that the reconstruction formula (7.43) remains valid for all ρ. Theorem 7.18 shows that the reconstruction OVD $\{R(x) : x \in \mathbb{X}\}$ defined as $|R\rangle = \mathbf{F}^{-1}|P\rangle$ is the optimal choice for the dual frame.

Equation (7.34) shows that $\Delta_\tau(R) = \operatorname{Tr}(\mathbf{F}^{-1})$. We will minimize the quantity

$$\operatorname{Tr} \mathbf{F}^{-1} = \sum_{k=1}^{d^2} \frac{1}{\lambda_k}, \tag{7.48}$$

where $\lambda_1, \ldots, \lambda_{d^2} > 0$ denote the eigenvalues of $\mathbf{F}$. These eigenvalues satisfy the constraint

$$\sum_{k=1}^{d^2} \lambda_k = \operatorname{Tr} \mathbf{F} = \sum_{x \in \mathbb{X}} \tau(x) \operatorname{Tr} |P(x)\rangle\langle P(x)| \le \sum_{x \in \mathbb{X}} \tau(x) = d,$$

since $\operatorname{Tr} |P(x)\rangle\langle P(x)| = \operatorname{Tr} P(x)^2 \le 1$. We know that the identity operator I is an eigenvalue of $\mathbf{F}$:

$$\mathbf{F}I = \sum_{x \in \mathbb{X}} \tau(x) |P(x)\rangle = I.$$

Thus we in fact take $\lambda_1 = 1$ and then $\sum_{k=2}^{d^2} \lambda_k \le d-1$. Under this latter constraint it is straightforward to show that the right-hand side of (7.48) takes its minimum value if and only if $\lambda_2 = \cdots = \lambda_{d^2} = (d-1)/(d^2-1) = 1/(d+1)$, or equivalently,

$$\mathbf{F} = 1 \cdot \frac{|I\rangle\langle I|}{d} + \frac{1}{d+1}\left(\mathbf{I} - \frac{|I\rangle\langle I|}{d}\right). \tag{7.49}$$

Therefore, by Theorem 7.22, $\operatorname{Tr} \mathbf{F}^{-1}$ takes its minimum value if and only if F is a tight rank-one IC-POVM. The minimum of $\operatorname{Tr} \mathbf{F}^{-1}$ comes from (7.49). $\qquad\square$

7.5 The Cramér–Rao Inequality

The Cramér–Rao inequality belongs to estimation theory in mathematical statistics. Assume that we have to estimate the state ρ_θ, where $\theta = (\theta_1, \theta_2, \ldots, \theta_N)$ lies in a subset of $\mathbb{R}^N$. There is a sequence of estimates $\Phi_n : \mathcal{X}_n \to \mathbb{R}^N$. In mathematical statistics the $N \times N$ **mean quadratic error matrix**

$$V_n(\theta)_{i,j} := \int_{\mathcal{X}_n} (\Phi_n(x)_i - \theta_i)(\Phi_n(x)_j - \theta_j)\, d\mu_{n,\theta}(x) \qquad (1 \le i, j \le N)$$

is used to express the efficiency of the nth estimation and in a good estimation scheme $V_n(\theta) = O(n^{-1})$ is expected. Here, $\mathcal{X}_n$ is the set of measurement outcomes and $\mu_{n,\theta}$ is the probability distribution when the true state is ρ_θ. An **unbiased estimation scheme** means

$$\int_{\mathcal{X}_n} \Phi_n(x)_i \, d\mu_{n,\theta}(x) = \theta_i \qquad (1 \le i \le N)$$

and the formula simplifies:

$$V_n(\theta)_{i,j} := \int_{\mathcal{X}_n} \Phi_n(x)_i \, \Phi_n(x)_j \, d\mu_{n,\theta}(x) - \theta_i \theta_j \,.$$

(In mathematical statistics, this is sometimes called the covariance matrix of the estimate.)

The mean quadratic error matrix is used to measure the efficiency of an estimate. Even if the value of θ is fixed, for two different estimates the corresponding matrices are not always comparable, because the ordering of positive definite matrices is highly partial. This fact has inconvenient consequences in classical statistics. In the state estimation of a quantum system the very different possible measurements make the situation even more complicated.

Assume that $d\mu_{n,\theta}(x) = f_{n,\theta}(x) \, dx$ and fix θ. $f_{n,\theta}$ is called the **likelihood function**. Let

$$\partial_j := \frac{\partial}{\partial \theta_j}.$$

Differentiating the relation

$$\int_{\mathcal{X}_n} f_{n,\theta}(x) \, dx = 1,$$

we have

$$\int_{\mathcal{X}_n} \partial_j f_{n,\theta}(x) \, dx = 0.$$

If the estimation scheme is unbiased, then

$$\int_{\mathcal{X}_n} \Phi_n(x)_i \partial_j f_{n,\theta}(x) \, dx = \delta_{i,j}.$$

Combining these, we conclude that

$$\int_{\mathcal{X}_n} (\Phi_n(x)_i - \theta_i) \partial_j f_{n,\theta}(x) \, dx = \delta_{i,j}$$

for every $1 \le i, j \le N$. This condition may be written in the slightly different form

$$\int_{\mathcal{X}_n} \left((\Phi_n(x)_i - \theta_i)\sqrt{f_{n,\theta}(x)} \right) \frac{\partial_j f_{n,\theta}(x)}{\sqrt{f_{n,\theta}(x)}} \, dx = \delta_{i,j}.$$

Now the first factor of the integrand depends on i while the second one depends on j. We need the following lemma.

Lemma 7.27 *Assume that u_i, v_i are vectors in a Hilbert space such that*

$$\langle u_i, v_j \rangle = \delta_{i,j} \qquad (i, j = 1, 2, \ldots, N).$$

Then the inequality

$$A \geq B^{-1}$$

holds for the $N \times N$ matrices

$$A_{i,j} = \langle u_i, u_j \rangle \quad \text{and} \quad B_{i,j} = \langle v_i, v_j \rangle \qquad (1 \leq i, j \leq N).$$

The lemma applies to the vectors

$$u_i = (\Phi_n(x)_i - \theta_i)\sqrt{f_{n,\theta}(x)} \quad \text{and} \quad v_j = \frac{\partial_j f_{n,\theta}(x)}{\sqrt{f_{n,\theta}(x)}}$$

and the matrix A will be precisely the mean square error matrix $V_n(\theta)$, while in place of B we have

$$\mathbf{I}_n(\theta)_{i,j} = \int_{\mathcal{X}_n} \frac{\partial_i(f_{n,\theta}(x))\partial_j(f_{n,\theta}(x))}{f_{n,\theta}^2(x)} \, d\mu_{n,\theta}(x).$$

Therefore, the lemma tells us the following:

Theorem 7.28 *For an unbiased estimation scheme the matrix inequality*

$$V_n(\theta) \geq \mathbf{I}_n(\theta)^{-1} \tag{7.50}$$

holds (if the likelihood functions $f_{n,\theta}$ satisfy certain regularity conditions).

This is the classical **Cramér–Rao inequality**. The right-hand side is called the **Fisher information matrix**. The essential content of the inequality is that the lower bound is independent of the estimate Φ_n but depends on the classical likelihood function. The inequality is called classical because on both sides classical statistical quantities appear.

Example 7.29 Let F be a measurement with values in the finite set $\mathcal{X}$ and assume that $\rho_\theta = \rho + \sum_{i=1}^n \theta_i B_i$, where the B_i are self-adjoint operators with $\mathrm{Tr}\, B_i = 0$. We want to compute the Fisher information matrix at $\theta = 0$.

Since

$$\partial_i \operatorname{Tr} \rho_\theta F(x) = \operatorname{Tr} B_i F(x)$$

for $1 \le i \le n$ and $x \in \mathcal{X}$, we have

$$\mathbf{I}_{ij}(0) = \sum_{x \in \mathcal{X}} \frac{\operatorname{Tr} B_i F(x) \operatorname{Tr} B_j F(x)}{\operatorname{Tr} \rho F(x)}.$$

$\square$

The essential point in the quantum Cramér–Rao inequality compared with Theorem 7.28 is that the lower bound is a quantity determined by the family Θ. Theorem 7.28 allows us to compare different estimates for a given measurement but two different measurements are not comparable.

As a starting point we give a very general form of the quantum Cramér–Rao inequality in the simple setting of a single parameter. For $\theta \in (-\varepsilon, \varepsilon) \subset \mathbb{R}$ a statistical operator ρ_θ is given and the aim is to estimate the value of the parameter θ close to 0. Formally ρ_θ is an $m \times m$ positive semidefinite matrix of trace 1 which describes a mixed state of a quantum mechanical system and we assume that ρ_θ is smooth (in θ). Assume that an estimation is performed by the measurement of a self-adjoint matrix A playing the role of an observable. (In this case the positive operator-valued measure on $\mathbb{R}$ is the spectral measure of A.) A is an unbiased estimator when $\operatorname{Tr} \rho_\theta A = \theta$. Assume that the true value of θ is close to 0. A is called a **locally unbiased estimator** (at $\theta = 0$) if

$$\frac{\partial}{\partial \theta} \operatorname{Tr} \rho_\theta A \Big|_{\theta=0} = 1. \tag{7.51}$$

Of course, this condition holds if A is an unbiased estimator for θ. To require $\operatorname{Tr} \rho_\theta A = \theta$ for all values of the parameter might be a serious restriction on the observable A and therefore we prefer to use the weaker condition (7.51).

Example 7.30 Let

$$\rho_\theta := \frac{\exp(H + \theta B)}{\operatorname{Tr} \exp(H + \theta B)}$$

and assume that $\rho_0 = e^H$ is a density matrix and $\operatorname{Tr} e^H B = 0$. The Fréchet derivative of ρ_θ (at $\theta = 0$) is $\int_0^1 e^{tH} B e^{(1-t)H} \, dt$. Hence the self-adjoint operator A is locally unbiased if

$$\int_0^1 \operatorname{Tr} \rho_0^t B \rho_0^{1-t} A \, dt = 1.$$

(Note that ρ_θ is a quantum analogue of the **exponential family**; in terms of physics ρ_θ is a **Gibbsian family** of states.) $\square$

Let $\varphi_\rho[B, C] = \text{Tr}\, \mathbb{J}_\rho(B)C$ be an inner product on the linear space of self-adjoint matrices. $\varphi_\rho[\,\cdot\,,\,\cdot\,]$ and the corresponding superoperator $\mathbb{J}_\rho$ depend on the density matrix ρ; the notation reflects this fact. When ρ_θ is smooth in θ as already assumed above, we have

$$\frac{\partial}{\partial\theta}\text{Tr}\,\rho_\theta B\Big|_{\theta=0} = \varphi_{\rho_0}[B, L] \tag{7.52}$$

for some $L = L^*$. From (7.51) and (7.52) we have $\varphi_{\rho_0}[A, L] = 1$, and the Schwarz inequality yields:

Theorem 7.31

$$\varphi_{\rho_0}[A, A] \geq \frac{1}{\varphi_{\rho_0}[L, L]}. \tag{7.53}$$

This is the **quantum Cramér–Rao inequality** for a locally unbiased estimator. It is instructive to compare Theorem 7.31 with the classical Cramér–Rao inequality. If $A = \sum_i \lambda_i E_i$ is the spectral decomposition, then the corresponding von Neumann measurement is $F = \sum_i \delta_{\lambda_i} E_i$. Take the estimate $\Phi(\lambda_i) = \lambda_i$. Then the mean quadratic error is $\sum_i \lambda_i^2 \text{Tr}\,\rho_0 E_i$ (at $\theta = 0$) which is precisely the left-hand side of the quantum inequality provided that

$$\varphi_{\rho_0}[B, C] = \tfrac{1}{2}\text{Tr}\,\rho_0(BC + CB).$$

Generally, we want to interpret the left-hand side as a sort of generalized variance of A. To do this it is useful to assume that

$$\varphi_\rho[B, B] = \text{Tr}\,\rho B^2 \quad \text{if} \quad B\rho = \rho B.$$

However, in the non-commutative setting the statistical interpretation seems to be rather problematic and thus we call this quantity the quadratic cost functional.

The right-hand side of (7.53) is independent of the estimator and provides a lower bound for the quadratic cost. The denominator $\varphi_{\rho_0}[L, L]$ appears to play the role of Fisher information here. We call it the **quantum Fisher information** with respect to the cost function $\varphi_{\rho_0}[\,\cdot\,,\,\cdot\,]$. This quantity depends on the tangent of the curve ρ_θ. If the densities ρ_θ and the estimator A commute, then

$$L = \rho_0^{-1}\frac{d\rho_\theta}{d\theta}\Big|_{\theta=0} = \frac{d}{d\theta}\log\rho_\theta\Big|_{\theta=0},$$

$$\varphi_0[L, L] = \text{Tr}\,\rho_0^{-1}\left(\frac{d\rho_\theta}{d\theta}\Big|_{\theta=0}\right)^2 = \text{Tr}\,\rho_0\left(\rho_0^{-1}\frac{d\rho_\theta}{d\theta}\Big|_{\theta=0}\right)^2.$$

The first formula justifies calling L the **logarithmic derivative**.

A **coarse-graining** is an affine mapping sending density matrices into density matrices. Such a mapping extends to all matrices and provides a positive and trace-preserving linear transformation. A common example of coarse-graining sends a density matrix ρ_{12} of a composite system $\mathbb{M}_{m_1} \otimes \mathbb{M}_{m_2}$ into the (reduced) density matrix ρ_1 of component $\mathbb{M}_{m_1}$. There are several reasons to assume complete positivity for a coarse graining and we do so. Mathematically a coarse-graining is the same as a state transformation in an information channel. The terminology coarse-graining is used when the statistical aspects are focused on. A coarse-graining is the quantum analogue of a statistic.

Assume that $\rho_\theta = \rho + \theta B$ is a smooth curve of density matrices with tangent $B := d\rho/d\theta$ at ρ. The quantum Fisher information $F_\rho(B)$ is an information quantity associated with the pair (ρ, B). It appeared in the Cramér–Rao inequality above and the classical Fisher information gives a bound for the variance of a locally unbiased estimator. Now let α be a coarse-graining. Then $\alpha(\rho_\theta)$ is another curve in the state space. Due to the linearity of α, the tangent at $\alpha(\rho)$ is $\alpha(B)$. As is usual in statistics, information cannot be gained by coarse graining, therefore we expect that the Fisher information at the density matrix ρ in the direction B must be larger than the Fisher information at $\alpha(\rho)$ in the direction $\alpha(B)$. This is the monotonicity property of the Fisher information under coarse-graining:

$$F_\rho(B) \geq F_{\alpha(\rho)}(\alpha(B)). \tag{7.54}$$

Although we do not want to have a concrete formula for the quantum Fisher information, we require that this monotonicity condition must hold. Another requirement is that $F_\rho(B)$ should be quadratic in B. In other words, there exists a non-degenerate real bilinear form $\gamma_\rho(B, C)$ on the self-adjoint matrices such that

$$F_\rho(B) = \gamma_\rho(B, B). \tag{7.55}$$

When ρ is regarded as a point of a manifold consisting of density matrices and B is considered as a tangent vector at the foot point ρ, the quadratic quantity $\gamma_\rho(B, B)$ may be regarded as a Riemannian metric on the manifold. This approach gives a geometric interpretation of the Fisher information.

The requirements (7.54) and (7.55) are strong enough to obtain a reasonable but still wide class of possible quantum Fisher informations.

We may assume that

$$\gamma_\rho(B, C) = \text{Tr } B\mathbb{J}_\rho^{-1}(C)$$

for an operator $\mathbb{J}_\rho$ acting on all matrices. (This formula expresses the inner product γ_ρ by means of the Hilbert–Schmidt inner product and the positive linear operator $\mathbb{J}_\rho$.) In terms of the operator $\mathbb{J}_\rho$ the monotonicity condition reads as

$$\alpha^* \mathbb{J}_{\alpha(\rho)}^{-1} \alpha \leq \mathbb{J}_\rho^{-1} \tag{7.56}$$

for every coarse graining α. (α^* stands for the adjoint of α with respect to the Hilbert–Schmidt product. Recall that α is completely positive and trace preserving if and only if α^* is completely positive and unital.) On the other hand the latter condition is equivalent to

$$\alpha \mathbb{J}_\rho \alpha^* \leq \mathbb{J}_{\alpha(\rho)}. \tag{7.57}$$

It is interesting to observe the relevance of a certain quasi-entropy:

$$\langle B\rho^{1/2}, f(\mathbb{L}_\rho \mathbb{R}_\rho^{-1})B\rho^{1/2}\rangle = S_f^B(\rho\|\rho),$$

see (7.9) and (7.11), where $\mathbb{L}_\rho$ and $\mathbb{R}_\rho$ are in (7.10). When $f : \mathbb{R}^+ \to \mathbb{R}$ is matrix monotone (we always assume $f(1) = 1$),

$$\langle \alpha^*(B)\rho^{1/2}, f(\mathbb{L}_\rho \mathbb{R}_\rho^{-1})\alpha^*(B)\rho^{1/2}\rangle \leq \langle B\alpha(\rho)^{1/2}, f(\mathbb{L}_{\alpha(\rho)} \mathbb{R}_{\alpha(\rho)}^{-1})B\alpha(\rho)^{1/2}\rangle$$

due to the monotonicity of the quasi-entropy, see Theorem 7.7. If we set

$$\mathbb{J}_\rho = \mathbb{J}_\rho^f := f(\mathbb{L}_\rho \mathbb{R}_\rho^{-1})\mathbb{R}_\rho,$$

then (7.57) holds. Therefore,

$$\varphi_\rho[B, B] := \operatorname{Tr} B\mathbb{J}_\rho(B) = \langle B\rho^{1/2}, f(\mathbb{L}_\rho \mathbb{R}_\rho^{-1})B\rho^{1/2}\rangle$$

can be called a **quadratic cost function** and the corresponding monotone **quantum Fisher information**

$$\gamma_\rho(B, C) = \operatorname{Tr} B\mathbb{J}_\rho^{-1}(C)$$

will be real for self-adjoint B and C if the function f satisfies the condition $f(x) = xf(x^{-1})$, see (7.14). This is nothing but a monotone metric, as described in Sect. 7.2.

Example 7.32 In order to understand the action of the operator $\mathbb{J}_\rho$, assume that ρ is diagonal, $\rho = \sum_i p_i E_{ii}$. Then one can check that the matrix units E_{kl} are eigenvectors of $\mathbb{J}_\rho$, namely

$$\mathbb{J}_\rho(E_{kl}) = p_l f(p_k/p_l)E_{kl}.$$

The condition $f(x) = xf(x^{-1})$ gives that the eigenvectors E_{kl} and E_{lk} have the same eigenvalues. Therefore, the symmetrized matrix units $E_{kl} + E_{lk}$ and $iE_{kl} - iE_{lk}$ are eigenvectors as well.

Since

$$B = \sum_{k<l} \operatorname{Re} B_{kl}(E_{kl} + E_{lk}) + \sum_{k<l} \operatorname{Im} B_{kl}(iE_{kl} - iE_{lk}) + \sum_i B_{ii} E_{ii},$$

we have

$$\gamma_\rho(B, B) = 2 \sum_{k<l} \frac{1}{p_k f(p_k/p_l)}|B_{kl}|^2 + \sum_i \frac{1}{p_i}|B_{ii}|^2.$$

In place of $2\sum_{k<l}$, we can write $\sum_{k\neq l}$.   □

Any monotone cost function has the property $\varphi_\rho[B, B] = \mathrm{Tr}\,\rho B^2$ for commuting ρ and B. The examples below show that this is not so in general.

Example 7.33 The analysis of matrix monotone functions leads to the fact that among all monotone quantum Fisher informations there is a smallest one which corresponds to the (largest) function $f_{max}(t) = (1 + t)/2$. In this case

$$F_\rho^{\min}(B) = \mathrm{Tr}\,BL = \mathrm{Tr}\,\rho L^2, \qquad \text{where} \qquad \rho L + L\rho = 2B.$$

For the purpose of a quantum Cramér–Rao inequality the minimal quantity seems to be the best, since the inverse gives the largest lower bound. In fact, the matrix L has been used for a long time under the name of the **symmetric logarithmic derivative**. In this example the quadratic cost function is

$$\varphi_\rho[B, C] = \tfrac{1}{2}\mathrm{Tr}\,\rho(BC + CB)$$

and we have

$$\mathbb{J}_\rho(B) = \tfrac{1}{2}(\rho B + B\rho) \qquad \text{and} \qquad \mathbb{J}_\rho^{-1}(C) = 2\int_0^\infty e^{-t\rho}Ce^{-t\rho}\,dt$$

for the operator $\mathbb{J}_\rho$. Since $\mathbb{J}_\rho^{-1}$ is the smallest, $\mathbb{J}_\rho$ is the largest (among all possibilities).
There is a largest among all monotone quantum Fisher informations and this corresponds to the function $f_{min}(t) = 2t/(1 + t)$. In this case

$$\mathbb{J}_\rho^{-1}(B) = \tfrac{1}{2}(\rho^{-1}B + B\rho^{-1}) \qquad \text{and} \qquad F_\rho^{\max}(B) = \mathrm{Tr}\,\rho^{-1}B^2.$$

It is known that the function

$$f_\alpha(t) = \alpha(1 - \alpha)\frac{(t - 1)^2}{(t^\alpha - 1)(t^{1-\alpha} - 1)}$$

is matrix monotone for $\alpha \in (0, 1)$. We denote by F^α the corresponding Fisher information. When X is self-adjoint, $B = i[\rho, X] := i(\rho X - X\rho)$ is orthogonal to the commutator of the foot point ρ in the tangent space (see Example 3.30), and we have

$$F_\rho^\alpha(B) = -\frac{1}{\alpha(1 - \alpha)}\mathrm{Tr}\left([\rho^\alpha, X][\rho^{1-\alpha}, X]\right). \tag{7.58}$$

Apart from a constant factor this expression is the **skew information** proposed by Wigner and Yanase some time ago. In the limiting cases $\alpha \to 0$ or 1 we have

$$f_0(t) = \frac{1 - t}{\log t}$$

and the corresponding quantum Fisher information

$$\gamma_\rho^0(B, C) = K_\rho(B, C) := \int_0^\infty \text{Tr } B(\rho + t)^{-1} C(\rho + t)^{-1} \, dt$$

will be named here after Kubo and Mori. The **Kubo–Mori inner product** plays a role in quantum statistical mechanics. In this case $\mathbb{J}$ is the so-called **Kubo transform** $\mathbb{K}$ (and $\mathbb{J}^{-1}$ is the inverse Kubo transform $\mathbb{K}^{-1}$),

$$\mathbb{K}_\rho^{-1}(B) := \int_0^\infty (\rho + t)^{-1} B(\rho + t)^{-1} \, dt \quad \text{and} \quad \mathbb{K}_\rho(C) := \int_0^1 \rho^t C \rho^{1-t} \, dt \,.$$

Therefore the corresponding generalized variance is

$$\varphi_\rho[B, C] = \int_0^1 \text{Tr } B \rho^t C \rho^{1-t} \, dt \,.$$

All Fisher informations discussed in this example are possible Riemannian metrics of manifolds of invertible density matrices. (Manifolds of pure states are rather different.) □

A Fisher information appears not only as a Riemannian metric but as an information matrix as well. Let $\mathcal{M} := \{\rho_\theta : \theta \in G\}$ be a smooth m-dimensional manifold of invertible density matrices. The **quantum score operators** (or **logarithmic derivatives**) aredefined as

$$L_i(\theta) := \mathbb{J}_{\rho_\theta}^{-1}\big(\partial_{\theta_i} \rho_\theta\big) \qquad (1 \le i \le m),$$

and

$$Q_{ij}(\theta) := \text{Tr } L_i(\theta) \mathbb{J}_{\rho_\theta}\big(L_j(\theta)\big) \qquad (1 \le i, j \le m)$$

is the **quantum Fisher information matrix**. This matrix depends on a matrix monotone function which is involved in the superoperator $\mathbb{J}$. Historically the matrix Q determined by the symmetric logarithmic derivative (or the function $f_{max}(t) = (1 + t)/2$) first appeared in the work of Helstrøm. Therefore, we call this the **Helstrøm information matrix** and it will be denoted by $H(\theta)$.

Theorem 7.34 *Fix a matrix monotone function f to induce quantum Fisher information. Let α be a coarse-graining sending density matrices on the Hilbert space $\mathcal{H}_1$ into those acting on the Hilbert space $\mathcal{H}_2$ and let $\mathcal{M} := \{\rho_\theta : \theta \in G\}$ be a smooth m-dimensional manifold of invertible density matrices on $\mathcal{H}_1$. For the Fisher information matrix $Q^{(1)}(\theta)$ of $\mathcal{M}$ and for the Fisher information matrix $Q^{(2)}(\theta)$ of $\alpha(\mathcal{M}) := \{\alpha(\rho_\theta) : \theta \in G\}$, we have the monotonicity relation*

$$Q^{(2)}(\theta) \le Q^{(1)}(\theta). \tag{7.59}$$

(*This is an inequality between m × m positive matrices.*)

Proof: Set $B_i(\theta) := \partial_{\theta_i} \rho_\theta$. Then $\mathbb{J}^{-1}_{\alpha(\rho_\theta)} \alpha(B_i(\theta))$ is the score operator of $\alpha(\mathcal{M})$. Using (7.56), we have

$$\sum_{ij} Q^{(2)}_{ij}(\theta) a_i \overline{a_j} = \operatorname{Tr} \mathbb{J}^{-1}_{\alpha(\rho_\theta)} \alpha\left(\sum_i a_i B_i(\theta)\right) \alpha\left(\sum_j \overline{a_j} B_j(\theta)\right)$$

$$\leq \operatorname{Tr} \mathbb{J}^{-1}_{\rho_\theta}\left(\sum_i a_i B_i(\theta)\right)\left(\sum_j \overline{a_j} B_j(\theta)\right)$$

$$= \sum_{ij} Q^{(1)}_{ij}(\theta) a_i \overline{a_j}$$

for any numbers a_i. □

Assume that F_j are positive operators acting on a Hilbert space $\mathcal{H}_1$ on which the family $\mathcal{M} := \{\rho_\theta : \theta \in \Theta\}$ is given. When $\sum_{j=1}^{n} F_j = I$, these operators determine a measurement. For any ρ_θ the formula

$$\alpha(\rho_\theta) := \operatorname{Diag}(\operatorname{Tr} \rho_\theta F_1, \ldots, \operatorname{Tr} \rho_\theta F_n)$$

gives a diagonal density matrix. Since this family is commutative, all quantum Fisher informations coincide with the classical $\mathbf{I}(\theta)$ in the right-hand side of (7.50) and the classical Fisher information on the left-hand side of (7.59). Hence we have

$$\mathbf{I}(\theta) \leq Q(\theta). \tag{7.60}$$

A combination of the classical Cramér–Rao inequality in Theorem 7.28 and (7.60) yields the **Helstrøm inequality**:

$$V(\theta) \geq H(\theta)^{-1}.$$

Example 7.35 In this example, we want to investigate (7.60) which is equivalently written as

$$Q(\theta)^{-1/2} \mathbf{I}(\theta) Q(\theta)^{-1/2} \leq I_m.$$

Taking the trace, we have

$$\operatorname{Tr} Q(\theta)^{-1} \mathbf{I}(\theta) \leq m. \tag{7.61}$$

Assume that

$$\rho_\theta = \rho + \sum_k \theta_k B_k,$$

where $\mathrm{Tr}\, B_k = 0$ and the self-adjoint matrices B_k are pairwise orthogonal with respect to the inner product $(B, C) \mapsto \mathrm{Tr}\, B \mathbb{J}_\rho^{-1}(C)$.

The quantum Fisher information matrix

$$Q_{kl}(0) = \mathrm{Tr}\, B_k \mathbb{J}_\rho^{-1}(B_l)$$

is diagonal due to our assumption. Example 7.29 tells us about the classical Fisher information matrix:

$$\mathbf{I}_{kl}(0) = \sum_j \frac{\mathrm{Tr}\, B_k F_j \,\mathrm{Tr}\, B_l F_j}{\mathrm{Tr}\, \rho F_j}.$$

Therefore,

$$\mathrm{Tr}\, Q(0)^{-1}\mathbf{I}(0) = \sum_k \frac{1}{\mathrm{Tr}\, B_k \mathbb{J}_\rho^{-1}(B_k)} \sum_j \frac{(\mathrm{Tr}\, B_k F_j)^2}{\mathrm{Tr}\, \rho F_j}$$

$$= \sum_j \frac{1}{\mathrm{Tr}\, \rho F_j} \sum_k \left(\mathrm{Tr}\, \frac{B_k}{\sqrt{\mathrm{Tr}\, B_k \mathbb{J}_\rho^{-1}(B_k)}} \mathbb{J}_\rho^{-1}(\mathbb{J}_\rho F_j) \right)^2.$$

We can estimate the latter sum using the fact that

$$\frac{B_k}{\sqrt{\mathrm{Tr}\, B_k \mathbb{J}_\rho^{-1}(B_k)}}$$

is an orthonormal system and it remains so when ρ is added to it:

$$(\rho, B_k) = \mathrm{Tr}\, B_k \mathbb{J}_\rho^{-1}(\rho) = \mathrm{Tr}\, B_k = 0$$

and

$$(\rho, \rho) = \mathrm{Tr}\, \rho \mathbb{J}_\rho^{-1}(\rho) = \mathrm{Tr}\, \rho = 1.$$

By the Parseval inequality, we have

$$\left(\mathrm{Tr}\, \rho \mathbb{J}_\rho^{-1}(\mathbb{J}_\rho F_j) \right)^2 + \sum_k \left(\mathrm{Tr}\, \frac{B_k}{\sqrt{\mathrm{Tr}\, B_k \mathbb{J}_\rho^{-1}(B_k)}} \mathbb{J}_\rho^{-1}(\mathbb{J}_\rho F_j) \right)^2 \leq \mathrm{Tr}\, (\mathbb{J}_\rho F_j) \mathbb{J}_\rho^{-1}(\mathbb{J}_\rho F_j)$$

and

$$\mathrm{Tr}\, Q(0)^{-1}\mathbf{I}(0) \leq \sum_j \frac{1}{\mathrm{Tr}\, \rho F_j} \left(\mathrm{Tr}\, (\mathbb{J}_\rho F_j) F_j - (\mathrm{Tr}\, \rho F_j)^2 \right)$$

$$= \sum_{j=1}^{n} \frac{\mathrm{Tr}\,(\mathbb{J}_\rho F_j) F_j}{\mathrm{Tr}\,\rho F_j} - 1 \le n - 1$$

if we show that

$$\mathrm{Tr}\,(\mathbb{J}_\rho F_j) F_j \le \mathrm{Tr}\,\rho F_j.$$

To see this we use the fact that the left-hand side is a quadratic cost and it can be majorized by the largest one (see Example 7.33):

$$\mathrm{Tr}\,(\mathbb{J}_\rho F_j) F_j \le \mathrm{Tr}\,\rho F_j^2 \le \mathrm{Tr}\,\rho F_j,$$

because $F_j^2 \le F_j$.

Since $\theta = 0$ is not essential in the above argument, we have obtained that

$$\mathrm{Tr}\,Q(\theta)^{-1} \mathbf{I}(\theta) \le n - 1,$$

which can be compared with (7.61). This bound can be smaller than the general one. The assumption on the B_k's is not very essential, since the orthogonality can be attained by reparameterization.   □

Let $\mathcal{M} := \{\rho_\theta : \theta \in G\}$ be a smooth m-dimensional manifold and assume that a collection $A = (A_1, \ldots, A_m)$ of self-adjoint matrices is used to estimate the true value of θ.

Given an operator $\mathbb{J}$ we have the corresponding cost function $\varphi_\theta \equiv \varphi_{\rho_\theta}$ for every θ and the **cost matrix** of the estimator A is a positive definite matrix, defined by $\varphi_\theta[A]_{ij} = \varphi_\theta[A_i, A_j]$. The **bias** of the estimator is

$$b(\theta) = \big(b_1(\theta), b_2(\theta), \ldots, b_m(\theta)\big)$$
$$:= \big(\mathrm{Tr}\,\rho_\theta(A_1 - \theta_1), \mathrm{Tr}\,\rho_\theta(A_2 - \theta_2), \ldots, \mathrm{Tr}\,\rho_\theta(A_m - \theta_m)\big).$$

From the bias vector we form a **bias matrix**

$$B_{ij}(\theta) := \partial_{\theta_j} b_i(\theta) \qquad (1 \le i, j \le m).$$

For a locally unbiased estimator at θ_0, we have $B(\theta_0) = 0$.

The next result is the quantum Cramér–Rao inequality for a biased estimate.

Theorem 7.36 *Let* $A = (A_1, \ldots, A_m)$ *be an estimator of* θ. *Then for the above defined quantities the inequality*

$$\varphi_\theta[A] \ge \big(I + B(\theta)\big) Q(\theta)^{-1} \big(I + B(\theta)^*\big)$$

holds in the sense of the order on positive semidefinite matrices. (Here I denotes the identity operator.)

Proof: We will use the block matrix method. Let $X = [X_{ij}]_{i,j=1}^{m}$ be an $m \times m$ matrix with $n \times n$ entries X_{ij}, and define $\tilde{\alpha}(X) := [\alpha(X_{ij})]_{i,j=1}^{m}$. For every $\xi_1, \ldots, \xi_m \in \mathbb{C}$ we have

$$\sum_{i,j=1}^{m} \xi_i \bar{\xi}_j \operatorname{Tr}(X\tilde{\alpha}(X^*))_{ij} = \sum_{k=1}^{m} \operatorname{Tr}\left(\sum_i \xi_i X_{ik}\right)\alpha\left(\left(\sum_j \xi_j X_{jk}\right)^*\right) \geq 0,$$

because

$$\operatorname{Tr} Y\alpha(Y^*) = \operatorname{Tr} Y\alpha(Y)^* = \langle \alpha(Y), Y \rangle \geq 0$$

for every $n \times n$ matrix Y. Therefore, the $m \times m$ ordinary matrix M having the (i, j) entry $\operatorname{Tr}(X\tilde{\alpha}(X^*))_{ij}$ is positive. In the sequel we restrict ourselves to $m = 4$ for the sake of simplicity and apply the above fact to the case

$$X = \begin{bmatrix} A_1 & 0 & 0 & 0 \\ A_2 & 0 & 0 & 0 \\ L_1(\theta) & 0 & 0 & 0 \\ L_2(\theta) & 0 & 0 & 0 \end{bmatrix} \quad \text{and} \quad \alpha = \mathbb{J}_{\rho\theta}.$$

Then we have

$$M = \begin{bmatrix} \operatorname{Tr} A_1\mathbb{J}_\rho(A_1) & \operatorname{Tr} A_1\mathbb{J}_\rho(A_2) & \operatorname{Tr} A_1\mathbb{J}_\rho(L_1) & \operatorname{Tr} A_1\mathbb{J}_\rho(L_2) \\ \operatorname{Tr} A_2\mathbb{J}_\rho(A_1) & \operatorname{Tr} A_2\mathbb{J}_\rho(A_2) & \operatorname{Tr} A_2\mathbb{J}_\rho(L_1) & \operatorname{Tr} A_2\mathbb{J}_\rho(L_2) \\ \operatorname{Tr} L_1\mathbb{J}_\rho(A_1) & \operatorname{Tr} L_1\mathbb{J}_\rho(A_2) & \operatorname{Tr} L_1\mathbb{J}_\rho(L_1) & \operatorname{Tr} L_1\mathbb{J}_\rho(L_2) \\ \operatorname{Tr} L_2\mathbb{J}_\rho(A_1) & \operatorname{Tr} L_2\mathbb{J}_\rho(A_2) & \operatorname{Tr} L_2\mathbb{J}_\rho(L_1) & \operatorname{Tr} L_2\mathbb{J}_\rho(L_2) \end{bmatrix} \geq 0.$$

Now we rewrite the matrix M in terms of the matrices involved in our Cramér–Rao inequality. The 2×2 block M_{11} is the generalized covariance, M_{22} is the Fisher information matrix and M_{12} is easily expressed as $I + B$. We have

$$M = \begin{bmatrix} \varphi_\theta[A_1, A_1] & \varphi_\theta[A_1, A_2] & 1 + B_{11}(\theta) & B_{12}(\theta) \\ \varphi_\theta[A_2, A_1] & \varphi_\theta[A_2, A_2] & B_{21}(\theta) & 1 + B_{22}(\theta) \\ 1 + B_{11}(\theta) & B_{21}(\theta) & \varphi_\theta[L_1, L_1] & \varphi_\theta[L_1, L_2] \\ B_{12}(\theta) & 1 + B_{22}(\theta) & \varphi_\theta[L_2, L_1] & \varphi_\theta[L_2, L_2] \end{bmatrix} \geq 0.$$

The positivity of a block matrix

$$M = \begin{bmatrix} M_1 & C \\ C^* & M_2 \end{bmatrix} = \begin{bmatrix} \varphi_\rho[A] & I + B(\theta) \\ I + B(\theta)^* & Q(\theta) \end{bmatrix}$$

implies $M_1 \geq CM_2^{-1}C^*$, which reveals precisely the statement of the theorem. (Concerning positive block matrices, see Chap. 2.) $\square$

Let $M_\Theta = \{\rho_\theta : \theta \in \Theta\}$ be a smooth manifold of density matrices. The following construction is motivated by classical statistics. Suppose that a positive functional $d(\rho_1, \rho_2)$ of two variables is given on the manifold. In many cases one can obtain a Riemannian metric by differentiation:

$$g_{ij}(\theta) = \frac{\partial^2}{\partial \theta_i \partial \theta'_j} d(\rho_\theta, \rho_{\theta'})\Big|_{\theta = \theta'} \qquad (\theta \in \Theta).$$

To be more precise the positive smooth functional $d(\cdot, \cdot)$ is called a **contrast functional** if $d(\rho_1, \rho_2) = 0$ implies $\rho_1 = \rho_2$.

Following the work of Csiszár in classical information theory, Petz introduced a family of information quantities parametrized by a function $F : \mathbb{R}^+ \to \mathbb{R}$

$$S_F(\rho_1, \rho_2) = \langle \rho_1^{1/2}, F(\Delta(\rho_2/\rho_1))\rho_1^{1/2} \rangle,$$

see (7.9); F is written here in place of f. $(\Delta(\rho_2/\rho_1) := L_{\rho_2} R_{\rho_1}^{-1}$ is the relative modular operator of the two densities.) When F is matrix monotone decreasing, this quasi-entropy possesses good properties, for example it is a contrast functional in the above sense if F is not linear and $F(1) = 0$. In particular, for

$$F_\alpha(t) = \frac{1}{\alpha(1-\alpha)}(1 - t^\alpha)$$

we have the relative entropy $S_\alpha(\rho_1, \rho_2)$ of degree α in Example 7.9. We have

$$\frac{\partial^2}{\partial t \partial u} S_\alpha(\rho + tB, \rho + uC) = -\frac{1}{\alpha(1-\alpha)} \cdot \frac{\partial^2}{\partial t \partial u} \mathrm{Tr}\, (\rho + tB)^{1-\alpha}(\rho + uC)^\alpha$$

$$=: K_\rho^\alpha(B, C)$$

at $t = u = 0$ in the affine parametrization. The tangent space at ρ is decomposed into two subspaces, the first consists of self-adjoint matrices of trace zero commuting with ρ and the second is $\{i[\rho, X] : X = X^*\}$, the set of commutators. The decomposition is essential both from the viewpoint of differential geometry and from the point of view of differentiation, see Example 3.30. If B and C commute with ρ, then

$$K_\rho^\alpha(B, C) = \mathrm{Tr}\, \rho^{-1} BC$$

is independent of α and it is the classical Fischer information (in matrix form). If $B = i[\rho, X]$ and $C = i[\rho, Y]$, then

$$K_\rho^\alpha(B, C) = -\frac{1}{\alpha(1-\alpha)} \mathrm{Tr}\, ([\rho^{1-\alpha}, X][\rho^\alpha, Y]).$$

Thus, $K_\rho^\alpha(B, B)$ is exactly equal to the **skew information** (7.58).

7.6 Notes and Remarks

As an introduction we suggest the book by Oliver Johnson, *Information Theory and The Central Limit Theorem*, Imperial College Press, 2004. The Gaussian Markov property is popular in probability theory for single parameters, but the vector-valued case is less popular. Sect. 7.1 is based on the chapter T. Ando and D. Petz, Gaussian Markov triplets approached by block matrices, Acta Sci. Math. (Szeged) **75**(2009), 329–345.

Classical information theory is covered in the book I. Csiszár and J. Körner, *Information Theory: Coding Theorems for Discrete Memoryless Systems*, Cambridge University Press, 2011. Shannon entropy appeared in the 1940s and it is sometimes said that the von Neumann entropy is its generalization. However, it is a fact that **von Neumann** introduced quantum entropy in 1925. Many details can be found in the books [67, 73]. The f-entropy of Imre **Csiszár** is used in classical information theory (and statistics) [35], see also the chapter F. Liese and I. Vajda, On divergences and informations in statistics and information theory, IEEE Trans. Inform. Theory **52**(2006), 4394–4412. The quantum generalization was extended by Dénes Petz in 1985, for example see Chap. 7 in [67]. The strong subadditivity of the von Neumann entropy was proved by E. H. Lieb and M. B. Ruskai in 1973. Details on f-divergence are in the chapter [49]. Theorem 7.4 is from the chapter K. M. R. **Audenaert**, Subadditivity of q-entropies for $q > 1$, J. Math. Phys. **48**(2007), 083507. The quantity $(\mathrm{Tr}\, D^q - 1)/(1 - q)$ is called the q-entropy or the **Tsallis entropy**. It is remarkable that strong subadditivity is not true for the Tsallis entropy in the matrix case (but it holds for probability). Useful information can be found in the chapters [38] and S. **Furuichi**, Tsallis entropies and their theorems, properties and applications, Aspects of Optical Sciences and Quantum Information, 2007.

A good introduction to CCR-algebras is the book [69]. This subject is far from matrix analysis, but the quasi-free states are really described by matrices. The description of the Markovian quasi-free state is from the chapter A. Jenčová, D. Petz and J. Pitrik, Markov triplets on CCR-algebras, Acta Sci. Math. (Szeged), **76**(2010), 111–134.

Section 7.4 on optimal quantum measurements is from the chapter A. J. Scott, Tight informationally complete quantum measurements, J. Phys. A: Math. Gen. **39**(2006), 13507. MUBs have a large literature. They are commutative quasi-orthogonal subalgebras. The work of Scott motivated the chapter D. Petz, L. Ruppert and A. Szántó, Conditional SIC-POVMs, arXiv:1202.5741. It was shown by M. Weiner [83] that the existence of d MUBs in $\mathbb{M}_d(\mathbb{C})$ implies the existence of $d + 1$ MUBs.

The quasi-orthogonality of non-commutative subalgebras of $\mathbb{M}_d(\mathbb{C})$ also has an extensive literature; a summary appears in the chapter D. Petz, Algebraic complementarity in quantum theory, J. Math. Phys. **51**(2010), 015215. The SIC POVM is constructed in 6 dimensions in the chapter M. Grassl, On SIC-POVMs and MUBs in dimension 6, http://arxiv.org/abs/quant-ph/0406175.

Section 7.5 is taken from Sects. 10.2–10.4 of D. Petz [73]. Fisher information first appeared in the 1920s. For more on this subject, we suggest the book of Oliver Johnson cited above and the chapter K. R. Parthasarathy, On the philosophy of Cramér–Rao–Bhattacharya inequalities in quantum statistics, arXiv:0907.2210. The general quantum matrix formalism was initiated by D. **Petz** in the chapter [71]. A. Lesniewski and M. B. Ruskai discovered in [62] that all monotone Fisher informations are obtained from a quasi-entropy as a contrast functional.

7.7 Exercises

1. Prove Theorem 7.2.
2. Assume that $\mathcal{H}_2$ is one-dimensional in Theorem 7.14. Describe the possible quasi-free Markov triplet.
3. Show that in Theorem 7.6 condition (iii) cannot be replaced by

$$D_{123}D_{23}^{-1} = D_{12}D_2^{-1}.$$

4. Prove Theorem 7.15.
5. The Bogoliubov–Kubo–Mori Fisher information is induced by the function

$$f(x) = \frac{x-1}{\log x} = \int_0^1 x^t \, dt$$

and

$$\gamma_D^{\mathrm{BKM}}(A, B) = \mathrm{Tr}\, A(\mathbb{J}_D^f)^{-1}B$$

for self-adjoint matrices. Show that

$$\gamma_D^{\mathrm{BKM}}(A, B) = \int_0^\infty \mathrm{Tr}\,(D + tI)^{-1}A(D + tI)^{-1}B \, dt$$

$$= -\frac{\partial^2}{\partial t \partial s} S(D + tA \| D + sB)\Big|_{t=s=0}.$$

6. Prove Theorem 7.16.
7. Show that

$$x \log x = \int_0^\infty \left(\frac{x}{1+t} - \frac{x}{x+t}\right) dt$$

and deduce that the function $f(x) = x \log x$ is matrix convex.
8. Define

$$S_\beta(\rho_1\|\rho_2) := \frac{\text{Tr } \rho_1^{1+\beta}\rho_2^{-\beta} - 1}{\beta}$$

for $\beta \in (0, 1)$. Show that

$$S(\rho_1\|\rho_2) \leq S_\beta(\rho_1\|\rho_2)$$

for density matrices ρ_1 and ρ_2.

9. The functions

$$g_p(x) := \begin{cases} \frac{1}{p(1-p)}(x - x^p) & \text{if} \quad p \neq 1, \\ \\ x \log x & \text{if} \quad p = 1 \end{cases}$$

can be used for quasi-entropy. For which $p > 0$ is the function g_p matrix concave?

10. Give an example for which condition (iv) in Theorem 7.12 does not imply condition (iii).

11. Assume that

$$\begin{bmatrix} A & B \\ B^* & C \end{bmatrix} \geq 0.$$

Prove that

$$\text{Tr } (AC - B^*B) \leq (\text{Tr } A)(\text{Tr } C) - (\text{Tr } B)(\text{Tr } B^*).$$

(Hint: Use Theorem 7.4 in the case $q = 2$.)

12. Let ρ and ω be invertible density matrices. Show that

$$S(\omega\|\rho) \leq \text{Tr } (\omega \log(\omega^{1/2}\rho^{-1}\omega^{1/2})).$$

13. For $\alpha \in [0, 1]$ let

$$\chi_\alpha^2(\rho, \sigma) := \text{Tr } \rho\sigma^{-\alpha}\rho\sigma^{\alpha-1} - 1.$$

Find the value of α which gives the minimal quantity.

Bibliography

1. Ando T (1979) Generalized Schur complements. Linear Algebra Appl 27:173–186
2. Ando T (1979) Concavity of certain maps on positive definite matrices and applications to Hadamard products. Linear Algebra Appl 26:203–241
3. Ando T (1987) Totally positive matrices. Linear Algebra Appl 90:165–219
4. Ando T (1988) Comparison of norms $|||f(A) - f(B)|||$ and $|||f(|A - B|)|||$. Math Z 197:403–409
5. Ando T (1989) Majorization, doubly stochastic matrices and comparison of eigenvalues. Linear Algebra Appl 118:163–248
6. Ando T (1994) Majorization and inequalities in matrix theory. Linear Algebra Appl 199:17–67
7. Ando T (2009) Private communication
8. Ando T, Hiai F (1994) Log majorization and complementary Golden-Thompson type inequalities. Linear Algebra Appl 197/198:113–131
9. Ando T, Hiai F (2011) Operator log-convex functions and operator means. Math Ann 350:611–630
10. Ando T, Li C-K, Mathias R (2004) Geometric means. Linear Algebra Appl 385:305–334
11. Ando T, Petz D (2009) Gaussian Markov triplets approached by block matrices. Acta Sci Math (Szeged) 75:265–281
12. Appleby DM (2005) Symmetric informationally complete-positive operator valued measures and the extended Clifford group. J Math Phys 46:052107
13. Audenaert KMR, Aujla JS (2007) On Ando's inequalities for convex and concave functions. Preprint, arXiv:0704.0099
14. Audenaert K, Hiai F, Petz D (2010) Strongly subadditive functions. Acta Math Hungar 128:386–394
15. Aujla JS, Silva FC (2003) Weak majorization inequalities and convex functions. Linear Algebra Appl 369:217–233
16. Bendat J, Sherman S (1955) Monotone and convex operator functions. Trans Am Math Soc 79:58–71
17. Besenyei Á (2012) The Hasegawa-Petz mean: properties and inequalities. J Math Anal Appl 339:441–450
18. Besenyei Á, Petz D (2012) Characterization of mean transformations. Linear Multilinear Algebra 60:255–265
19. Bessis D, Moussa P, Villani M (1975) Monotonic converging variational approximations to the functional integrals in quantum statistical mechanics. J Math Phys 16:2318–2325
20. Bhatia R (1996) Matrix analysis. Springer, New York

F. Hiai and D. Petz, *Introduction to Matrix Analysis and Applications*,
Universitext, DOI: 10.1007/978-3-319-04150-6,
© Hindustan Book Agency 2014

21. Bhatia R (2007) Positive definite matrices. Princeton University Press, Princeton
22. Bhatia R, Davis C (1995) A Cauchy-Schwarz inequality for operators with applications. Linear Algebra Appl 223/224:119–129
23. Bhatia R, Kittaneh F (1998) Norm inequalities for positive operators. Lett Math Phys 43:225–231
24. Bhatia R, Parthasarathy KR (2000) Positive definite functions and operator inequalities. Bull Lond Math Soc 32:214–228
25. Bhatia R, Sano T (2009) Loewner matrices and operator convexity. Math Ann 344:703–716
26. Bhatia R, Sano T (2010) Positivity and conditional positivity of Loewner matrices. Positivity 14:421–430
27. Birkhoff G (1946) Tres observaciones sobre el algebra lineal. Univ Nac Tucuman Rev Ser A 5:147–151
28. Bourin J-C (2004) Convexity or concavity inequalities for Hermitian operators. Math Ineq Appl 7:607–620
29. Bourin J-C (2006) A concavity inequality for symmetric norms. Linear Algebra Appl 413:212–217
30. Bourin J-C, Uchiyama M (2007) A matrix subadditivity inequality for $f(A + B)$ and $f(A) + f(B)$. Linear Algebra Appl 423:512–518
31. Calderbank AR, Cameron PJ, Kantor WM, Seidel JJ (1997) Z_4-Kerdock codes, orthogonal spreads, and extremal Euclidean line-sets. Proc Lond Math Soc 75:436
32. Choi MD (1977) Completely positive mappings on complex matrices. Linear Algebra Appl 10:285–290
33. Conway JB (1978) Functions of one complex variable I, 2nd edn. Springer, New York
34. Cox DA (1984) The arithmetic-geometric mean of Gauss. Enseign Math 30:275–330
35. Csiszár I (1967) Information type measure of difference of probability distributions and indirect observations. Studia Sci Math Hungar 2:299–318
36. Donoghue WF Jr (1974) Monotone matrix functions and analytic continuation. Springer, Berlin
37. Feller W (1971) An introduction to probability theory with its applications, vol II. Wiley, New York
38. Furuichi S (2005) On uniqueness theorems for Tsallis entropy and Tsallis relative entropy. IEEE Trans Inf Theor 51:3638–3645
39. Furuta T (2008) Concrete examples of operator monotone functions obtained by an elementary method without appealing to Löwner integral representation. Linear Algebra Appl 429:972–980
40. Hansen F, Pedersen GK (1982) Jensen's inequality for operators and Löwner's theorem. Math Ann 258:229–241
41. Hansen F, Pedersen GK (2003) Jensen's operator inequality. Bull Lond Math Soc 35:553–564
42. Hansen F (2008) Metric adjusted skew information. Proc Natl Acad Sci USA 105:9909–9916
43. Hiai F (1997) Log-majorizations and norm inequalities for exponential operators, In: Linear operators, vol 38. Banach Center Publications, Warsaw, pp 119–181, 1994, Polish Acad Sci Warsaw
44. Hiai F, Kosaki H (1999) Means for matrices and comparison of their norms. Indiana Univ Math J 48:899–936
45. Hiai F, Kosaki H (2003) Means of Hilbert space operators, vol 1820. Lecture Notes in Math-Springer, Berlin
46. Hiai F, Kosaki H, Petz D, Ruskai MB (2013) Families of completely positive maps associated with monotone metrics. Linear Algebra Appl 439:1749–1791
47. Hiai F, Petz D (1993) The Golden-Thompson trace inequality is complemented. Linear Algebra Appl 181:153–185
48. Hiai F, Petz D (2009) Riemannian geometry on positive definite matrices related to means. Linear Algebra Appl 430:3105–3130
49. Hiai F, Mosonyi M, Petz D, Bény C (2011) Quantum f-divergences and error correction. Rev Math Phys 23:691–747

50. Hida T (1960/1961) Canonical representations of Gaussian processes and their applications, Mem Coll Sci Univ Kyoto Ser A Math 33(1960/1961):109–155
51. Hida T, Hitsuda M (1993) Gaussian processes. Translations of mathematical monographs, vol 120. American Mathematical Society, Providence
52. Horn A (1962) Eigenvalues of sums of Hemitian matrices. Pacific J Math 12:225–241
53. Horn RA, Johnson CR (1985) Matrix analysis. Cambridge University Press, Cambridge
54. Ivanović ID (1981) Geometrical description of quantal state determination. J Phys A 14:3241
55. Klyachko AA (1998) Stable bundles, representation theory and Hermitian operators. Selecta Math 4:419–445
56. Knuston A, Tao T (1999) The honeycomb model of $GL_n(\mathbb{C})$ tensor products I: proof of the saturation conjecture. J Am Math Soc 12:1055–1090
57. Kosem T (2006) Inequalities between $\| f(A + B)\|$ and $\| f(A) + f(B)\|$. Linear Algebra Appl 418:153–160
58. Kubo F, Ando T (1980) Means of positive linear operators. Math Ann 246:205–224
59. Lax PD (2002) Functional analysis. Wiley, New York
60. Lax PD (2007) Linear algebra and its applications. Wiley, New York
61. Lenard A (1971) Generalization of the Golden-Thompson inequality $\mathrm{Tr}(e^A e^B) \geq \mathrm{Tr} e^{A+B}$. Indiana Univ Math J 21:457–467
62. Lesniewski A, Ruskai MB (1999) Monotone Riemannian metrics and relative entropy on noncommutative probability spaces. J Math Phys 40:5702–5724
63. Lieb EH (1973) Convex trace functions and the Wigner-Yanase-Dyson conjecture. Adv Math 11:267–288
64. Lieb EH, Seiringer R (2004) Equivalent forms of the Bessis-Moussa-Villani conjecture. J Stat Phys 115:185–190
65. Löwner K (1934) Über monotone matrixfunctionen. Math Z 38:177–216
66. Marshall AW, Olkin I (1979) Inequalities: theory of majorization and its applications. Academic Press, New York
67. Ohya M, Petz D (1993) Quantum entropy and its use. Springer, Heidelberg (2nd edn, 2004)
68. Petz D (1988) A variational expression for the relative entropy. Commun Math Phys 114:345–348
69. Petz D (1990) An invitation to the algebra of the canonical commutation relation. Leuven University Press, Leuven
70. Petz D (1985) Quasi-entropies for states of a von Neumann algebra. Publ. RIMS. Kyoto Univ. 21:781–800
71. Petz D (1996) Monotone metrics on matrix spaces. Linear Algebra Appl 244:81–96
72. Petz D (1986) Quasi-entropies for finite quantum systems. Rep Math Phys 23:57–65
73. Petz D (2008) Quantum information theory and quantum statistics. Springer, Berlin
74. Petz D, Hasegawa H (1996) On the Riemannian metric of α-entropies of density matrices. Lett Math Phys 38:221–225
75. Petz D, Temesi R (2006) Means of positive numbers and matrices. SIAM J Matrix Anal Appl 27:712–720
76. Petz D (2010) From f-divergence to quantum quasi-entropies and their use. Entropy 12:304–325
77. Reed M, Simon B (1975) Methods of modern mathematical physics II. Academic Press, New York
78. Schrödinger E (1936) Probability relations between separated systems. Proc Cambridge Philos Soc 31:446–452
79. Suzuki M (1986) Quantum statistical Monte Carlo methods and applications to spin systems. J Stat Phys 43:883–909
80. Thompson CJ (1971) Inequalities and partial orders on matrix spaces. Indiana Univ Math J 21:469–480
81. Tropp JA (2012) From joint convexity of quantum relative entropy to a concavity theorem of Lieb. Proc Am Math Soc 140:1757–1760
82. Uchiyama M (2006) Subadditivity of eigenvalue sums. Proc Am Math Soc 134:1405–1412

83. Weiner M (2013) A gap for the maximum number of mutually unbiased bases. Proc Am Math Soc 141:1963–1969
84. Wielandt H (1955) An extremum property of sums of eigenvalues. Proc Am Math Soc 6:106–110
85. Wootters WK, Fields BD (1989) Optimal state-determination by mutually unbiased measurements. Ann Phys 191:363
86. Zauner G (1999) Quantendesigns—Grundzüge einer nichtkommutativen Designtheorie, Ph.D. thesis, University of Vienna
87. Zhan X (2002) Matrix inequalities, vol 1790. Lecture Notes in MathSpringer, Berlin
88. Zhang F (2005) The Schur complement and its applications. Springer, New York

Index

Symbols
$A \circ B$, 68
$A \sigma B$, 197
$\langle \cdot, \cdot \rangle$, 4
$A : B$, 196
$A \# B$, 190
A^*, 3
A^t, 3
$B(\mathcal{H})$, 10
$B(\mathcal{H})^{sa}$, 10
$E(ij)$, 2
$G_t(A, B)$, 206
$H(A, B)$, 197, 207
$H^\perp$, 5
I_n, 1
$L(A, B)$, 207
M/A, 59
$[P]M$, 61
AG(a, b), 187
$\Delta_p(Q)$, 305
$\mathbb{J}_D$, 151
Φ, 241
$\Phi_p(a)$, 242
$M_f(A, B)$, 212
$\mathrm{Tr}\, A$, 3
Tr_1, 277
$\|A\|$, 9
$\|A\|_p$, 245
$\|A\|_{(k)}$, 245
$\mathbb{M}_n^{sa}$, 10
$\mathbb{M}_n$, 1
χ^2-divergence, 287
$\det A$, 3
ℓ_p-norms, 242
$\mathcal{H}$, 4
$\mathcal{P}_n$, 188
$\ker A$, 6

$\mathbb{P}_n$, 217
$\mathrm{ran}\, A$, 6
$\sigma(A)$, 15
$\|\|A\|\|$, 243
$\|\| \cdot \|\|$, 243
$a \prec_w b$, 230
$a \prec_{w(\log)} b$, 231
$m_f(A, B)$, 203
$s(\Lambda)$, 233
$v_1 \wedge v_2$, 41
2-positive mapping, 86

A
Absolute value, 29
Adjoint
    matrix, 3
    operator, 10
Ando, 222, 268
Ando and Hiai, 263
Ando and Zhan, 269
Annihilating
    polynomial, 15
Antisymmetric tensor product, 41
Arithmetic-geometric mean, 202
Audenaert, 183, 320
Aujla and Silva, 183

B
Baker–Campbell–Hausdorff
    formula, 109
Basis, 5
    Bell, 38
    product, 37
Bernstein theorem, 110
Bessis–Moussa–Villani conjecture, 131

F. Hiai and D. Petz, *Introduction to Matrix Analysis and Applications*,   327
Universitext, DOI: 10.1007/978-3-319-04150-6,
© Hindustan Book Agency 2014